Bytelines

A Life Story

JIM O'DONNELL

To order additional copies of this book, contact:
Xlibris Corporation
1-888-795-4274
www.Xlibris.com
Orders@Xlibris.com
53229

Bytelines

CONTENTS

IN HOG SIGNO VINCES.

IN THIS SIGN THOU SHALT CONQUER

FOREWORD

On July 13, 2006, I set out to write this memoir for my children and grandchildren. Having grandchildren late in life means that they will never really get to know you. This was the case for me and for my children. I never knew my grandparents as they lived and died in Ireland before I had a chance to go there. I might have gone over in 1950 and seen my maternal grandfather Patrick Quinn. He was the only grandparent living when it was possible, after the war, to visit Ireland. Regretfully, I didn't go when I had the chance, and he died in 1951.

My own children never got to know my father as he died in 1969, before our youngest son was born. Even I never really got to know much about him as he didn't share much of himself with us. He was kindly and generous, but only seemed comfortable with his friends from the Old Country.

A little family background may be useful. My parents were James (1896) and Margaret Quinn (1903) O'Donnell. They were born in what is now Northern Ireland. They had three children: me (1932), Patricia (1935), and Maureen (1941). Beyond that, you'll have to read and learn about us, the only branch of our O'Donnell clan extant outside of the Old Country.

In 1957, I married Catharine Marie "Kitty" Carlin, of Moorestown, New Jersey. We met while working at RCA in the summer of 1956. We have four boys: Thomas (1963), Christopher (1965), Patrick (1968), and James (1970). Now you will know about whom I speak when I mention Kitty or my boys.

I decided to write more rather than less about me and my family because I've found that you can never anticipate what readers find of interest—reading is such a personal experience. I've tried to check my family-related facts with relatives and with my sisters. I've also used the

Internet extensively to verify dates and such outside of the family. I suppose that I'll have to list references and credits before I'm finished. I've included lots of pictures that have come from many sources. Pictures can certainly help bring to life a time long gone, the time of my life.

August 7, 2008

BYTE

Definition: In computer science, a *byte* (pronounced "bite") is a unit of measurement of information storage, most often consisting of eight bits. In many computer architectures, it is a unit of memory addressing.

NOTATION

Listed here are some notations I'll be using to avoid endless repetition of much-used words and phrases:

Br.: A religious brother
Fr.: A religious priest
Msgr.: An honorific for a priest or bishop
Sr.: A religious sister (nun)
St.: Saint (male and female)
CIP: Catholic Institute of the Press
IRL: The Republic of Ireland
NIR: Northern Ireland, a part of the United Kingdom
Ireland: Inclusive of IRL and NIR or before division
NYC: New York City
NY: The state—NYC and upstate
OPA: U.S. Office of Price Administration, during WWII
RC: Roman Catholic
WWI & WWII: The world wars

ABOUT NEW YORK CITY

In New York City, you have to recognize the distinction between county (state) and borough (city). Within the city, it's the borough of "the Bronx," which has its own courts and borough hall. From the perspective of the state, it is Bronx County, hence it also has a county courthouse and jail. This distinction holds true for the city's other boroughs—see table below. In 1898, the independent city of Brooklyn and the various cities that now comprise Queens were incorporated into what was then called Greater New York, an appellation that has fallen by the wayside over the years. However, the Catholic Diocese of Brooklyn has retained its independence from the Archdiocese of New York

Borough (City)	County (State)	Catholic Diocese
Manhattan	New York	New York
The Bronx	Bronx	New York
Staten Island	Richmond	New York
Brooklyn	Kings	Brooklyn
Queens	Queens	Brooklyn

PART 1

Early Years (1932-1943)

Irish Background

Our extended O'Donnell family and friends all came from Ireland. As we will journey back and forth to our native soil, it may help to know something about Ireland and then Northern Ireland. During the long British occupation of Ireland, the Catholic faith was suppressed in favor of Protestant sects. The Church of Ireland was the established church and was the beneficiary of the Catholic churches and monasteries seized in the time of Henry VIII. Through the efforts in the Parliament of Daniel O'Connell and others, Ireland regained "home rule," and the Catholic Church was "emancipated" in 1829. This meant that Catholics could freely practice their faith, have churches, and have schools. So prior to that time, scant records exist for Catholic births, marriages, and deaths. Such records as existed were maintained by the various Protestant sects, most notably the Church of Ireland. However, in 1856, civil registration was mandated, and records of our family can be found starting about then. Some years ago, the government funded collection of the individual parish records and their entry into computerized databases. In Northern Ireland, the Catholic parish records are held in a beautiful new records center near the Cathedral of St. Patrick in the town of Armagh. However, the Public Records Office of Northern Ireland in Belfast, known as PRONI, is the main repository for all records in the province. Access to records in neither place is easy as they are unavailable online.

As a consequence of the 1916 Easter Rebellion and its aftermath, in 1920, a settlement was reached, which created the Irish Free State of twenty-six counties. Great Britain retained sovereignty over six of the nine counties in Ulster Province, and they form what we know today as Northern Ireland (NIR). Included among the six is the county of Tyrone. The total package is now called the United Kingdom (of Great Britain and Northern Ireland). After World War II (WWII), the Irish Free State became totally independent of the United Kingdom and became the Irish Republic (IRL). So while my parents were born in Ireland, pre-NIR, they carried British passports on their journeys to North America.

Although Donegal is the historic home of the O'Donnell clan (see appendices A and B), my immediate ancestors on both sides of the family came from County Tyrone. In early times, the county was divided into baronies, parishes, and townlands. In Tyrone, there were three baronies—north, west, and south. Our family areas in and around the town of Dungannon were in the Tyrone south barony that borders

County Armagh to the southeast. In very recent times, NIR was divided into twenty-six administrative units, one of which is Dungannon. A nice historic article on Tyrone and Dungannon appears as appendix C.

Historically, parishes were significant administrative units within baronies and counties and were determined by the Church of Ireland. Later, Catholic parishes did not always follow the established names and boundaries, creating much confusion for researchers. In some cases, the Catholic parish first adopted the established name, but later renamed the parish, thus having two names. Within parishes were townlands, similar to American tracts or housing estates within a township. Townlands, however, have no administrative or political function but do serve to delineate election districts. Villages, towns, and cities round out the political divisions of NIR.

My Father

The O'Donnell ancestral homeland was Tyrconnell, now County Donegal in the IRL. Our branch of the family was probably allied with the O'Neill clan and settled in their bailiwick Tyrowen, now County Tyrone, presumably named for Owen O'Neill. As mentioned, Tyrone is now in NIR but has a long border with County Donegal to the northwest.

The earliest known homestead of the family is that of my great-grandfather Edward and his wife, Mary Casey. Their home was at what's now known as 15 Derrylappen Road, Derrylappen Townland, in the Brantry District of southeast Tyrone. Historically, the area was within the barony of South Tyrone and today is in the Dungannon administrative unit. The area is along the border with County Armagh and features the locally popular Brantry Lough (a lake or pond).

The family lands lay within Eglish parish, which was established in 1834, shortly after Catholic emancipation. The Eglish Church of St. Patrick is today a fine-looking building on Eglish Road. Alongside the church are the parish burying grounds, old and new, where the ancestors were interred. Unfortunately, the earliest family graves are unmarked. The parish today has 230 families. Derrylatinee public school serves the parish now as it did in my father's time, but in a fine new building.

When we visited my great-grandfather's homesite in 1990, we found a prosperous farm with lots of livestock and a newish farmhouse. The original house had been torn down many years before. The farm has come down through the James O'Donnell branch of the family, James being the oldest son and my grandfather Edward, the youngest. The current James, my second cousin who was born the same year as I, did not choose to stay with the ancestral farm, so his younger brother Sean and his wife, Sadie Hagan, and large family are the present occupants. James took over the farm of his wife Christina Hughes's family. Their family is also large and very prosperous and own one of the largest kitchen suppliers in Ireland.

My grandfather Edward and his wife, Catherine Donnelly, had a farm at what's now known as 31 Gort Road, Gort Townland, the Brantry, Dungannon district. Catherine's family was from the Galbally area, northwest of the Brantry district, but not all that far away. Her father had the unusual name of Murtha, and was known as Murty. Her mother was Catherine McQuaide. At Gort, my grandparents had nine children, of whom eight survived to be adults. My father, James, was second, born in 1896 and the oldest male. His father was forty-five, and his mother twenty-nine. My future godfather was Francis "Frank" IV, born in 1900. At early ages, my father and his sister Cassie were sent to live with their uncle Canon Francis Donnelly who was parish priest (pastor) in Carrickmore parish, some miles to the northwest of Dungannon. Aunt Cassie served as housekeeper while my father drove for his uncle and perhaps others, somehow earning enough to finance his voyage to North America.

The story is that my father went to Belfast with his funds for the voyage and found that the ship was delayed or had already left. Stranded in Belfast, he fell among gamblers and lost his stake. He had to return to Carrickmore to rebuild his nest egg and eventually sailed to Canada. We don't know where he started from, where he landed, or from what ship. It's possible that he arrived in Halifax, Nova Scotia, as that was the equivalent of New York's Ellis Island. In any event, his destination was the farm of a family named Marshall. It was located near Montreal, and we tried, unsuccessfully, to find it when we traveled to Canada in the summer of 1949. We don't know how long he spent there until he made his way to New York, via the border crossing at Rouses Point, NY. He was finally in the USA, but as an illegal.

Dates for my father's arrival in New York City (NYC) are elusive. The only concrete piece of information we have is his Irish Republican Army (IRA) transfer dated March 10, 1926. In that document, Captain James O'Donnell is transferred from the Tyrone Brigade to the foreign reserve list in the USA. His departure date is given as May 1925, and his address abroad is given as 505 Columbus Avenue. This address is on the west side of Manhattan, near Eighty-fifth Street, and may have been the home of his younger brother, Francis, or of a friend from Carrickmore.

His first job in NYC was probably as a clerk in one of the grocery chain stores owned by earlier Irish immigrants. Prominent among them was Butler Stores where you worked long hours for low wages. These stores were small and carried loose stock, burlap bags of beans, coffee, and the like. Butter was sold out of tubs, and cigarettes were sold one at a time or by the pack. These stores were hard hit by the 1930s recession.

Dad was an inveterate storyteller who loved to sit with friends and swap tales. He also loved poetry and music. We have the text of two poems that he wrote, apparently during his stay in Carrickmore. They were intended to be sung to some Irish melodies popular in the 1920s. (See appendix J for the text of one of them.) My father had a violin that he would play from time to time and an Autoharp that I never saw him use in any way. To the world outside family and friends, he was known as a bagpiper in the County Tyrone Pipe Band, about which there will be later. My father loved Charlie Chaplin, and I can remember seeing *The Gold Rush* with him, probably on television. Many of the Irish turned against Chaplin when he became involved with Communist causes, but mostly because he "cradle-robbed" the seventeen-year-old aspiring actress Oona O'Neill for his fourth wife. Detractors failed to consider that she had been estranged from her sodden father, the great playwright Eugene O'Neill. He disinherited her when she married. Oona and Charlie had a good long marriage and had eight children. After Chaplin's death, she became reclusive and died in Switzerland, at age sixty-five, from the ravages of the family disease of alcoholism.

My Mother

Margaret Teresa Quinn was born on March 10, 1903, at home on the farm in Cornamaddy Townland near Pomeroy, County Tyrone. She was the fourth of eight children born to Patrick Quinn and Mary Clarke and was the second girl. Patrick was forty-five at the time, and Mary was twenty-nine. Patrick's parents were Michael and Mary Quinn, and he had undoubtedly spent his whole life in Cornamaddy, except for a sojourn in America as will be seen later. Mary Clarke's parents were Patrick Clarke and Brigid Quinn, of Lurgylea, a nearby townland. Both townlands, Cornamaddy and Lurgylea, lay between Pomeroy town and Cappagh village, then in the Dungannon district. Later Pomeroy became part of the Cookstown administrative district. However, such districting is ever-changing, and Northern Ireland has recently been realigned to reduce the number of districts and make them more representative of population distribution. The most likely arrangement would put Dungannon and South Tyrone in a new "West" district which would include Omagh, Cookstorm, and County Fermanagh.

At an early age, Mother helped out on the farm and attended Altmore School, quite a walk in bad weather. She often spoke of walking to school and of her teacher Ms. McComiskey who was schoolmistress for many years. We know that Mother had a cooking class there, possibly a first step toward her future career. The Cornamaddy Quinns were in Pomeroy parish (RC) and attended mass in Altmore Chapel, which was nearer than Assumption, the parish church in Pomeroy. Altmore School and the chapel were near the Shields estate, which is just outside the village of Cappagh. The Shields were a well-to-do Catholic family who probably supported both the chapel and the school. One of their sons emigrated to the USA and became a general who served the Union during the Civil War. Later, he became senator of several Western states in the days when senators were appointed by the governor. The Shields's home was designated a historic site and converted into a beautiful inn. We stayed there in 1968. Sadly, the building was firebombed during the sectarian strife of the 1970s.

Mother's social life included barn dances and occasional trips to Pomeroy on market day. She also spoke of a Boyle who was her boyfriend and who was to own a popular pub in Cappagh. Mother probably left school at sixteen, in 1919, to help out on the farm. Fortunately, being on the farm, her family was not ravaged by the worldwide flu epidemic that raged during and after World War I (WWI). Other than helping around the home, I don't know what else she did until she left for the USA. Mother, at age nineteen, arrived at Ellis Island on September 26, 1922, via the Anchor Line *Tuscania* out of Londonderry, NIR. This was the second *Tuscania* as the original had been torpedoed and sunk in 1918.

On this voyage, Mother was accompanied by her first cousin Mary Quinn whose father, Joseph Quinn, had moved from Ireland to Middlesbrough, England. There he had married a Mary Quinn, and young Mary was the eldest child. She and Mother were of the same age. I don't know for sure where Mary embarked, but she probably came over to Ireland and got on the ship with Mother in Londonderry. When released from Ellis Island, Mother and Mary set out for Philadelphia where they had relatives awaiting their arrival. Their aunt Mary Ann Quinn had emigrated years before, went to Philadelphia, and eventually married John Cartin. The Cartins became a large Philadelphia clan and easily accommodated the newcomers.

The Quinn family's connection to Philadelphia included my grandfather Patrick, Mary Ann's brother. He had lived in the area, working in the iron and steel industry until he returned to Ireland on the death of his father, Michael Quinn. As the oldest of Michael's children, Patrick inherited the Cornamaddy farms. The only other capable male in the family who could have acquired them was the aforementioned Joseph, who had already settled in England. There was another brother, Francis, who was senior to Joseph, but a cripple for whatever reason. He never married and stayed on the farm until his death.

Mother and Mary came to Philadelphia and worked jobs in the kitchen of Girard College, essentially a prep school for boys. Girard College was founded in 1848 by the terms of the will of financier and banker Stephen Girard. He was a French businessman who figured prominently in the American Revolution. Originally for "poor white male orphans," the courts have, since Mother's time, changed admission policies to include motherless boys as well as fatherless boys between six and eighteen, other races, and now girls. I guess Mother picked up some institutional cooking tips at the college. We believe Mother only stayed in Philadelphia for a year or so. Perhaps she left about the time her cousin Mary entered the convent, where she became Sister Florina. She remained a nun in the Philadelphia area until her death.

At some point, perhaps in 1924, Mother went to visit Gallagher cousins in NYC. There she met Mary, a second cousin on the Quinn side. Mary was born in 1908 in NYC, so she knew her way around even though she was only in her teens. Mary was a fun person, and she and Mother hit it off right away. So much so that Mother moved to NYC. There she entered domestic service with well-to-do families. She often spoke of these families, particularly the Tracys, and enjoyed being in service with her own room and in the heart of the great city. It was quite a wonderful change for a farm girl. Eventually she became a cook and traveled with her families to their summer homes in the Hamptons and Nantucket.

We Become a Family

My parents met somewhere in the great city, but I know not where or how. Perhaps it was at one of the dances held by the Irish societies, most likely that of the County Tyrone society. In any case, they were married on April 28, 1931, in St. Francis de Sales Church on East Ninety-sixth Street, Manhattan, which is just off Park Avenue. Mother was twenty-eight, and my father was thirty-seven. The best man was Frank O'Donnell, younger brother of the groom, and the bridesmaid was Mary Quinn, younger sister of the bride. Officiating was Fr. A. V. Storm. Frank and Mary eventually became my godparents. There is a picture that shows my parents in those days, probably about to leave by car for their honeymoon at Niagara Falls.

I was born on Wednesday, April 13, 1932, on the isle of Manhattan. To be more exact, I was born to Margaret Quinn and James O'Donnell at Woman's Hospital, located at 141 W 109th Street in Manhattan. Woman's Hospital is now part of St. Luke's Hospital, whose address is given as 114th Street and Amsterdam Avenue. This is an area that might have been fairly nice in those days as it lies just west of Central Park and southwest of Morningside Park, not far from Columbia University. Riverside Park, along the Hudson River, was only a few blocks to the west. Now, of course, it abuts or is part of Harlem, which has encroached from the north. In 1979, St. Luke's merged with Roosevelt Hospital, becoming St. Luke's-Roosevelt Hospital Center. Roosevelt Hospital itself is located on West Fifty-eighth Street.

According to my original birth certificate, no. 110JJ, my parents were living at 294 Cypress Avenue in the Bronx. This, my first home, was between East 139th and East 140th Streets and just south of St. Mary's Park. I was christened in the parish Church of St. Luke, 623 East 138th Street, on May 15, 1932. The church and the school are still there, and the present pastor is Msgr. Gerald Ryan. Of course, this formerly Irish-Jewish area has long ago been given over to Hispanic people.

We lived on Cypress Avenue through September 1934. This part of the Bronx was known as Mott Haven and was among the first settled in the post-Civil War period as the city moved upward into what was then Westchester County. In 1932, the housing stock was old, originally 2½-story brownstone townhouses from the late 1800s that gave way to five-story walk-ups known as "old law" tenements. On the main streets, the tenement ground floors were usually occupied by small shops. They were typically centrally steam-heated by coal-fired furnaces. It was fun to watch the coal trucks unload into the basement via openings in the sidewalk. I remember that the coal trucks had chain drives. Around the fringes of Mott Haven were industrial sites, rail yards, bridge approaches, and gasworks. Young as I was, I have some vague memories of life on Cypress Avenue.

At that time, my father was a route man for the Ward Baking Company. He had an electric van with a noisy chain drive and made bread deliveries around the area. He would come by and give me a ride on the van, which probably had a steering tiller instead of a wheel. It was quite a thrill! I've tried to find out who made the vans for Ward, without success. Mack, however, was a leading supplier of electric chain drive vehicles in the 1920-30s.

As near neighbors, we had Rose and Frank McElroy. My father knew Frank from his time in Carrickmore, NIR. We became very close to that family and later often went to the shore (Rockaway Beach) with them. They had a daughter, Rose, who was near my age, and later, Maureen who would be near my sister Patricia's age. On Cypress Avenue, they had a dog that I can remember seeing from my high chair. He was known to me as Bow-wow and fascinated me no end. Years later, I got my own dog, but more about that later.

To Ireland and Then on the Move

In the summer of 1933, my parents took me to Ireland, probably to wow their parents. I would have been the first O'Donnell grandchild. I remember nothing of the trip, but many pictures show me with a dog, whose I don't know. I never thought to ask what boat we sailed on or to what port. I'd love to know. Apparently it was a happy trip, since it had been twelve years since my mother had seen her parents.

Upon our return to the Bronx, things went bad when my father lost his job. Before leaving for Ireland, my father had been assured that his job would be there for him upon his return—not so. My sister Patricia believes that his job actually went to one of his friends, Frank McElroy, whom he knew in Carrickmore, NIR. Ward Baking Company had been established in New York City in 1849, and Tip Top bread was its featured product. Ward had a large factory and a day-old store not far from where we lived. From 1929, the company started to go bad as the founding Ward family left the business. Eventually Ward was taken over by Continental Baking, makers of Wonder Bread.

With no job and few prospects in the Depression of 1934, we started to make a series of relocations that took us from the Bronx to Long Island City, over to New Jersey, back to the Bronx, then to Washington Heights and, finally, to Riverdale in 1943. (For the complete listing of residences, see appendix D.) The first stop on this long trail was to 675 East 135th Street, where we lived from September 1934 to August 1935. I suspect that this short move was made to take advantage of a free month's rent common in those hard times. I have no recollection of life on 135th Street, although the aforementioned memories of the McElroys and their Bow-wow may have occurred here, rather than there.

At Christmas 1934, we made a visit to my godfather Uncle Frank O'Donnell and his wife, Nell Daly, of County Cork. I believe they were living on Columbus Avenue in Manhattan. It is my earliest memory of them. Frank was working in the shipping department of R. H., Macy's flagship store on West Thirty-fourth Street. Nell was to become a hairdresser in one of the city's upscale salons on Fifth Avenue. They had a surprise gift for me, a fire truck with siren and lights. Nell was not too happy as I chased it around the room, bumping into furniture and the Christmas tree. Thus, my first impression of Aunt Nell was that she was a fussbudget. This proved to be true, but was later mitigated by her wonderful stories told in her very precise speaking voice.

I recall, at the age of two or three, going with my parents to various social events, typically weddings, wakes, and funerals. I have a strong recollection of taking exciting ferry trips, probably to New Jersey or to Staten Island. I also remember sipping drinks at a party and getting a bit tipsy, much to the amusement of the indulgent adults. So this is how it starts?

A major family event did occur while on 135th Street, namely the birth of my sister Patricia Margaret on March 17, 1935, also at Woman's Hospital in Manhattan. I don't seem to remember anything about her arrival in our home or how we were cared for when Mother was away.

In August 1935, we moved to 34-43 Thirty-fourth Street in Astoria, Long Island City, where we stayed until May of 1936. There we had hoped to make a living by taking over and running a defunct coffee shop. Bad idea as we could not undefunct it. The few memories of Astoria include visits to the shop's mysterious basement where unused ice cream shop chairs and tables were stacked. I also recall looking out the back window to watch rain fall on the small garden. But mostly I remember moviemaking as Long Island City was where many films were shot pre-Hollywood. I got into some street scenes while I was supposed to be watching my baby sister who would gleefully speed away from me on her tricycle. Mother was mighty unhappy with my babysitting. I also remember that some older kids held me by the heels so I could reach down into a catch basin to recover their ball. Scary, but that's life in the big city.

New Jersey and School

In May of 1936, we moved to 195 Valley Road in Clifton, New Jersey. This was close to where my uncle Mike Quinn operated a general store, also on Valley Road. Uncle Mike (1898-1972) was my mother's brother and the oldest of her siblings. He married Kathleen Donaghy (1900-1961) in Paterson, New Jersey, in 1928. Margaret Quinn, my mother, was maid of honor. Aunt Kathleen, whose maiden name is sometimes spelled as Donohue, was from the Cookstown area of County Tyrone, not far from Pomeroy. They went on to have three surviving children: Patricia (1930), Barry (1933), and Eileen (1936). Our family has remained close to the Quinn family up to the present day.

When we moved to Clifton, it was a modest city with an industrial belt along U.S. 46. It was to grow mightily during WWII with house workers in the many nearby defense plants. Valley Road was the main connector between the cities of Paterson, the Passaic County seat, and Montclair in Essex County. Valley Road is so named because it lies along the base of a ridge that extends from Paterson to the southwest. The part of the ridge nearest Paterson and opposite where we lived is called Garrett Mountain. This remote part of Clifton was officially known by the weird name of Albion Place but was called "bean town" by the locals. Uncle Mike was to become very active in Clifton's civic affairs. Among other posts, he served as health commissioner and had a shiny gold badge to prove it.

The motivation for this move was a job. My father was to be a route man or distributor for Mi-Oun Cakes and Pies, an Astoria, Long Island, firm. We have a picture of him with his truck, but I'm not sure where it was taken. At 195, we lived in a multifamily block of apartments, but I don't remember much from that time. I do remember that we played in a nearby sandbox. One day another kid angered me so that I smacked him on the head with my digging tool, which happened to be a heavy long-handled sieve. He started to bleed and scream, so I ran home and hid under the bed while I waited for the police. I'm not sure, but I do believe that a policeman came and sternly reprimanded me and probably my mother as well. I was four and a half years old and supposedly should have been able to control my anger. My sister Pat was one and a half.

Sometime after this, my first criminal act, in the summer or early fall of 1936, we moved a few blocks north to 127 Valley Road. This was a two-family house right on the road, facing the school, Clifton's PS5. There was a school crossing right outside, so from our front window, I could watch the children on their way to school. We were also closer to Uncle Mike, Aunt Kathleen, and family who lived over his store. I started kindergarten in the fall term of 1937, at five, and remember singing a song in the 1937 Christmas party to which parents were invited. I know that I enjoyed kindergarten immensely, particularly the kindergarten blocks. I must have behaved myself because I was advanced to grade 1B for the winter-spring term of 1938. My teacher was Constance Spangenberg. I was promoted to grade 1A for the 1938 fall term, where my teacher was Hetah Dunn. Other than their names on my report cards, I remember nothing about those teachers. (For a full listing of my schools, see appendix E.)

Life was difficult during the winter of 1937-38. Our flat had no central heating. The only heat came from a big coal stove in the kitchen, so that's where we stayed in cold weather, which was often in that horrible mid-Depression winter. We took baths once a week using a gas-fired water heater to heat the water and the bathroom. When the gas heater was off, which was most of the time, you did not tarry in the bathroom. I can remember Mother coming in to talk while I was bathing, probably because it was warm there at bath time. She was often very down, almost in tears, and lonely. She missed New York and her many friends there. She hated that my dad was always late getting home, usually well past dinner, so that they had very little time together.

For me, there were many things of interest along Valley Road. A major attraction was the arrival of our upstairs neighbors who probably owned the building. Apparently they had something like a traveling rodeo and showed up in late fall with horses and wagons that they stored for the winter in a barn behind our building. They did other interesting stuff that entertained my cousin Barry and me until the cold winter set in. Barry, a year younger than I, was my constant companion in those days. Another attraction was a Tin Lizzie owned by a neighbor. We used to admire his impeccably kept Model T Ford that sat proudly in his side yard.

Come summer, there were outdoor movies at the school, which were also an occasion for mischief. We used to gather and dry cattails, which we called punk. These could be lit and "smoked" during the films to, presumably, keep mosquitoes away. They could also be the cause of "accidental" burns to your neighbors. Best of all was story hour at Quinns'. Their neighbors, the Avatos, ran a shoe store. Cosmo Avato was a friend of my older, by two years, cousin Patricia Quinn. Cosmo would, after dark, read ghost stories to Patricia, Barry, and me. They all lived there, but I had a scary walk home. Cosmo eventually became a doctor on the West Coast.

Another significant event in my life occurred in the summer of 1938. Uncle Mike took Barry and me to swimming lessons at Barbour Pond on top of Garrett Mountain. I became a reasonably competent swimmer for my age, but Barry never became comfortable in the water. In September of 1938, at age forty-two, my father received his U.S. citizenship certificate at Paterson, the Passaic County seat. This was an important family event. Strangely, it was granted by the Department of Labor, not Justice as was my mother's some years later. Dad's home address was stated as 115 Valley Road, perhaps Uncle Mike's address. His height was recorded as 5 feet 9 inches, and his weight 159 pounds. There's a mysterious handwritten note on the certificate: "Entry at New Ross." I found a New Ross in Canada, near Halifax, but it's neither a seaport nor a border crossing point. The only other New Ross I turned up is in Ireland. So what does it mean?

I guess the cake business did not work out because in November of 1938, we were on the move again. But I must tell you of a small tragedy that accompanied that move. Either for Christmas or for my birthday, I was given a puppy. I believe it was a Boston terrier that we called Buster. So as we went off in the cake truck, Buster was left behind. I tearfully left Clifton wondering what was to become of him. I never did find out.

Back to the Bronx

November of 1938 finds us back in Mott Haven, at 683 East 138th Street, Bronx. I resumed my first grade, second half or 1B as it was known, at PS65. The school was located a few blocks north on 141st Street. I remember going there with my mother when she signed me up, well after the term had started. My first day of school was marvelous. There was such a rich school's smell of books, paper, pencils, and chalk that I remember to this day. I was very happy at PS65. However, I didn't enjoy PS65 for long as I was soon transferred to St. Luke's, our parish school. I spent the next three and a half years at St. Luke's where I made my first Communion and my Confirmation. I remember that the Dominican nuns at St. Luke's were tough and had to be. Both schools today serve a predominantly Latino population. PS65 is now called Mother Hale School, and appropriately enough, the principal is Ms. Fern Cruz.

One of the great events of my early school years was a first trip to the library. Our class walked over to the public library on Alexander Avenue, a fairly long walk. Along the way, we passed some of the early Mott Haven row houses. We probably also passed the infamous building where the janitor had killed a little girl and put her body in the furnace. Anyway, the library was a revelation to me. It was a beautiful building, which still stands. Best of all was the card catalog, but I'm not sure that we were allowed to use it. My love affair with libraries started there.

St. Luke's Church was about a block west on East 138th Street, but the school fronted on 139th Street between Cypress and St. Anne's Avenues. I only have one memory of life at St. Luke's school, and it is an embarrassing one. The school gave a medal for the outstanding scholar of the marking period. You were to proudly wear the "gold" medal on your school uniform for the period. While in the second grade in 1939, I won the medal and promptly lost it in the schoolyard. Because the nuns had stressed how important it was to safeguard the medal, I was in tears. To my rescue came a cute blonde classmate named Ann Sheridan. She told me not to cry, that we'd find the medal, and we did, thanks to Ann. I've always remembered her name because of the bubbly and popular screen actress of the same name. Sadly, Ann (the actress), a Texan, died in 1967 at age fifty-two. She had appeared in some ninety-three films and TV series. Happily, St. Luke's is still a viable parish of the Archdiocese of New York, and Msgr. Gerald Ryan is now the pastor.

There was another local girl to whom I was strongly attracted at about the same time. Her name was Sandra, and she was the daughter of a furrier who had a shop around the corner on Cypress Avenue. I had probably met her while at public school. Anyway, I would look in the shopwindow to see her or go around the back of the shop to look in their windows. They kept Sandra well out of sight. I suppose they didn't want her associating with one of the Irish ruffians of the neighborhood.

Our street was one of the most important in Mott Haven as it ran from the Harlem River to the East River. It was very wide and had streetcar tracks down the middle. The neighborhood was mostly Irish and Jewish, and something was happening all the time, and often to me. Our tenement building had a short "stoop" to the main floor's landing. Alongside the stoop were small shops on each side, usually owned by

Jews. Similar buildings and shops lined our north side of the street from Cypress Avenue on the west, to Jackson Avenue on the east. For some reason, many of the Jewish shopkeepers adopted me, and I ran errands for them. They were always kind and generous to me. I got my first hamburger when I went to the White Tower on Bruckner Boulevard for one of the shopkeepers. He gave me money for two burgers and gave me half of one as my reward. It was so delicious I've never forgotten it.

There were no supermarkets in those days. The A&P store across 138th Street was no bigger than Sammy's little shop in our building. Around the corner on Cypress Avenue, a Jewish shopkeeper acquired a few storefronts and was consolidating them to form a prototypical supermarket. Somehow, I got involved helping him to clean out the accumulation of decades. Once, I lifted a pile of empty burlap bags, and a mouse ran out. The shopkeeper saw it and yelled, "Jimmy, look out, it's a mice!" I later found a half dollar under another pile. My conscience made me ask if I could have it and was told yes. But when I left to go home to dinner, I was jumped by bigger kids and relieved of the fruits of my labor. How did they know? Such was life in the South Bronx.

My good relations with the Jewish shopkeepers were sorely tested by the actions of the local rowdies. Their favorite trick was to push open the shop's door and throw in a "stink bomb," anything that would burn and create a stink. Old rolls of film seemed to be very effective. Much as I tried to divert my rowdy friends from my friendly Jewish shopkeepers, I became an accessory before and after the fact.

I got my punishment one day, but it was undeserved. As I entered our building's front hall, I was grabbed and beaten by the police. Apparently some rowdy had thrown a rock through the big window of a shop on our corner. The police were called and told that a blond kid had done it. Checking around, they found that an appropriate blond kid, me, lived at 683, so they waited for me. My screams attracted the attention of some neighbors who convinced the police that I was a good kid and would never have done such a thing. The police left, angry and unapologetic. My suspicion of police motives and behavior may go back to that day in 1939. But then, again, perhaps not. I do remember that I often joined the local rowdies in hurling insults through the basement windows of the local police station. It was located nearby on Alexander Avenue and home of the Fortieth Precinct, an outpost of the infamous "Fort Apache."

One of the bad things about the Depression era's city life was the scorn for people on "relief." I don't think we were on relief, but we probably should have been. I remember a family named Flanagan, who lived in our building, who was mocked for being on relief. Mr. Flanagan just lay about their apartment drunk most of the time. One day in the street, Mrs. Flanagan smacked me for bad-mouthing her and her family. I don't remember if I really was one of the culprits, but I was certainly handy. And I punched her back.

Whitey the Cop

Our street in those days was busy with many horse-drawn commercial vehicles. I remember the iceman, the vegetable man, and the merry-go-round man, all with horse and wagon. The poor iceman had to cut a chunk of ice, carry it up five flights of stairs, and dole it out as he went. I don't think any of the twenty-plus apartments in our building had refrigerators. We lived on the fourth floor and had a less than state-of-the-art icebox. In the winter, we used a box cut into an outer wall as our refrigerator. If it got too cold overnight, the milk would explode out of the bottle. There were only heavy glass bottles then, and they had to be returned for credit. People would often yell down from the window to a peddler what it was they wanted, and he would bring it up to them. While deliveries were made, the horses were left alone in the street.

This brings me to the story of Whitey, the local cop on the beat. Whitey spent much of the day in a saloon next to our building. He was generally a figure of ridicule in the neighborhood. One day, a horse bolted west on 138th Street, dragging his vegetable wagon behind. Having been alerted by the commotion, Whitey bolted out of the saloon and down the street after the horse. He was able to catch the horse and bring it to a halt, much to the amazement of all the onlookers. Whitey's stock rose.

Sometime later, I was busted by this very Whitey. It happened this way. It was a dull day on the street, and some of the rowdies decided it would be fun to throw things at the neon sign over our local saloon. The idea, of course, was to break the sign and run. Stupid me, I was caught in the act when Whitey appeared in the doorway. He grabbed me and, knowing I was a local, asked me to take him to my parents. I was very much afraid that my poor father, who was home for some reason, would be arrested and taken to jail. Even worse, I knew we were poor, and I was very much afraid they would try to make my father pay for the sign. When we got to our apartment, my father and Whitey had a few drinks and took turns severely chastising me. Thankfully, that seemed to end the matter. But my rogue friends and I were not averse to stealing pennies from unattended newspaper stands in a time when a penny would buy a cigarette and a match.

Oddly enough, the local saloon was to further figure in my young life. For it was there in 1941 that I saw my first television program. It happened to be the Louis-Conn fight from the Polo Grounds won by Joe Louis, "the Brown Bomber." The set was a large console that projected the image upward to be viewed on a large mirror. Television broadcasting and development was suspended for the war years, but made a big comeback afterward, initially on tiny eight- to ten-inch screens.

A Happening Place

There was a lot to do and explore in and around 138th Street. We had windows that faced directly on the street. One day, we heard a commotion and rushed to the front window to see police chasing a car toward the East River and firing shots. Our windows were protected by grilles so that kids could not fall out. They also prevented us from leaning out far enough to follow the action. However, later that day, Mother took us for a walk in that direction. It was local wisdom that whooping cough could be helped by a trip to the gasworks that lay on the river at the east end of 138th Street, an area known as Hunt's Point. Apparently my sister was ill and needed exposure to the gas that leaked from the gasworks. When we got there, we saw the very car, with bullet holes, up against the bulkhead at the end of the street. The criminals had chosen the wrong street on which to escape. The car was big and black, very much like those you'd see in the gangster movies of the day.

From the gasworks, you could look across the East River to an airport then called North Field, now LaGuardia (LGA). The field was set up for planes to land on the water and taxi up a ramp to the terminal. It was the age of the Pan American Flying Clippers, which were able to fly all over the world because they could land on water, long before major land-based airports came into being. If you were there at the right time, you could see the European Clipper splash down and climb up the ramp. It was a beautiful sight. Most of the clippers were lost overseas in the early days of the war, particularly the exotic China Clipper lost at Wake Island. After the war, land-based aircraft started to serve the overseas market, using bases developed for the ferry service whereby we shipped warplanes to Europe and elsewhere.

Another fascinating place was the Oak Point railroad yard of the New York Connecting Railroad (NYCRR) near the East River. This railroad was established in the 1920s to allow freight to be exchanged among the railroads serving the city. Chief among these were the New York Central (NYC); the New York, New Haven, and Hartford Railroad, known simply as the New Haven; and the Pennsylvania Railroad (PRR). Freight cars from the PRR crossed the Hudson through a tunnel onto Manhattan and then went on to their Sunnyside yards in Queens through an East River tunnel. Once in Queens, cars bound for New England had to be barged across to the Bronx through a treacherous stretch of water known as Hell Gate. This was neither safe nor satisfactory, so the NYCRR built a magnificent high tied arch bridge over Hell Gate and a long trestle across Randall's and Ward's Islands on into the Bronx. Thus, freight cars could be moved directly by rail from the PRR to the NYC and New Haven tracks in the Bronx. Although originally built for freight, Amtrak now uses the Hell Gate Bridge for its high-speed New York-Boston passenger service.

Naturally, we inquisitive kids had to explore the nearby rail yard. We would marvel at all the different railroad names on the cars and climb on them until chased by the "bulls," as we called the railroad police. Surely it was a dangerous place to play, and there were stories of kids electrocuted by the overhead wires. Another dangerous place where we played was on the roofs of the low industrial buildings below 138th Street. Once on the roof, we usually found places to jump off, often into piles of sawdust blown out of woodworking shops. We sometimes found treasure troves of paper pads discarded by businesses whose addresses or names had changed. Paper was scarce and expensive, so we carried our loot home and used the back sides for scribbles and homework. People in the businesses below had access to the roof via trapdoors or skylights. Many times we'd be chased when an irate foreman suddenly popped up seemingly out of nowhere.

As kids, we often went on outings to Randall's Island, which lay at the junction of the East and Harlem Rivers, near Hell Gate. Access to the island was by way of a walkway over the Bronx Kill on the new Triborough Bridge, which connected Queens and the Bronx to each other and to Manhattan. Today it seems a terribly long and difficult walk with lots of ramps and steps, but it didn't seem to bother us then. Once on the island, we'd spread a blanket and picnic, play ball, and play cards, whatever. But our favorite thing to do was to watch the trains cross the nearby Hell Gate Bridge. My sister Pat and I would count the cars and argue as to who had the correct count. Often, friends would join us on the outings and count cars, but they were never as competitive as Patricia and I. Later a very nice stadium (Downing Stadium) was built on the island, and that's where my high school played its home football games. More on that later.

Another memory of life in Mott Haven was snow, and lots of it. Because 138th Street was an important street with a streetcar line, snow was quickly removed and pushed to the sidewalk, and there piled up in mountains. Walkways were cut through by shopkeepers for their customers. One day while playing on top of such a pile, we dug down and found a car buried underneath. There were lots of good snowball fights from one mountain to another.

The streetcars were a danger to kids who loved to hop on back and ride them down the street. One kid fell off onto the adjacent track as a car came the other way. I don't think he lived to tell of it. A favorite stunt was to pull the pickup, called a trolley, off the wire so that the car couldn't move. The unhappy driver had to get off, put it back, and get out of there before they unhooked him again. In summer, the sides of the cars were removed and replaced with open grilles. This gave the local ruffians an opportunity to pelt the passengers with pebbles, trash, and whatever came to hand. Such fun!

Summer evenings were spent on the sidewalk until dark. Parents often came down to chat on the stoop. Popular sidewalk games were potsy, similar to hopscotch, and skully, which was played with soda bottle caps on a chalked layout of squares. The object of skully was to slide your cap through the sequence of squares and get to the home square first. Since empty caps were hard to control, we usually pressed an orange or lemon peel onto them for added weight.(For the full story on this very popular New York street game, see appendix F). One day while at play, I happened to be sitting at the curb when a car ran over my feet. I don't remember that I felt much of anything.

Street singers were a frequent source of free local entertainment. These men would not actually sing in the street, but in the enclosed area behind the buildings. It was kind of like singing in the shower; their voices sounded very powerful and full. Those who enjoyed the concert would throw down coins. Wise guys would throw down trash, old shoes, whatever. A street singer figures in the movie *Winterset*, from the Maxwell Anderson play. I've always liked that little-known film and am always reminded of our Bronx singers. A bizarre form of entertainment was occasioned by the arrival of the police and fire department to rescue people who got stuck in a bathtub or on the toilet. We would try to find out which apartment so that we could peek in until chased. I never did see anything. We lived in a "front" apartment, so our windows faced the street. In good weather, we, and others so situated, spent a lot of time hanging out the windows or on the fire escape where we could watch what was happening on the street, and usually something was.

Of course, with young children in the apartment, we had grilles fastened to the windows such that we could lean out without falling. Usually, we also had a window that faced the backyard and gave access to a clothesline strung from the window to a clothes pole that served all units. In addition to hanging out on the stoop or in the street, we also made use of the roof of the building. It was accessible via the stairs that served all the apartments. We kids did not often play up, there but it was a popular place to take family pictures. It also served as a place to hang wash to dry on the many wash lines that were strung between T-bars fastened to the roof. The roof was dangerous as there wasn't much to prevent you from plunging to your death. I don't remember how I felt then, but I now have a great fear of heights.

The Winds of War and a New Baby Sister

As 1939 ended, even we kids became aware that there was a war going on somewhere in the world. We didn't know exactly where or why or what it all meant. As we came into 1940, the nation began to prepare for war by beefing up our puny military. Men were drafted from among the Irish immigrants and went off to training in distant places, such as Louisiana. Business started to pick up, and my father was by then driving a taxi, probably for one of the Toals, Peter or Terence, whom he knew from his years in Carrickmore.

To boost business in Depression NYC, a World's Fair was opened in 1939 and was such a success that it was extended into 1940. It was called the World of Tomorrow. As a sign of the times, a bomb was exploded at the opening of the fair's British pavilion. Various known Irish Republican Army (IRA) men were taken in for questioning, some of them friends of the family. But the Irish-dominated police force never found the culprits.

I attended the fair several times, most notably when I went with my father on the closing day in the fall of 1940. The fair was a magical eye-opener to a street kid from the Bronx. I remember the wonderful Firestone Tire rain forest, the Beech-Nut and the Wonder Bread mini factories with free samples, and the astounding General Motors' vision of the future. The free Hostess cupcakes were baked fresh as you watched and tasted great. On the last day, people were stripping the buildings of ornaments for souvenirs. The hot dog stands were out of buns and were using hamburger buns, telling people to bite off the ends of the dogs.

But the major event of 1940 was the presidential campaign, Franklin D. Roosevelt versus Wendell Willkie. Roosevelt was criticized for moving toward American involvement in the war by aiding the British, specifically his so-called "lend-lease" arrangements wherein we swapped old destroyers to Britain for bases in the Caribbean and elsewhere. The Irish, generally disposed to the Democratic Party, were not fond of Roosevelt for two reasons: his support of Britain and his snubbing of Al Smith as a running mate in 1932. To the Irish, Al Smith, a Catholic and former New York governor, was their man. When Roosevelt adopted Smith's state programs for his "New Deal," but didn't offer Smith a place in his cabinet, the Irish were mighty unhappy. But where could they go? They realized Roosevelt would win and didn't want to waste their votes on Willkie, so they reluctantly voted the Democratic ticket. As the Roosevelt election cavalcade came down East 138th Street, I was pushed to the curb, handed a small American flag, and told to wave as the man went by. I did my bit for his reelection.

Coming into 1941, our family was augmented with the birth on January 9 of my sister Maureen Frances, also at Woman's Hospital. This time, I do remember that we had a woman to look after us while my mother was away. I don't remember anything about the woman, nor do I remember what my father was doing for us. I do remember that Maureen slept in a lower bureau drawer for several weeks or months until a crib was found for her. I remember pushing her pram and "minding" her when my mother went into various shops. I remember that the Chinese laundryman would come out and offer us lichee nuts, but we wouldn't take them. We helped Maureen learn some of the words to a popular song of the day, "Mairzy Doats." She would do her best to sing the tricky song for us.

After Maureen's arrival, a common outing was a family trip to nearby St. Mary's Park. We all wanted to push Maureen's baby carriage, but Mother said it was good for her back, so we rarely got to push. Many family pictures were taken in that park. Molly and Frank Donnelly were close family friends from Tyrone. Molly's brother was a professional photographer, and he took some of the better ones. I can remember buying "comic books" on the way to read in the park. They were my favorite early reading until I was able to go to the library by myself.

In 1940, my sister Patricia started kindergarten at PS65. She later transferred to St. Luke's for first grade. She also remembers from PS65 a Cypress Avenue furrier's daughter, probably the sister of the one I pursued. Now I wonder how a furrier made a living in those hard times. Poor Pat was to suffer many indignities in her young life. It started with me accidentally breaking her nose with a broom handle back in Clifton. On 138th Street, she was bitten on the finger by a bulldog when she and the dog went after her ball as it rolled into the street. Mother took her to nearby Lincoln Hospital for treatment. The woman whose dog it was bought Pat an ice cream cone. Fortunately, the dog was found to be free of rabies.

Summertime at Rockaway Beach

Rockaway Beach—a.k.a. Rockaway Park—had been developed by Robert Moses during the LaGuardia years of the 1930s, probably to relieve crowding at the older city beaches such as Coney Island and Riis Park. It had a beautiful boardwalk, with an art deco comfort station and gift shop every few blocks. These shops had wonderful stuff for the home and an excellent selection of toys and games. While the merchandise was high class, it was priced beyond our reach. Instead, we patronized the many shops a block or two off the boardwalk. Shore Front Parkway ran alongside the boardwalk and had a median beautifully planted with flowers. At Beach Ninety-sixth Street, there was an amusement park called Playland, with a roller coaster and many other rides. I loved to go there for the fun house and to play skeeball. Perhaps best of all was a nearby street lined with Irish bars and dancehalls. Whole families spent the evening there. Many places were open to the street to catch a passing breeze, so kids could run in and out while parents talked and danced. Is it any wonder that Rockaway Beach was called the "Irish Riviera"?

In the summertime, Mother loved to go to the beach, and the beach was Rockaway. I was never that fond of the beach myself. At first we went by subway to Penn Station where we transferred to the Long Island Railroad (LIRR) for the long trip to Rockaway Beach. I loved to stand in the first car as the LIRR train traversed the long trestle over Jamaica Bay. The cars would be packed as whole families relocated for a week or two at the beach. Later we went by car, either an idle taxicab or, after the war, the "green monster." The latter was a converted 1939 DeSoto cab that had been painted a shabby green. We used it for several years and sold it to a man who used it to ferry servicemen from Washington to Texas and California. By the time it was retired, it must have had a quarter of a million miles on it.

If we went to the beach for a week, we'd stay at a rooming house. We'd all bunk in one room and share kitchen and bath with several other families. These places were famous for their porches with dark green rocking chairs and sand, lots of sand everywhere. It was so hot in the rooms that evenings were usually spent on the boardwalk. There our parents would meet many friends from the Old Country, and we'd each get an ice cream cone and, rarely, something from the wonderful shops on the boardwalk. I recall one time a friend bought ice cream cones for my sister Pat and me. When we came to the steps down from the boardwalk, Pat tripped and fell. She and the cone went splat on the sidewalk. The man said, "Look what happened to the ice cream!" Poor Pat! Sadly, it's all gone now. As the Irish became more affluent, they abandoned Rockaway for more upscale places. The old rooming houses where we stayed have been torn down and replaced with new apartment blocks. Playland, itself, is a fading memory.

Apparently it was an Irish tradition that one went to the beach for "the waters" on the Feast of the Assumption, August 15. So in the summer of 1941, we made a day trip to the beach. There we met and sat with our friends the McCreeshes. It was an overcast day, and the surf was very rough. I was playing in the surf when next I know I was way out to sea. I was rescued by the lifeguards and carried onto the beach. There I was resuscitated, and I remember hearing the fearful screams of my mother. I remember that she rode with me in the ambulance to the hospital. Next, I remember coming to full consciousness in the hospital, with nurses asking why my pulse was racing. Otherwise, I don't remember the hospital stay. I do remember my father coming to take me home, and he may have carried me out to a borrowed car. He had bought me a kite, which we later took to St. Mary's Park for a test flight. So in my ninth year, I almost became a statistic.

A final word on Robert Moses. When I was a kid, he was chairman of New York City's parks and recreation commission. First he built parks like Rockaway Park, and then he built roads to get to the parks. He became chairman of New York State's parks and recreation commission. He then built the beautiful Jones Beach State Park out on Long Island and the Southern State Parkway to get to it. We did go to Jones Beach once in a while, but it was a long trip, and the surf seemed very rough. While it was beautiful, it was isolated; there was nothing but the beach itself and some picnic grounds. Moses then became chairman of something new, the city's Triborough Bridge and Tunnel Authority. He then built the bridge of that name to connect three boroughs—Bronx, Manhattan, and Queens. We used that bridge quite often, mostly on foot. Lastly, he became chairman of New York State's power commission. In that capacity, he evicted Seneca Indians from a part of their reservation to build Robert Moses's hydropower plant near Niagara Falls. No wonder he was known as the "power broker."

I Become a Real Yankee Fan and War Comes

I was just a kid standing on my stoop (from the Dutch *stoup*) one day when I was accosted by a big kid. He wanted to know what baseball team I liked. I didn't know any teams and wisely asked what my choices were. He told me there were the Giants, the Yankees, and the Dodgers. I knew from school that Americans were often called Yanks, so I said Yankees. He said I was OK and left. I probably avoided a beating had I chosen one of the other teams. I thought no more about this encounter and still knew nothing about baseball until I was over nine years old.

One of my last prewar memories is of the 1941 Dodger-Yankee baseball World Series. It grew out of my friendship of a shut-in neighbor Joe Greene. Joe was about my age but had an incurable disease, which would kill him not long after we moved away. Joe's father was a sandhog who worked in tunnels of which there are dozens in and around New York City. His was a really well-paid job in a time when many men were out of work. Consequently, my friend Joe had every toy imaginable. I loved to visit him and play with them, and I remember his toys even now. Joe's mother always treated me well and appreciated my coming to visit her son. Joe was wise beyond his years and was very knowledgeable about baseball, probably as a result of his father's interest. Joe and I listened to many of the 1941 series games together, and he explained what was happening to an ignorant Jimmy. My love affair with baseball grew out of my friendship with a dying boy.

Because he was shut-in, Joe listened to the radio quite a bit. We all did after school or after outdoor play, but Joe listened to ball games during the day and became quite knowledgeable for his age. One day when I was there, he had on one of the 1941 World Series games, and we became excited together as the Yanks blew the Dodgers away. I became a true Yankees fan and, in 1942, started to sneak off to Yankee Stadium where kids were let into the bleachers in the fourth or fifth inning. That was the last year of the great teams of the late 1930s, managed by Joe McCarthy and starring Joe DiMaggio. As center fielder, Joe had anchored the great Yankee outfield with Charlie "King Kong" Keller in left and Tommy "Old Reliable" Henrichs in right. After serving in the military, the threesome played together again in the late 1940s but were never quite the force they had been before the war. In those days, for reasons unknown to me, outfielders left their "mitts" on the grass behind the infield when they went to bat, a real no-no today.

A big event in the waning days of 1941 was the reelection campaign of Mayor Fiorello H. LaGuardia, of Italian-Jewish parentage. There was a campaign office in our block, and we kids were often allowed to hang out there. Consequently, we proudly wore our LaGuardia badges and ran occasional errands. We were very much aware of LaGuardia, that he was a good mayor and should be reelected. He ran as a "fusion" candidate, in that he was backed by all major parties: Democrat, Liberal, and Republican. I guess it was no surprise when in November he was reelected for a third term, a first in the history of Greater New York, that is, since 1898. Some years later, during a newspaper strike, LaGuardia famously read the "funnies" (Dick Tracy et al) on the radio. It was a bravura performance, well remembered by all who heard it. He was funnier than the funnies. He was not to live much longer. He died in 1947 at age sixty-five and is fondly remembered as Greater New York's finest mayor.

One of the things we did in those days was to take motoring trips to visit the Quinns in Clifton. I'm not sure how we got the vehicles we used until later when we had taxicabs to use. At first, we'd take the Weehawken ferry across the Hudson, and my father would always get lost looking for the Paterson Plank Road. Later, of course, we were able to use the George Washington Bridge and NJ Route 46. On these trips, my sister Pat and I would count gas stations or out-of-state license plates, with mutual claims of cheating. At the Quinns, we had a long-running dispute over the relative merits of New York versus New Jersey.

In the late 1930s, Uncle Mike bought some land out west on Valley Road on the south slope of Garrett Mountain. In 1940, he built a house there with help from in-laws and others. I can remember them working there at night by the light from auto headlights. He also built a house next door that he sold to people named Saams, some of whom lived there until recently. At some point, he sold off the remaining building lots. Cousin Eileen Quinn and her husband, Joe Dodd, still live in that house, although they have made substantial improvements.

Anyway, we were visiting there on Sunday, December 7, 1941. Uncle Mike listened to the evening news, and we heard that Japan had attacked the United States at Pearl Harbor and that we would be at war. We had neither heard of Japan nor Pearl Harbor. Cousin Patricia got out an atlas, and we looked up Japan. It looked like such a small country compared to the United States. We didn't think the war would last very long. But before long, we were at war with Germany and Italy as well. I remember on the drive home to New York wondering what lay in store for us. It was quite a while before I could find Pearl Harbor on a map.

The Christmas of 1941 was the last without rationing. As we had been doing for prior Christmases, Mother took us shopping at the Hub. This was the major shopping area of the South Bronx, located where Westchester Avenue forked off from Third Avenue at East 149th Street. It was quite a walk, and we all were tired and cold by the time we got home. The major attraction at the Hub was Hearn's department store. It had an extensive toy department, and we would show Mother what we wanted for Christmas. However, we got very few of them to go along with our clothes and fruit. I vaguely remember that there was another store at the Hub that featured electric trains. But they were priced out of reach for us at that time. Little did I know that this was to be our last Christmas in Mott Haven.

The Bronx at War

In the aftermath of the December 7 attack, "Remember Pearl Harbor" became the rallying cry, and the Bronx struggled to get on a wartime footing. I guess newsreel footage of the London blitz made air-raid concerns paramount. So in the absence of an air-raid warning system, fire trucks were scattered around so that their sirens could be used to warn of air raids. Constant practicing and sounding of alarms caused much annoyance and little protection. Air-raid shelter signs started to appear on buildings so designated.

Kids were issued ID tags, which we were to wear so that in case of disaster, our bodies could be identified. I still have my tag, which is made of plastic and a bit over an inch in diameter. It bears my name, birth date, and this identifier: "15-12371 N.Y.C." Who knows what that signified in a time before Social Security numbers became ubiquitous? Some years later, as an experiment, I tried to burn the tag to see if it would have survived a fire. Oddly enough, it was self-extinguishing, pretty good under the circumstances.

As part of the blossoming civil defense agency, an air-raid warden's service was developed, with armbands for the various rankings. At nine, I became a lowly messenger, but my job was mainly to help collect stuff for the war effort. This stuff included newspapers, old rubber articles, tin cans, and the tin foil from cigarette packages. In those days, real tin foil was used. We'd separate it from the paper backing and roll the foil into balls for collection. Mostly, we'd sit around the store used by the wardens and play chess. They had an inexpensive wooden set, a classic, and that's where I learned to play. I have a similar chess set that was owned by my uncle John Sander who loved to play.

Next on the wartime program were air-raid drills. When the warning was sounded, we had to take shelter and turn off all lights. Helmeted air-raid wardens would go around and check that all was well and that, at night, no light was escaping. We had a big dining table with a table cloth that dropped to the floor on all sides. We'd crawl under it and read by flashlight until we heard the "all clear." A major concern in those days was the German submarine force lurking off the coast to intercept ships leaving the port of New York for Britain. City lights created perfect silhouettes for the Germans to see the ships at night. So a "brownout" was established, and the city lights were dimmed for the duration. Also, places along the coast were required to have blackout curtains.

A lesser concern was coastal defense, so old gun batteries, many dating back the Civil War, were refurbished and manned during the war years. As we went to the Long Island beaches, we could see the gun emplacements on Brooklyn Heights and elsewhere. They were probably in the same spots that George Washington used against the British in 1776. In addition, bunkers, pillboxes, and armed patrols were installed on beaches that were deemed particularly vulnerable to invaders and saboteurs. We had armed guards at our favorite spot, Rockaway Beach, which is at the far southeastern edge of the city. It all seems pretty ridiculous now, but some German saboteurs were caught coming ashore on Long Island from a submarine. At the outset, our World War II planning seemed to rely upon what worked in the last war or wars. This has been the case ever since.

It's axiomatic that war is good for business. WWII lifted the United States out of the Depression, which hit bottom in the bleak year of 1937. All business sectors benefited, particularly transportation. American railroads had been on precipitous decline since WWI, many nearly bankrupt. The need to move vast quantities of men and war material gave them new life. Old rolling stock was restored to service, and all was well for American railroading until the end of the war. Shipping was in great demand, and there were terrible losses to German submarines in the early war years. The valor of the merchant marine sailors has never been fully recognized, even today. Passenger liners were converted to troopships and made speedy crossings to the various war fronts. Commercial aviation was in its infancy and did not play a major part in the war. However, commercial pilots opened dangerous routes across the Atlantic for military aircraft.

Most important to our family was the boom in the New York taxicab business. Sometime prior to the war, my father started driving a cab, probably for one or both of the Toal brothers (see appendix G). One of the brothers lent him money to buy his first cab, probably a 1939 DeSoto seven-passenger vehicle. But the medallion, or license, to run a cab is what he really bought. As business prospered, Dad would drive the cab during the day, and another man would take the night shift. After the war, he bought his first new DeSoto and gradually built up a small fleet.

In all our years on 138th Street, we only became really close to one family, the O'Regans, Charles and Mary. They had a daughter, Moira, around my sister Pat's age. She became a nun and ultimately librarian at my high school. They had other children, but I only know of Charles who was much younger than I. He later attended my high school and college and also became an engineer. The O'Regans often joined us on our trips to Randall's Island. We stayed in touch with them even after moving away. Oddly, my mother always addressed her friend as Mrs. O'Regan, never Mary. A few years ago, I had an e-mail from young Charles with a family update; I'm sure I saved it, but I've been unable to find it.

Radio Days and Nights

On his way home from work, my father usually picked up a copy of the *New York Journal-American*, a Hearst paper and notorious right-wing rag. However, it did have some very good columnists, particularly in the sports section. My father enjoyed the political columnists who were almost all anti-Roosevelt and anti-Britain. I read the paper myself after Dad finished and wondered how our government could have gone so wrong. The paper did not survive the arrival of television.

However, entertainment and news primarily came to us by radio. The period from the mid-1930s to the 1940s was known as radio's golden age. We had a wonderful monster RCA radio console, which on its legs stood about four feet high. For the technical crowd, it was TRF (tuned radio frequency), a pre-superheterodyne design that used huge capstan-driven tuning capacitors. My dad used it to listen to his favorite evening news programs and boxing from Madison Square Garden. We kids used it to listen to our favorite predinner shows, such as *Jack Armstrong, the All-American Boy* presented by Wheaties, "Breakfast of Champions." Other favorites were *Captain Midnight* and *Sky King*. These shows were all fifteen minutes in duration, and many offered stuff you could mail in for using box tops. Decoder badges and rings were popular gadgets that were incorporated into the story line. One of my favorites required that you buy Ovaltine and send in the foil seal to get your decoder. We drank a lot of chocolate milk in those days, usually made with a product called Bosco. I wanted Ovaltine to put in my milk, but Mother said no, it was too expensive. She did buy something called wheat germ and insisted that we use the horrible stuff on our cereal because it was so good for us.

Later in the evening, there were many more popular shows for the older-age group. I particularly enjoyed Carlton Morse's *I Love a Mystery*, known to fans as ILAM. It featured the crime-fighting adventures of Jack Packard, Doc Long, and Reggie York. I remember the distinctive voice of future Oscar winner Mercedes McCambridge who usually played the damsel in distress. This show ran in various time slots and for varying lengths of time well into the 1950s. Today there are a number of Web sites for devotees of ILAM.

Radio's "golden age" featured many well-written and well-acted mystery-suspense shows. Most famous were *Suspense, Inner Sanctum* with its creaking door, *Escape, Lights Out* written by Arch Oboler, and *The Green Hornet*. I remember that *Lights Out* was sponsored by Ironized Yeast, whatever that was. As a kid, I was scared the first times I heard some of these shows, particularly when my parents sent me to bed, and in the dark, I would hear the Green Hornet buzz off with his faithful servant Kato. Blue Coal sponsored the very popular Sunday-afternoon radio program *The Shadow*, which featured Orson Wells as the hero Lamont Cranston.

The program opened with the famous line "Who knows what evil lurks in the hearts of men? The Shadow knows!" Classic!

There were excellent dramatic shows, and we learned to appreciate them as we grew older. I'm thinking of the *Lux Radio Theater* and *The First Nighter* and, the heavyweight champion of all time, Orson Welles's *The Mercury Theater*. They did many great shows up into the 1950s, but most memorable was the 1938 broadcast of H. G. Wells's *The War of the Worlds*. In that program, New Jersey was attacked by Martians, and it was so realistic that it caused a panic. We lived in Clifton then, and I can remember the stir that it caused. As we grew older, we started to enjoy Dad's favorites, the comedians Jack Benny and Fred Allen, and then later, Bob Hope. I liked Fred Allen best. Bob Hope became famous for his travels to visit the troops around the world. In fact, Bob Hope put on a show for us at Keesler Air Force Base in 1952.

The aforementioned Blue Coal was one of the most widely advertised products of the time. It was a trademark of the Reading Coal Company, a subsidiary of the Reading Railroad. It was hyped to be clean burning anthracite coal mined in eastern Pennsylvania. Blue Coal local distributors moved the coal, daubed with blue paint, by truck from the rail lines to homes and factories. Since the trucks had chain drives, you could hear them coming from afar. They always seemed to be dripping dirty coal water on the streets. We loved to watch the coal go down the chute into the cellars of the buildings on our street. In other places, the coal had to be carried in canvas buckets to the coal bins.

While speaking glowingly of radio days, I can't neglect movies or films as they are now called by the cognoscenti. I went to the independent el cheapo movies most Saturdays. In Mott Haven, we had the Osceola, a 556-seat theater at 258 St. Ann's Avenue. Why a Bronx movie theater was named after a Seminole Indian chief is a mystery to me. In such theaters, we saw a double feature, usually an A movie and a B movie, cartoon, newsreel, and serial episode. The featured movies were mostly Westerns with Randolph Scott and other second-tier actors, with an occasional crime story. I remember seeing some of the Hepburn-Tracy movie previews, but don't think I ever saw one of their films as a youngster. The newsreels were a very popular way to get war news in pretelevision days. Of the serials, I only remember Dick Tracy, but there were many cliff-hangers. Only once in a great while did I get to go to one of the RKO or Loew's movie palaces as they were more expensive, maybe a dime or more. It was probably at one of them that I saw my early favorite movie,

1941's *They Died with Their Boots On*, starring Errol Flynn as George Armstrong Custer. This blockbuster had wonderful music and much U.S. Seventh Cavalry derring-do until the end, when he dies wearing his "arrow shirt." The big chain theaters were magnificent and featured elegant decor and much-better seating. After the war, the big theater chains were separated from their studio owners in a landmark court decision. Under independent owners, most of the great movie palaces were renamed, and some closed. The shabby Osceola Theater today survives, physically, as a shabby supermarket.

Rationing and Propaganda

The Office of Price Administration (OPA) was established in 1941 to administer price control that began that summer. Some eight thousand rationing boards were created with two hundred thousand volunteers to assist the eventual sixty thousand employees of the agency. Food rationing included restrictions on sugar and meat; clothing rationing restricted silk and nylon. Gasoline rationing began in May of 1942 on the East Coast, limiting use to five gallons per week. By the end of 1942, half of the nation's automobiles were issued an A sticker, allowing four gallons per week. The other half of automobiles had either a B sticker (supplementary allowance for war workers) or a C sticker (vital occupation such as doctor). Truckers had a T sticker for unlimited amounts. A black market developed in stolen or counterfeit stickers that were used in up to 30 percent of gasoline sales. By 1945, there had been 32,500 motorists arrested for using such false stickers, with 1,300 convictions and 4,000 gas stations closed.

To the individual, particularly to us kids, rationing was a complicated mess. There were ration books and ration tokens. I've not been able to find any sources that explain how it all worked. Maybe no one remembers. Anyway I do remember red and blue OPA tokens. Somehow I think the red ones were for meat and the blue ones for sugar. I still have a blue one. Our family was concerned not only about rationing of food and clothing but that my father would have ample gasoline for his taxicab. I believe that the OPA issued C or T stickers to cabs so that they had adequate supplies.

Moving to a wartime footing also involved government propaganda and control of the news. We usually didn't learn of battles and disasters until long after they occurred. It seems to me that we didn't get details, including pictures, of the disaster at Pearl Harbor until late in 1942. One of the major propaganda efforts of the war was to get women into the workforce to replace men going into the services. Thus, we had the famous Rosie the Riveter posters everywhere. The campaign was so successful that by the end of the war, women constituted one-third of the workforce. Another major effort was directed against spilling information that would help the German U-boats patrolling offshore. Posters with the slogan Loose Lips Sink Ships also appeared everywhere.

The Hudson River piers had always been readily accessible from the nearby streets. However by 1942, they had been enclosed by high fences to forestall spies and saboteurs. Wartime paranoia seemed justified when the SS *Normandie* burned at her Hudson River pier in February of 1942. The pride of the French Line had been seized by the U.S. government and was being converted for use as a troop ship. Initial thoughts of sabotage gave way to the reality that a workman's torch had started the fire that destroyed the ship. She lay at her last pier all during the war and for some time after and could be seen from the West Side Highway that overlooked the piers. Neither highway nor piers remain today.

Serious war news came over the radio, and people listened intently every evening to their favorite broadcasters on the various networks: Mutual, CBS, NBC Blue, and NBC Red. I believe there was a Dumont network as well. The newsreels shown between features in movie theaters mostly heralded victories and played down defeats. They also covered the sports scene, so we got to see our football and baseball heroes in action. Another facet of wartime was the constant promotion of war bonds. These were savings bonds designed to soak up excess cash so as to reduce upward price pressure. We were encouraged to buy war bonds by movie stars over the radio and in the movie newsreels. It was a successful program, and you could buy them on what amounted to an installment plan. Whole life insurance was also widely promoted and could be bought and paid for in weekly installments. A man would come around each week to collect the dime that was due on a policy. I had a paid-up $1,000 MetLife policy that my parents purchased for me. I eventually cashed it in and collected the face amount, plus accrued interest.

Bubble gum "trading cards" had historically featured baseball heroes, as football and basketball had not yet become significant. During the WWII, "war cards" appeared in our bubble gum packs. They were very well done and I liked them although other kids didn't. Wings' cigarette packs included a card that, as I recall, featured military aircraft. I was occasionally able to beg the card from someone who bought a pack of Wings. Any of these cards would be very valuable today, but I never kept those that I had. Bubble gum card packs were, at five cents, expensive in the Depression and early wartime years. I don't know whose bubble gum cards we had in the 1930s and 1940s. Topp's first successful product was Bazooka bubble gum in the late 1930s. It was packaged with a small comic on the wrapper. They didn't get into trading cards until 1950.

As I write this memoir, I was reminded of something that you don't think much about. These days we use an awful lot of paper products, most of which didn't exist or were not available in wartime. I can't remember when I first saw "bath tissue," which we knew as Kleenex until quite recently. It had to be well after the war. Another everyday product we take for granted is the common plastic "sponge." These didn't come into being until after WWII. There had been natural sponges, but they disappeared for the duration and were quite expensive in any case. Therefore, washing up was done with rags from old and discarded items of clothing. Clothes were washed in the kitchen sink using very strong yellow soap, scrub brush, and a long-vanished item called a washboard. Clothes were wrung out by hand and taken to the window that gave access to the clothes line.

On the Move Again

A final thought on our life in the South Bronx. During the NY World's Fair of 1963-64, we had a visit from Mother's sister Kitty. She had been a nurse in England and had gone through several husbands. She returned to Ireland and lived there with her last husband. His death may have been the occasion of her first and only visit to "America," meaning the United States. The family assembled in Riverdale to see Aunt Kitty, and some of us decided to take her to the fair. On the way, we made a detour through Mott Haven and down our old street. Our old building was still standing, but there seemed to be a gloomy pall over the whole neighborhood. Aunt Kitty could not believe that we had actually lived there for many, quite happy, years. I wondered myself.

In September of 1942, we left St. Luke's parish and moved to Incarnation Parish. Our new address was 503 West 175th Street in the Washington Heights area of upper Manhattan. I don't recall why we moved, possibly for a free month's rent, more likely because the South Bronx was going bad. There had been much talk about "spics" moving into the area—the spics were from Puerto Rico. During the move, I remember being very concerned because we took with us a steel clothes cabinet that didn't belong to us. Despite my scruples, I was to use that cabinet until I left New York in 1956 for my new life in New Jersey.

Washington Heights was appropriately named as it sat on a high bluff that overlooked the Hudson River to the west and the Harlem River to the east. We lived on the east end of West 175th Street, at Amsterdam Avenue. Thus, we overlooked the Harlem River and High Bridge, a bridge completed in 1848 to carry water from the Bronx reservoirs into the Manhattan reservoirs. I later learned that the bridge, as originally built, was supported by a continuous series of masonry arches like the Roman aqueducts. In the 1920s, a single steel central span replaced many of those arches so that larger ships could use the Harlem River. On top of the modified bridge was a walkway that permitted us to easily walk back into the Bronx and, in my case, to get to Yankee Stadium. However, it was a tough climb down from the heights to the bridge. Furthermore, a long unused ramp went from near our corner down to the unfinished Harlem River Drive. On snow days, we would sled down the ramp for a good part of the way toward the Polo Grounds at 155th Street. The trip back up the ramp took the rest of the evening, so only one trip per school day was possible.

In the fall of 1942, my sister Patricia and I were enrolled in Incarnation School—she in grade 2B, and I in grade 5B. The school dated from 1910 and was run by Sisters of Charity from Mount St. Vincent in Riverdale. It was conveniently located for us in the next block west on our street. We had to line up class by class in the street in front of the school and march in when the bell rang. We also had a lineup for the nine o'clock Sunday mass that we were obliged to attend, en masse. The sermon at the kiddie mass was aimed at instructing us in some esoteric aspect of our Catholic faith. Any fidgeting at mass was rewarded by a clip on the ear by Sister who was sitting close behind the class. Once we'd made our first Communion, we had to march up to the communion rail in orderly columns to receive the Eucharist. Sometimes parents attended the kiddie mass, but most adults avoided it. Shortly after we moved away, a wacko gunman fired into the kids lined up for school and killed a boy that had been in my class. I literally dodged a bullet. As I recall, his name was Mahoney.

A favorite thing I remember first seeing at Incarnation School was the weekly messenger. It was a little newspaper edited for schoolkids that was delivered to our desks. I seem to remember that there were two of them, one of which was exclusively for the Catholic schools. I recall reading about the opening of the Pennsylvania Turnpike, which had occurred before the war in 1940. The article told of how they used the abandoned railroad tunnels through the Allegheny Mountains. I was to enjoy the messengers again at my next school.

Sometimes my mother would send me to the store to pick up milk and groceries. She would insist that I go to a store owned by an Irishman, one Jimmy Slane. I wasn't happy as it was much farther than the nearby Jewish shops. One day I went there, paid, picked up my bag of groceries, and trekked home. When I got home and gave the change to Mother, she checked his arithmetic (always done on the bag) and found that he had shortchanged me. Back I was sent to Slane's to get the extra coins. I made such an anti-Irish fuss that I was never sent there again. My sister Patricia says that she had to go in my place. Poor Pat!

The neighborhood offered some amenities that were lacking in the South Bronx. A few blocks away in High Bridge Park was a huge public swimming pool complex. It had a large Olympic-sized adult pool and a large kiddie pool. The adult pool had islands you could swim to and climb on. The islands actually contained underwater lights for night swimming. I was eleven in the summer of 1943 and, therefore, was old enough to go to the pool by myself. I paid a small fee to check in whereupon I'd get a basket and key tag. I then changed in the locker room, put my clothes in the basket, and returned the basket to the front counter. The key tag was attached to an elastic band, which I wore while swimming. I know that Mother took my sisters to the pool, but I don't remember seeing them. While the pool was multiracial, things seemed to be mostly under control. However, we were all atwitter when a body was found in the pool one night.

The O'Brien Family

We became very friendly with the O'Brien family who lived upstairs, I believe, in our five-story building. Mr. O'Brien was deceased, having left Charlotte with four children. From oldest to youngest, they were Betty, Jeanie, Tommy, and Kenneth. Tommy was perhaps a year older than me, and Kenneth was close to my sister Pat's age. We were to remain friendly with the O'Briens long after we moved away. As with Mary O'Regan, Mother always spoke of Mrs. O'Brien, never Charlotte. Mrs. O'Brien, nee Cavanaugh, had a well-to-do brother who lived at Atlantic Highlands, New

Jersey. They often took a boat from Manhattan to Keansburg to visit him, and I guess he helped them out financially. Tommy and I became close friends, and he inspired me to take up tennis. Somehow I got a racket and learned to play with him at Fort Tryon Park on the far west side. There were many courts in the park, and they overlooked the Hudson River, and few players. So I started playing tennis in Manhattan at age eleven.

Tommy and I noticed that kids were playing baseball in the nearby park. To that point, I had never played ball other than stoop ball or curb ball. In those games, you threw a ball against the curb or a step and it rebounded into the "field." Your opponents tried to catch or quickly retrieve the ball as the number of bounces determined whether you "hit" a single, double, triple, or home run. The older kids played stick ball using chalked bases or manhole covers. Both games required a lively rubber ball called a spaldeen. The name was a corruption of Spalding, a well-known supplier of sports equipment. Because it was wartime, rubber balls became scarce and prized. It was tragic when a ball rolled into a storm sewer's catch basin. If at all possible, we retrieved the ball by dragging off the grate and climbing down into the catch basin. That was no fun. The older kids often played stick ball with a "half-ball," and that produced a lot of spin.

Anyway, Tommy O'Brien convinced me that we should form a baseball team to play the other teams in the park league. We spent a lot of time trying to figure out what to name the team. My suggestion of "Crusaders" was adopted, and we got team baseball caps with a *C* embroidered on them. I convinced Mother that I needed a bat, ball, and glove to play. So she took me to Davega's, the local outlet of a popular chain of the day. There I got a black Ernie Lombardi bat and a shapeless glove, all at rock-bottom prices. Now that we were fully equipped, we never played a game. I moved away shortly after.

Because Tommy was older and had older sisters, he was more worldly wise than I was. For one thing, he was very much into music, and the singer Peggy Lee was a great favorite of his. I started to pay attention to music thereafter, and we both became great fans of Nat King Cole as a pianist and as a singer. When visiting at the O'Briens', his sisters encouraged us to dance to the latest radio tunes as there were many music broadcasts. I was a most reluctant dancer. The O'Briens, particularly Tommy, were peculiar in that they misspoke in a sort of English cockney fashion. A "bottle" was a "bokkle," and they "bunked" into things that I bumped into.

Tommy was also an entrepreneur. He had a newspaper concession at a nearby Catholic church. On Sundays after mass, he sold the Brooklyn diocesan paper called the *Tablet*. Someone else had the concession for the New York archdiocesan paper, which I believe was called the *Catholic News*. Even after we moved away, I would take the subway down to help Tommy and pick up some cash. Many times we would leave from the church and go directly to Yankee Stadium over the High Bridge. We tried to go to all the Sunday-afternoon doubleheaders. There were many in those days as lights had not yet been installed at the ballpark. You paid for one ticket and saw two games, not like today. I remember being very tired during some long second games, especially if the Yankees were losing. The long walk home to Washington Heights was daunting.

I became an avid baseball autograph collector. After a game, I and others would go to the clubhouse exit to intercept the players and shove our autograph books in front of them. Some players were very sneaky and would go out another door to avoid us. I once caught Charley "King Kong" Keller, a star Yankee outfielder, sneaking out. He must have had a bad day as he took my autograph book and threw it into the street. Otherwise, the players were very courteous. I had autographs of all the Yankees of the 1940s, including, Hall of Famers Joe DiMaggio and Bill Dickey—the star catcher before Yogi Berra, who also wore number 8. We often walked with the players to the nearby Grand Concourse Hotel where many stayed. I specifically remember walking and talking with star reliever Joe Page and journeyman outfielder Johnny Lindell, together, one day. On the downside, I remember a late game with the Tigers when their second baseman Jimmy Bloodworth hit a homer to win the game. We saw a man come out of the clubhouse and start walking toward the hotel. Nobody recognized him, but I decided to ask for his autograph anyway. To my shock, he obligingly signed "Jimmy Bloodworth." I didn't know if I should tear that page out of my book. I didn't, and it stayed in my book until the book itself disappeared sometime in the 1950s.

With Tommy O'Brien, I joined the Cub Scouts of Incarnation Parish. I bought my Cub Scout manual and quickly made progress through the ranks and into the bottom level of the associated Boy Scout Troop. I happily bought my scout manual and remember learning my knots. I wish I still had my manuals; they were classics. We went on hikes over the nearby George Washington Bridge and down to the west bank of the Hudson. There we made campfires and cooked whatever we had carried with us, usually potatoes for roasting. They were known as "mickies," presumably because the Irish were known as "micks." Some of the scouts sneaked off to smoke cigarettes they had brought with them. I later started sneaking smokes on scout hikes along the Hudson.

With Family Friends and Off Again

During the year we spent in Washington Heights, we continued to travel to Rockaway Beach, but now we went in my father's cab, not by train. We often went at the same time as the McElroys, Frank and Rose, and family. Frank was a pal from Dad's days in Carrickmore. One year, we jointly rented the top floor of a bungalow and erected a blanket to divide the area. During our stay, I picked up a bad infection in a finger on my left hand. It was treated with smelly poultices. On the boardwalk at night, young Rose McElroy and I always got vanilla ice cream cones. My sister Patricia and Rose's younger sister Maureen always got chocolate cones. Rose and I, "the Vanillas," always walked together as did the Chocolates. Eventually Rose and Frank had a son called Frankie but he was too young to play with us.

We visited back and forth with the McElroys for many years. One time, when the McElroys were visiting us at home, Rose and I decided to go into the bathroom and play "doctor." Her mother must have heard us because she came charging in to rescue Rose. The McElroys moved into a house in the East Bronx area known as Castle Hill. They joined a very nice private beach club on Castle Hill Boulevard, not far from the East River. I believe we got to go swimming there on our few visits. About ten years ago, I traced the McElroys to an apartment in the East Bronx. By phone, I spoke to Rose who told me that Frank was too ill to speak to me. It's too bad that we had not continued the family relationship.

Other family friends we often saw at Rockaway Beach were the McCreeshes, Barney and Molly, and family. I don't remember the connection, but they too were from the Old Country. Barney ran a saloon in the Bay Ridge section of Brooklyn, and they had an apartment upstairs. We often visited there and enjoyed climbing out the back window onto the roof of the saloon below. Their oldest was Arthur, a year or so older than me. Consequently, Molly always called me Junior, which I hated. Their daughter Mary was a year or two older than my sister Pat. A younger son, also Bernard, was known as Teenie. Arthur was very bright and got a scholarship to the top Latin school in the city, a Jesuit school called St. Regis Prep. For some years, I was in touch with Arthur and even visited him and his wife at their home in Maryland. But now I've totally lost touch with all of the McCreeshes.

While living in Washington Heights, I made my first solo trip to the Quinns in Clifton. I got a bus to Paterson at the bus station at our (east) end of the George Washington Bridge. The bus took me to the center of the city that was and probably still is a rather gloomy place. There I got a bus that ran from Paterson to Montclair on Valley Road. I would stay with the Quinns for a week or two and return with my parents when they came out to visit. The Quinns' property ran up the side of Garrett Mountain in a series of terraces covered with grapevines. Even better, there were delicious huckleberries that we picked and took to Aunt Kathleen who made them into what has always been my favorite pie. Most people haven't heard of huckleberries or, if they have, think they are just blueberries. I decided to research this issue and found the following:

Wild blue mountain huckleberries are related to domestic, commercially raised blueberries and can be used interchangeably in recipes. However, the fruit of the wild berries is very special. It's much more flavorful, possessing that unique character and magical taste that only wild berries enjoy. Some of the best picking aficionados say is where there's heavy brush and it's terribly steep. There is only one sure way to tell one from the other, blueberries have a large number of tiny soft seeds, whereas huckleberries have ten rather large bony seeds.

In later years, Uncle Mike sold off most of the hillside to a firm that was quarrying rock out of the mountain.

In those days, Cousin Barry had a .22-caliber rifle that we took turns shooting at bottles and cans. Barry and I also hiked back into town (Albion Place) and visited our favorite spots along the way. A favorite was a pistol range where we would collect the many spent shells (brass) and speculate on the types of weapons from which they came. In those days, there were cow pastures along Valley Road. Another favorite was the drugstore soda fountain just off Valley Road on Fenner Avenue in Albion Place. They had great shakes and sundaes. Fenner Avenue was an important street, which connected Valley Road and Clifton's Broad Street. Across Valley Road from Fenner was the entrance to Mountain Park Road, which led to the top of Garrett Mountain and to Lambert Castle. Garrett Mountain Park overlooks the city of Paterson and has many hiking and biking trails.

Cousin Patricia Quinn came to stay with us from time to time, but never Barry; he just never liked New York. I remember Patricia being with us at Rockaway Beach. On a rainy day, we went to a favorite toy store and bought picture puzzles to take back to our room or rooms. Our favorite picture puzzles in those days featured triumphant war scenes, where our boys were thrashing the dastardly Japs on land and sea. We had recently beaten the Japs at Midway and thought the war was well on toward victory. While it was to drag on for another three years, the war became less and less a factor in our daily lives. There was one exception however. Someone invented a horrible butter substitute called "oleo" for short. We were told that real Americans should use this vegetable stuff so that real butter could go to the poor Russians and to our boys overseas. The American dairy industry would not allow the stuff to be sold colored yellow, like butter. So it was sold in whitish bricks, later sticks, with a separate packet of coloring so you could roll your own, so to speak. Presumably, the purchaser was to melt down the butter, mix in the coloring, and recast the result into sticks or bricks as he wished. This weird situation prevailed for many years until, finally, oleo could be sold looking something like butter. It never did taste nearly as good as butter, so who needed it anyway?

Even in 1943, our Washington Heights neighborhood was going bad as spics were moving in. A big kid, a spic from the next block, made me pay him protection money. I recall the amount was ten cents. For that I got a "token," which entitled me to protection from whomever, mostly him I suppose. Now the neighborhood and Incarnation Church, in particular, predominantly serve families from the Dominican Republic. It's good to know the parish is still functional. However, we were soon on our way to Riverdale.

But before we move on, I'd like to mention two strange events that occurred either in Washington Heights or in Mott Haven, I'm not sure which. In the first of these, my father smacked me for the first and only time. My sister Pat and I were jumping on the bed and making a racket. It was early in the morning, and I'm sure my mother insisted that he go into our room and put an end to the ruckus so they could get some sleep. My father was no disciplinarian, so we were surprised to see him come into the room, smack me, and walk out. I don't recall that he said anything.

The other event was rather sad. Mother's pal and second cousin Mary McAleer (nee Gallagher) had a brother Frankie who was in the submarine service in the Pacific. While home on wartime leave, he had acquired a baseball glove, which was intended for me. He arrived unannounced one day at our door and rang the bell and said who he was. However,

Mother had gone out and told me that I was not to open the door for anyone until she returned. Through the door, I dutifully told the visitor that Mother did not want me to open the door for anyone. He went away disappointed, and after the fact, everyone was upset about this abortive visit. I suppose I took the heat for being such a dummy. Although we often saw his sister Mary at family affairs, I never met Frankie and never got the glove. I believe that Frankie died an early death after release from his heroic service in the navy.

While in Washington Heights in February of 1943, Mother was finally granted her citizenship certificate by the U.S. District Court for the Southern District of New York. The document states that she was thirty-nine years of age, white, married, five feet three inches tall, weighed 135 pounds, and was formerly Irish. Close enough I suppose, though her passport was British. For some reason, these naturalization documents cite the number of years since independence. In Mother's case, the figure given was 167, at least that is correct, assuming 1776 is the year of independence.

My O'Donnell grandparents, Edward (1851-1942) and Catherine Donnelly (1867-1941) in 1933. They lived in Gort, Brantry District, near the town of Dungannon in what is now Northern Ireland.

My Quinn grandparents, Patrick (1858-1951) and Mary Clarke (1874-1944). Their home was in Cornamaddy, near the town of Pomeroy in what is now Northern Ireland.

My O'Donnell grandfather's homestead in Gort, Brantry, 1968. Left to right: aunts Maggie and Cassie, Chris, Kitty, Tommy, two cousins, and their friend. Our rental Ford Cortina is at the right.

Uncle Joe Quinn shows Kitty in grandfather Quinn's homestead that consisted of attached house and barn, 1968. My mother grew up here with her parents and seven siblings. In summer, the house and the barn accommodated lots of visitors from the United States of America. The house has since been torn down and the site used for a new house for one of Sean Quinn's daughters.

Wedding of Uncle Mike Quinn to Kathleen Donohue at St. Joseph's, Paterson, New Jersey, January 25, 1928. Uncle Mike was twenty-nine, and Aunt Kathleen was twenty-seven. The best man was the bride's brother Peter. The bridesmaid was Margaret Teresa Quinn. The priest was John Cartin of Philadelphia, a Quinn cousin.

My parent's wedding day, April 28, 1931, at St. Francis de Sales, Manhattan. My aunt Mary Quinn (later Sander) was bridesmaid, and uncle Frank O'Donnell was best man. Mother was twenty-eight, and Dad was thirty-four.

Mother arrived in NY from Londonderry on the Anchor Line's *Tuscania* on September 26, 1922, at age nineteen. The ship was the second of that name, the first having been sunk during the war. Mother may well have been on its maiden voyage.

Mother is seated at the far left during her cooking class. Standing right is her teacher Ms. McComisky. Many of the girls were related to Mother.

Girard College was where my mother worked after arriving in Philadelphia in 1922. She had an aunt who worked there and arranged a job for her in food service. The "college" was actually a boarding school for gifted boys from impoverished families.

My parents are off to honeymoon at Niagara Falls, 1931. I don't know whose car they used.

My father (seated right) and some friends. Standing in the center is Frank McElroy, and to his left is Eddie O'Connor who was married to Mother's cousin Brigid Clarke.

My godmother Aunt Mary Quinn and her husband, John Sander, wedding day June 23, 1934.

Kitchen of my O'Donnell grandfather's house, 1970. *Left to right:* aunts Cassie and Maggie and uncles Eddie and John. The house was built right on the bare ground, so it was very damp, and the wallpaper would fall off the walls every spring.

With the Sean O'Donnells at the Derrylappen location of my great-grandfather's homestead, 1999. Sean, a second cousin, raises livestock and had a big mushroom operation until the prices collapsed.

The center of the town of Dungannon, called the Diamond, 1991. Dungannon is the administrative and educational hub for South Tyrone, Northern Ireland. It is a heavily Protestant town of some twenty thousand in an otherwise highly Republican area. Our Donnelly and O'Donnell families have lived in and around the town for generations.

Church of St. Patrick at Eglish, 1999. The church has served many generations of O'Donnells since 1834. *Left to right:* Margaret Murphy (Eddie's wife), Kitty, and first cousin Eddie O'Donnell.

Old Eglish graveyard, 1999. Family legend has it that my great-grandfather Edward O'Donnell is buried under the shrub in the center of the picture.

My great-uncle Canon Francis Donnelly (1860-1940) was the parish priest (pastor) of St. Columcille in Carrickmore and is buried in the churchyard. My father and aunt Cassie O'Donnell lived with him for a time.

Derrylatinee school, 1999. My father went to this "public" school near Eglish, but in a much-earlier building.

Original Donnelly homesite in Ballynahaye, near Galbally, 1999. It now faces a hill with a cell phone tower, probably in Clonavaddy townland.

Church of St. Patrick, Aughnagar, Killeeshil Parish, 1999. My Donnelly forbears are buried in this churchyard. The church dates from 1862. There is no trace of a headstone for my great-grandfather Murtagh Donnelly.

Interior of Aughnagar church with second cousin Patrick Donnelly, of Altaglushan near Galbally, County Tyrone, Northern Ireland, 1999.

Kitty and I visited my O'Donnell grandfather's home in 1999. It is still in the family but is no longer occupied and is showing signs of abandonment.

Church of the Immaculate Conception, Altmore, Pomeroy Parish, 1968. This austerely beautiful chapel was badly damaged during subsequent sectarian strife, but has been restored. It dates from 1870. It was Mother's family church.

Altmore House, 1968. We stayed in this historic home and inn with Tom and Chris who can be seen playing on the right. It was the home of the well-to-do Shields family who gave the United States a general who later became a senator for three states. The Shields family probably supported the nearby chapel and Altmore school, which was attended by my mother's family. The building was subsequently destroyed during the sectarian strife.

Cappagh village and distant mountain, 1968. Cappagh lies on the main road between the O'Donnell and Quinn homesites and was the home of several O'Donnell and Clarke relatives. The big black car was in service with cousin James Clarke. He often drove my mother and my sisters on their visits.

Town of Pomeroy on fair day, 1968. I toured the pubs with Uncle Joe Quinn while the ladies looked at the merchandise. When I visited in 1962, I drove into Pomeroy and asked around for the home of "Yankee Pat." I was promptly sent to the right place a few miles from town. Of course, it was Uncle Joe's place by then. That's Kitty in green at the right of the picture.

My official portrait at age two, 1934.

Arthur McCreesh and I in St. Mary's Park, Mott Haven, the Bronx, 1935.

Chain drive as used on Mack electric trucks in the 1920-30s. Trucks with this type of drive made a distinctive ratcheting sound that could be heard for some distance.

An electric delivery van of the 1920-30s. My father drove one similar to this for the Ward Baking Company until we went to Ireland in 1933.

Clifton's PS5 in Albion Place. This is where I started school in 1937. The wing on the far left was added since my time. In summer, outdoor movies were shown in the yard behind the school.

My father with his cake truck, circa 1938. The venture was a failure, so we moved back to the Bronx.

Birthday party behind Quinn's store on Valley Road, circa 1938. Standing are Aunt Kathleen holding Cousin Eileen, Kathleen's brother Pete Donohue, Marion Avato, and Mother. Cousin Barry and I are seated at the near end of the table on the right. My sister Pat is getting off her chair on the left.

St. Luke's Church on East 138th Street in Mott Haven, the
Bronx. Here is where I was baptized, confirmed, and made
my first Communion.

FDR comes down our street during the 1940 presidential
election campaign. I was given a flag to wave as he went by.

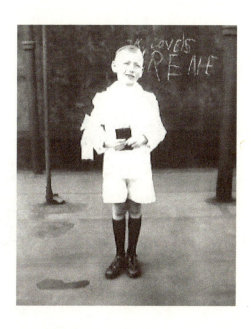

On the roof for a photo shoot after first Communion in 1939.

World-famous Hell Gate Bridge over the East River. It provided a railroad link to allow for freight to be exchanged among the major railroads of the northeast. From Randall's Island, we would count the freight cars on the long trains as they crossed the bridge.

Streetcars ran up and down East 138th Street in our time. An elevated IRT subway line crosses in the background.

This horse and wagon were still in service on East 138th Street in 1948. Runaway horses spiced up otherwise uneventful days on the street.

SAN FRANCISCO-HAWAII *Overnight!*

Via **PAN AMERICAN · TO THE ORIENT**

This ad promotes Pan American Airways Clipper service to the orient. We could see the European Clipper arrive at North Field (now LaGuardia) from the foot of East 138th Street. The flying boats did not survive the war.

It may not look it, but this was the Hub of the Bronx well into the 1950s. The building with the water tower is Hearn's department store where we did our Christmas shopping in the late 1930s and early 1940s.

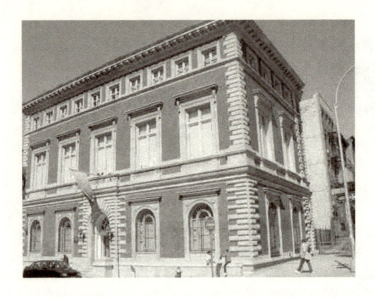

The Mott Haven Public Library on Alexander Avenue. My destiny as a library rat was revealed here, circa 1939.

Wartime symbols. Top left is our ID issued early in 1942 so that our bodies could be identified in the event of an air raid and fire. It is charred due to an experiment I conducted on the material. The blue object is a rationing token, value of one whatever. Below is my air force dog tag that's made of durable steel.

Catharine Carlin, circa 1936.

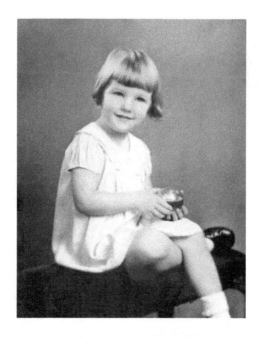

Catharine Carlin, circa 1937.

The Carlin sisters, circa 1938. *From the left:* Marietta, Agnes, Patricia, and Catharine.

Our family in St. Mary's Park, circa 1939. A friend of the family who was a professional photographer took this one.

Girls on the taxicab at Quinn's, circa 1942. *Left to right:* Eileen Quinn, my sisters Patricia and Maureen, and Patricia Quinn.

The social center of the South Bronx was the Concourse Plaza Hotel pictured here in the late 1940s. Many baseball players stayed here in its heyday. My future high school was just a few blocks south on the Grand Concourse, the center of middle-class Jewish life in the Bronx.

Joe McCarthy was my first and all-time favorite Yankee manager. A Hall of Famer, he served as Yankee manager from 1931 to 1945. The Yankees had some bad years during the war, so Joe was fired in 1946. He then went to the Boston Red Sox.

Mother and Uncle Mike at 636 Valley Road, Clifton, late 1940s.

The 1939-40 World's Fair newspaper dated the last day of the fair, October 27, 1940. My dad and I were there on that last day. It was eerie.

This is what they've done to my old Mott Haven movie theater, the Osceola.

Confirmation Day, 1941.

Rockaway Beach, "the Irish Riviera," in its heyday in the 1940s.

My sister Pat and I loved the fun house at Rockaway's Playland.

We never ventured on Playland's roller coaster. Sadly, Playland was torn down many years ago. Developers wanted the land for expensive high-rise housing.

Built to carry water from upstate into Manhattan, Highbridge also served as a pedestrian crossing over the Harlem River. That's Washington Heights in the background. We lived there for a year, 1942-43. This is a picture of the bridge as built in the 1840s. In the 1920s, to accommodate larger ships, a steel span replaced the masonry arches at the center of the bridge.

Incarnation School, Washington Heights, Manhattan, dates from 1910. Note the separate doors for boys and girls.

I'm a newly minted Cub Scout, 1943.

Rosie the Riveter encouraged women to go into "war work."

Art deco ad for the SS *Normandie* of the French Line. In 1942, she burned at her pier in New York while being converted to serve as a troopship.

PART 2

Riverdale Years (1943-1950)

We Make a Big Move

I don't recall having heard anything about moving to Riverdale or about a house. Seemingly out of the blue, in October of 1943, at age eleven, I walked into a house at 437 West 261st Street and was told that it was ours. Wow! It was a simple 2½-story brick structure with garage tucked underneath, accessible via the backyard. The front door opened into a short hall with a coat closet and a glass door into the living room. Off the living room were the stairs to the second floor, a "sun parlor" across the front of the house and a glass door to the back hall. Off the back hall were a pantry, stairs to the basement, powder room, kitchen, and a large dining room than ran all the way across the back of the house.

The second floor had three bedrooms, a linen closet, a full bath with separate tiled shower, and an enclosed stairway to the third floor. The front bedroom ran the width of the house and overlooked 261st Street. This was usually my parents' room. Another large bedroom ran across the back of the house and overlooked the backyard and adjacent empty lots. My sisters usually shared this room. Along the upstairs hall was a small bedroom that was mine until I moved to the more spacious room on the third floor. Room assignments varied, as you will see. I later learned that my parents had paid some $8,000 for the house, and there was a mortgage that was later paid off. One curious element of the house was the escutcheon plates on the walls in each room and down the halls. It turned out they covered fittings for gaslights. I guess the 1920s builders were not convinced that electric lighting was here to stay.

The government promoted "victory" gardens during WWII. Although we had a very small twenty-five-foot-wide lot, I did have room for a victory garden in our backyard. There I planted and raised radishes, string beans, and tomatoes and perhaps, other vegetables that I've forgotten. The soil must have been very fertile because I had bumper crops. I particularly enjoyed the string beans, which I picked off the vine and ate raw. Very few ever made it into the kitchen. In order to get his taxicab into the garage, my father had to back up close to my garden. The car's exhaust would blacken some of my tomatoes. I recall that we ate them anyway.

Our street ran from Broadway and Van Cortlandt Park west to the Hudson River. Broad north-south avenues named after scientists crossed the section from Broadway to Riverdale Avenue. We lived between Delafield and Liebig Avenues. West of Riverdale Avenue was wooded with hills down to the Hudson River. It was an area of large estates and the College of Mount St. Vincent, which included a Sisters of Charity motherhouse. The overall area was known locally as Jew Hill, probably because of a Jewish old-age home situated there. Our part of North Riverdale featured single family and two family homes, mostly built in the 1920s. It was heavily Irish American and mostly Catholic. However, on the hills west of Riverdale Avenue and south of 259th Street, there were apartment blocks that had a more mixed population, including Jews. These buildings were erected just before the war and had elevators, shortly to be of benefit to me. My mother was to live there for a time.

For transportation, we had a bus on Riverdale Avenue that ran to the IND subway at West 207th Street in northern Manhattan. On Broadway, also known as the Albany Post Road, we had a choice of streetcars. The Yonkers cars ran from Getty Square in Yonkers to the IRT subway station at 241st Street. They used an electric pickup that worked off a power line buried in the street between the tracks. The NYC streetcar, known as the C car, ran from the city line at 262nd Street and Broadway all the way across the Bronx. Thus, it connected with many of the elevated subway lines that served the Bronx. It used an overhead electric pickup called a trolley. I was to ride the C car for years on my way to high school. It was OK except that it labored up the steep hill on Kingsbridge Road as it traveled east from Broadway. Sometimes the cars ran off the tracks on the way down the hill and tied up traffic for hours. The hills in Yonkers were so steep and the turns so tight that the streetcar lines had "runaway" tracks to stop out of control cars. (See appendix I for more background information on Riverdale.)

Welcome to St. Margaret's

My sister Patricia and I were quickly enrolled in the parish school St. Margaret of Cortona. The school had eight classrooms and was staffed by Sisters of Charity from the nearby Mount St. Vincent motherhouse. The schools we had previously attended were large and offered simultaneous A and B grades. Because we had started school in the spring term, in the fall of 1943, we were in the B grade. Because St. Margaret's had only an A grade, we were left back a term. I wasn't happy about this, but with so many things happening at once, I had to let it slide. So Patricia went into the third grade, and I started over in the sixth grade. Since the term was well under way, you felt as though you were under a microscope when you walked into class that first day with everyone checking you out.

St. Margaret's School was in a large stone building that fronted on 260th Street and ran back some distance. You entered through the front door and went up a steep flight of stairs to the classrooms on the floor above. On the entry level were the auditorium dead ahead, a large library to the left, and a large unused room on the right. The latter was used as a cloakroom for school events. On a landing, part way up the stairs, was the school office where I was to spend a lot of time in later years. There was a back stairway that was mostly used by the nuns to quickly get to their residence on the top floor. The back stairs were also used by students assigned to take the erasers outside to beat the chalk out of them. There was also a side entrance that gave the nuns access to an elevator up to their residence. The school had a basement, part of which was set up as a clubroom. Infrequent school dances were held there.

The parish church was a fine stone building that dated from 1891. The handsome interior had exposed wooden trusses supporting a slate roof. Shortly after arriving at St. Margaret's, I was recruited to be an altar boy and served as one until I went into the Air Force in 1951 at age nineteen. After learning the Latin mass responses, I spent many happy hours in the church, sacristy, and rectory. An altar boy might be assigned to any of the daily or Sunday masses. If you served an early daily mass, you would be allowed to come late to class after a fine breakfast in the rectory with the priest. Often we had to serve visiting priests, some of whom would be quite old and feeble. One old timer kept us in smothered laughter as he fumbled and mumbled his way through mass. One visiting priest dropped his handkerchief on the altar steps. It was filthy and looked as though he used it to clean his car. The other altar boy and I couldn't take our eyes off it as we wondered if we should remove it. I don't remember what we did, but I suspect we let it lie there. The two boys serving an ordinary mass competed for the job of ringing the chimes during the offertory. Our church had chimes, not a bell. With a rubber mallet, you were to strike the three chimes in a particular order: *bing, bang, bung*; then *bung, bang, bing*; then, lastly, *bing, bung, bang*. Everyone in the congregation knew the prescribed order. If you slipped up by hitting them in the wrong order or two at once, you never heard the end of it.

During all my years at St. Margaret's, the pastor was Msgr. Joseph Doyle, and he was assisted by Fr. James Dale. Fr. Dale was a rather reserved man, and among other things, he directed the altar boys. A later and welcome addition was Fr. Bartholomew Kilcoyne who came to us from Cardinal Hayes High School where he was known as Black Bart. Fr.

Kilcoyne was a hell-raiser who became what today would be called a youth minister. He took over the altar boys from Fr. Dale and often chided us for taking advantage of the system. It was not a false charge.

Serving mass for Fr. Kilcoyne could be an adventure. He was fast, very fast, and you had to get your Latin responses out quickly, or he might turn around and give you the "evil eye." One cold morning, he sent me off the altar down to the basement to tell the janitor, whom he called Creeping Moses, to put more coal on the fire. He also did not like much wine in his chalice, saying, "How would you like to drink a bottle of wine before breakfast?" He often took us on trips in his car. He was from Brewster, New York, and took us up to his family home there and to nearby Candlewood Lake for swimming. I also recall a trip to Palisades Park, a huge resort across the river in New Jersey. Fr. Kilcoyne was not averse to swatting an unruly kid, something Fr. Dale would never do. Saintly Msgr. Doyle remained quite aloof from the hurly-burly of parish life.

One memorable afternoon, a school assembly was called. I don't recall whether it was just my class or several classes. In any case, we were seated in the auditorium awaiting events when Fr. Kilcoyne strode up the aisle to the front of the room amid a chorus of boos. Now I and mostly everyone else liked Fr. Kilcoyne, even though he could be tough at times. I joined in the booing just for the hell of it. Fr. Kilcoyne was, apparently, furious and demanded that the culprits stand up and that we would stay until they did so. No one stood. My conscience bothered me, so eventually I stood up. Fr. Kilcoyne said something like "O'Donnell, don't be a hero, you were not the only one, so sit down!" The assembly then proceeded as though nothing had happened.

Adventures in the sacristy, the priest's dressing room, during high mass or novena always included making a lot of smoke in the thurible, the device used by the priest to cense the altar, crucifix, and the casket at a funeral. We got maximum smoke by putting candle wax in with the burning charcoal. Hopefully, when the thurifer opened the door to go out on the altar, he'd be followed by a cloud of smoke. Speaking of smoke, one day when I leaned over to get something in the sacristy, I didn't notice that lighted candles were sitting on the floor. We wore a loose-fitting surplice (white smock) over our cassocks. Mine was set on fire by the candles. Fortunately Fr. Dale noticed right away and put it out. I don't recall how he did it. I had another unusual encounter with Fr. Dale. One day, he told me that he often wondered why I was such a diligent altar boy when I didn't seem very religious. I told him that I just loved the pageantry of the church services. He said he did too.

Somehow or other my friend Bob Foran became lead altar boy. As such, he served as "master" during high mass and for other major parish events. The master worked closely with the priests and was almost a subdeacon. Neither I nor any of the other altar boys felt confident enough to challenge Bob Foran for this position. As I became a senior boy, I usually opted for the job of thurifer. He's the one that carries the thurible full of hot coals in processions or comes out onto the altar for a censing ceremony. Usually the thurifer and his incense-bearing assistant got to hang out off the altar in the sacristy. Hanging out was something that I was good at. I remained an altar boy right up through college. Bob and I served the principal mass at 11:00 a.m. every Sunday for years and years.

Within the parish boundaries of St. Margaret's were a monastery of the Passionist Fathers and the aforementioned Sisters of Charity motherhouse at the College of Mount St. Vincent. Altar boys from our parish served at both places on special occasions, and were treated like royalty before and after. Especially enjoyable were religious extravaganzas at what we called the Mount. This beautiful campus overlooked the Hudson River and the New Jersey Palisades beyond. On special occasions at the Mount, we would be part of processions of bishops, scores of priests, nuns, and scholastics, all preceded by altar boys—us—carrying cross, candles, and thurible. Afterward, there were goodies to be had in the dining room with its marvelous view of the river.

Much less interesting was an assignment to serve during a "mission" service. Once a year, a group of priests, generally from one of the missionary orders, came to St. Margaret's to preach a "mission." During Sunday masses and at evening services, the visiting priests gave hellfire and damnation sermons intended to keep the faithful from falling into sin. Since the parishioners who attended the services were mostly already on the path of righteousness, it now seems like foolishness. The priests should have gone door to door to find and preach to the parish shirkers. I hated drawing a "mission" assignment and having to listen to the extended hellfire preaching. I have to admit that the mission priests were very good preachers and a nice change from the usual bland Sunday sermons.

The Mount campus had many attractive buildings and features that drew us, neighborhood kids. I particularly remember the beautiful trees in the fall. There were horse chestnut trees with huge nuts that we used in a game. Adjacent to the college tract was Mount St. Vincent

Academy, a girls' high school. My sister Patricia attended the academy for two years until the school moved north of Yankers. Another nice feature of the Mount was a large goldfish pond just off Riverdale Avenue. It featured a decorative bridge and benches for contemplation and was great for ice-skating. Eventually, the nuns decided that they didn't want the neighborhood kids on their property, so a watchman was engaged to keep us away. We then had to adopt a new ice-skating venue, a frozen stream up on Jew Hill.

In those days there was a Mount St. Vincent railroad station on the main line of the New York Central Railroad that ran along the east side of the Hudson River. However, it was a rare occurrence when a passenger boarded or left a commuter train at our station. It was at the far western end of our street and down a steep hill that was treacherous in winter. Since Riverdale was the last part of the city to get plowed or salted, we would usually be able to ride our sleds down the hill for several days after a storm. In good weather, we liked to hike down to the river and watch the trains go by. We'd stand on the bridge over the tracks and watch the signal lights. They'd change from green when a train was coming, long before we'd see it or hear it. That was our signal to rush down and put pennies on the tracks to be flattened. I enjoyed seeing the famous long-haul passenger trains of the day, particularly the 20th Century Limited. The limited went to Chicago where the passengers transferred to the Santa Fe Super Chief for the run to Los Angeles. There was even a radio program on which famous passengers were interviewed as they boarded the train at Grand Central Terminal. Sadly, the New York Central is no more, known to railroad buffs as a "fallen flag" line. The Central became part of the failed Penn-Central and then part of an unprofitable Conrail. Now the tracks along our part of the Hudson are owned by Metro-North, a publicly operated commuter rail service. Beyond the Westchester suburbs, the tracks are part of CSX Transportation.

School Days

Sixth grade with Sr. Marie Rita went by in a hurry. Sister was unusual in that she seemed to favor boys over girls. She must have had brothers. My class had two girls for every boy, so a new boy was welcome. It was a large class of some forty-eight students, and I felt very uncomfortable at first. However, I quickly became involved in the usual class projects and homework. I remember that there was a lot of memorization, particularly the Catechism questions and answers. At that point in my life, I just went along with the program. Geography was one of my all-time favorite subjects. Class assignments included making, on poster board, a large map of the continent under study and applying materials to symbolize the products of each country. It was a challenge to find stuff that would do. I remember removing the brass band from a pencil and using it to represent gold. I know I did a map for South America, but I'm not sure about any of the other continents. Perhaps they were assigned to other students.

After school one day, a classmate asked if I played football. I said I hadn't but would try. He led me to a field where I met a group of boys, some from my class and some from the fifth grade. Among the group were the brothers Garrison, John and Charlie. John was in my class and Charlie a year behind. I learned that they were avid footballers; in fact, John went on to Columbia University where he played on the varsity and had a career in the U.S. Army. Anyway, on that fall day, we played tackle football without any equipment other than the ball, which Charlie had just gotten. During a pileup on top of Charlie and the ball, the ball exploded and ended that day's play. I guess it was a cheap wartime football. In fact, lots of toys began to be made of less sturdy stuff than before the war.

Many of the boys and girls in my class lived nearby in North Riverdale. Others lived farther away in Central Riverdale or in ritzy Fieldston. Since we had no telephone and I had no bicycle, I had to walk to a classmate's house to find out what was happening. Bicycles were not to be had during the war, and I needed one very badly to hang out with the guys. I convinced Dad that I could get a paper route and earn some money if I had a bike. Lo and behold, one day a bike appeared. It was old and had been busted up and repaired at a bike shop in Yonkers. I believe it was an old balloon-tired Rollfast, but I was glad to have it. It served me for many years, and I've never had another bike.

I became very close friends with the classmates who lived nearest me, specifically Bob Foran, Fritz Wiener, and John Collier. After class, I rarely saw my more distant classmates, with a couple of exceptions. From time to time, we'd go to visit Roy Capaldo whose family owned, what seemed to me, a mansion down near Broadway. They had an amazing twelve-cylinder Packard that was right out of a Hollywood movie. The whole family dressed beautifully and drove off to church in their magnificent chariot with Mr. Capaldo at the wheel, Panama hat on head. They also had a Hollywood telephone on a telephone table, the kind of phone with the hook on the side for the receiver. Later on, we were to be really dazzled by Roy's beautiful sister whose name I've forgotten. Roy went to Cardinal Hayes High School with me and then earned an engineering degree at Manhattan College (1954). In college, he joined the ROTC and graduated into the U.S. Air Force where he became a pilot. Roy retired from the service and moved from California to Colorado where other members of his family had relocated.

Another exception was John Gallagher who lived on West 259th Street and Liebig Avenue. We occasionally had one-on-one visits back and forth, but John was never a part of our group activities. This was perhaps due to his distance from the core group or perhaps because he had better things to do. John had a peculiar voice, and since my little sister Maureen was usually first to the door, she would let me know that Squeaky was there. John's family owned a Dodge-Plymouth car dealership on Broadway and West 240th Street. The Gallaghers had a spacious home that featured all the latest prewar architecture and technology and was a marvel to me. John had an older brother Charlie, who graduated from Manhattan College in 1952 with an engineering degree. I don't remember meeting him and believe he was working in his father's dealership. John had a younger sister, Regina, whom I dated a few times while in college. After the war, the Gallaghers were the first to have television. Our gang then tried to ingratiate themselves with John in order to get an invitation to see this marvel. They had a huge tabletop set with perhaps a ten-inch screen. In the mid-1940s, there was almost no programming, just the ever-popular test pattern. John went to Manhattan Prep and to college, but I don't recall where. When I last saw him, he was living in Rockland County, NY. My sister Patricia told me that Regina Gallagher married someone from Manhattan College and has eleven children.

Two nonclassmates were to figure prominently in my school days; they were Tommy O'Rourke and Jimmy Conway, especially Tommy. They were one class behind me. Tommy lived along my route to school, but Jimmy lived some distance away and was to join us in later years. Tommy was a prodigious reader who consumed several books a day. Admittedly they were not classic novels, but his speed was amazing. He introduced me to many authors that I came to love and to the semiprivate Riverdale Neighborhood House library where we'd go together to pick up the new juvenile books and classic detective stories. Among our favorite juveniles were the Jerry Todd and Poppy Ott books authored by Leo Edwards, a pen name. The books were funny and had totally outrageous titles such as *Poppy Ott and the Stuttering Parrot* (1926) and *Jerry Todd and the Flying*

Flapdoodle (1934). Tom and I both loved the English seaside stories by Arthur Ransome: *Swallows and Amazons* (1930), *Coot Club* (1934), and *The Big Six* (1940).

Other series books were very popular during the 1930s through the 1950s. My favorites were the Hardy Boys by Franklin W. Dixon, Tom Swift by Victor Appleton, and Bomba the Jungle Boy by Roy Lockwood. Popular girls' series featured Nancy Drew by Carolyn Keene and the Bobbsey Twins by Laura Lee Hope. I was later to learn that these authors were creations of the fabulous Stratemeyer Syndicate. Edward Stratemeyer and, later, his daughters created the characters and the plots. The actual task of writing was then farmed out to a stable of writers who completed the book. The earlier books had an edge that made them more interesting and challenging than later versions. Edward died in 1930; and after WWII, his daughters decided to rewrite, update, and reissue many of the books. Critics say that they deleted minor characters and subplots that had made the books enjoyable in the first place. Sic transit gloria mundi!

As we got older and more mobile, Tommy O'Rourke and I joined and haunted the Yonkers Public Library. It was quite a ride on our bikes, but we were up to the challenge. In contrast to our small local library, the Yonkers library was big and spread out over three floors. Tommy and I made a major discovery one day on our way to the library, the Alicat Bookshop on South Broadway. This place was a revelation to me, a used bookshop that had all of our favorite juveniles and detective stories, in hard cover for ten cents or so. The books were stacked literally from floor to ceiling, and we were allowed to rummage freely through them by the forgiving owner. I later took my wife, Kitty, for a visit to the shop. I loved it so. The bookshop, which opened in 1936, is still going and has become a publisher. The output consists of "chapbooks," small books of poetry, essays, and the like. Popular crime novelist Charles Willeford had a book of poetry published by Alicat Press.

In addition to being a prodigious reader, Tommy O'Rourke was a daredevil and very accident-prone. My sisters and I well remember the day he fell through the ice on the goldfish pond at the Mount and had to walk home, soaking wet, in the dead of winter. On another occasion, we were exploring a "haunted house," and Tommy decided, on a dare, to jump out the second-floor window. He caught his elbow on a hook on the outside of the window and ripped out a piece of his elbow. Tommy's parents, Timothy and Geraldine, were also Irish-born but much stricter than mine. I'm sure he caught hell often. In later years, his parents and mine played cards together, probably the whist and euchre they had learned back in Ireland.

Probably in 1944 when I turned twelve, my father took me for the first time to the Ringling Bros. and Barnum & Bailey Circus, which played each spring in New York's Madison Square Garden. I enjoyed walking through the areas where the animals were kept, but I really did not enjoy the circus acts. However for many years thereafter, we went because the circus and my birthday coincided, and the circus was my birthday treat. I did enjoy the famous clown Emmett Kelly, but I realized my father enjoyed the show more than I did. Come to think of it, that's probably why we went.

I developed various schemes for earning money, some benign and one not so benign. I became interested in magic and magic tricks. I sent away for tricks that arrived with the necessary how-to instructions. Once I had become somewhat proficient, I did shows from time to time and charged a penny admission. I usually got my sisters and their friends to come. I wasn't very good, and they were not impressed, so I quickly abandoned magic as a moneymaker. One day, I set up a table at the curb and set out various articles, each with a number that corresponded to a numbered slip in a jar. For a penny or nickel, I don't remember which, a kid could draw one of the slips from the jar. If it matched one of the items on the table, it was his. A prized article on the table was a spaldeen, a high-bouncing pink ball that had practically disappeared during the war. One day, a kid kept paying and drawing for the spaldeen. He eventually became irate and accused me of cheating, saying there was no slip in the jar for it. Of course, he was right, but I denied it and closed up shop for the day. I always felt bad about my behavior that day, and I don't recall if I ever gave him back his money or gave him the spaldeen, probably neither.

I Became a Paperboy

My nearest classmate was Robert "Bob" P. Foran who lived upstairs in a duplex around the corner on Delafield Avenue. He had an older brother, Arthur, who seemingly eased Bob's way in life. Their father, also Arthur, had an extensive and expensive U.S. stamp collection that I got to see once in a while. I became fascinated and decided to take up that hobby. I was never able to accumulate some of the rare and expensive stamps held by Mr. Foran. Among his holdings were the famous *Graf Zeppelin* airmail stamps of the 1920s. Anyway, Bob had a newspaper route. He delivered the *Bronx Home News*, a once-popular local evening paper that died years ago. In its heyday, it was famous for its front-page "plunge column" that featured local murders and suicides and particularly deaths by falling from the upper floors of Bronx buildings. I started to help him as he had a fairly large route. He later arranged for me to get my own paper route.

I undertook to deliver the *Yonkers Herald-Statesman*, an evening newspaper, all over North Riverdale. I had an extensive and hilly route that was difficult to cover on the bike I had, but it was doable. I had seventy-five papers to deliver, six days a week, but never on Sunday. On Fridays, I collected 30¢ for the paper and usually got a tip of 5¢, sometimes 10¢. On Saturdays, I went to the bare back room of a candy store in Yonkers to pay my paper bill of 25¢ per customer. Thus, I was making 5¢ per customer, per week, plus tips. In total, I was making about $7 per week, a rather princely sum at the time. I opened a bank account in Yonkers to hold my growing nest egg.

I started to spend money on two collections, stamps and books. Inspired by President Roosevelt, often shown with his stamp collection, and Arthur Foran's U.S. collection, I started to collect American stamps. I often made trips downtown to Gimbel's, which had the city's best-known stamp dealership. I not only bought stamps and albums, but the definitive Scott's stamp catalogs as well. The latter were big and expensive, but could be bought at a discount when they became replaced by a later edition. I developed a good U.S. collection, except that I never did acquire the really high-end stuff that was in Arthur's collection. At the same time, I started collecting the famous boys' series books and then went on to the juveniles of Tunis, Heyliger, and Ransome. To shelve my books, I bought a bookcase kit from Sears, Roebuck and had it delivered to the house. It had three shelves and side cupboards with doors. I unpacked and built the unit in my attic room. I then stained, sanded, and varnished it as though I knew what I was doing. It turned out very well and served me for many years. I'm sure I took it with me when I left home, but can't recall what happened to it.

Since few of my friends were interested or had money to spend, I usually went downtown by myself. My favorite thing to do was to stop at Nedick's for a hot dog and an orange drink; they were the best. At the Nedick's I patronized, you ate while standing at the window where you could watch the shoppers going into and out of Macy's across the street, or perhaps it was Gimbel's. In those days, a Catholic organization known as the Legion of Decency rated films, and each year, at mass, we promised to observe its rules.

We promised never to see a film listed as C for "condemned." I assumed that a file was given a C rating because of sexual license or nudity. While walking down Forty-second Street one day, I noticed a film playing that I knew was rated C. I decided to risk my immortal soul and see *Miss Tatlock's Millions.* It was a charming film with an outstanding cast and contained none of the expected scenes of depravity, so I couldn't understand why it had been given a C rating. I later learned that it got a C rating because it made light of some aspect of Catholic politics or theology. I plan to see it again and try to detect its subversive message.

My *Herald-Statesman* newspaper customers included residents of the high-rise apartment blocks west of Riverdale Avenue. Because the buildings were interconnected, I was able to ride up the elevator in one building and go down the stairs in the next, dropping off papers as I went. I had one really strange customer whose house was set way back on Riverdale Avenue. I had to leave my bike, walk up a steep set of wooden stairs, and hike up a trail in the woods to the house. By then, their dogs would have alerted the residents, two elderly women, to my imminent arrival. They always waited until I came to the door to go up a ladder into the attic and get a pickle jar of coins. They would then count out thirty cents for the paper and a nickel for me. I guess it was a pleasant weekly interlude for them, but for me, it was time wasted as I tried to be home on time for dinner.

The blizzard of 1947 struck a day or two after Christmas. That day I waited at the usual place for my papers to be dropped off. As the afternoon wore on, I climbed into an open car at the by then closed gas station that was my pickup point. I waited for a long time, to near dark, before I realized that the papers would not be coming at all. I had a hard time getting out of the car as snow had piled up against the door. Now my bicycle and I had to get home through deep snow. Some twenty-four-plus inches of snow finally fell during that storm, a record for the NYC. Nothing was moving in the city, but my Yonkers papers were dropped off at the usual place the following day, two days' worth. It was during school recess, so I had all day to get around and deliver them. When I went to collect the following Friday, many people, mostly those who lived in the apartment blocks, refused to pay for the paper that hadn't been delivered on time. The city's worst blizzard was no excuse. It was a hard lesson, but worse was to come.

On the following New Year's Day, 1948, the city's transport workers went on strike. Mercurial Mike Quill, a Kerry man whose brogue got thicker every year, led the union. When the schools reopened, there was no public transportation. I was attending Cardinal Hayes High School (Hayes) at the time. We had recently gotten our first telephone on a party line that soon thereafter became a private line. Our number was KIngsbridge 9-8926, and Mother retained that number until she left Riverdale. I asked my mother to use our new telephone to call the school to explain that I would be absent due to the strike. When she reluctantly did so, she was connected to the infamous Jabbo, Fr. Jablonski who was dean of discipline. She was told "Get him in here!" in no uncertain terms.

Jabbo had no idea what conditions were like in the nether regions of the city. Having no car, no subways, no buses, and no streetcars, how could I get there? I called the New York Central Railroad to find out when a train was due at Mount St. Vincent station and learned I was too late for the morning local trains on that line, but that I might be able to get a train on the Putnam line. The latter was about two miles away across Van Cortlandt Park, and I had never been there before. So off I went with my school bag and mushed through deep snow to the station. A steam train arrived shortly after I did and dropped me off at a station near Yankee Stadium, the last stop for steam-driven trains entering the city. I then trudged up the hill to school, arriving about midmorning. I don't remember being welcomed by Jabbo, nor do I remember having to do this again, perhaps because the strike was quickly settled. The Putnam division was abandoned, along with so many other rail lines, years ago.

Other Schoolmates

A classmate that I always felt close to in many ways was John Augustine Collier. John lived several blocks east on Spencer Avenue near West 262nd Street. His parents were from the Pennsylvania anthracite coal region around Pittston. He had two sisters and a brother, but I only remember a cute younger sister, Catherine, who was in my sister Patricia's class at St. Margaret's. I don't recall much about John's parents, but I think that his father worked for the telephone company. If so, they were relatively well-off as his father would have had a job during the Depression. They had an interesting house with a separate study that impressed me. His father was usually seen back there reading a newspaper. Once I was invited to go with the family when they went to visit his grandparents in Pennsylvania. There I met his grandfather, who was a locomotive driver for the Lehigh Valley Railroad, a so-called coal road. I believe that we were given a short ride in his diesel locomotive. I also remember that John and I went out one night to a nearby taproom. While we were there, drinking Iron City Beer, a shift of miners came in, and their faces were covered with coal dust. It was memorable.

Another close classmate was Ferdinand P. Wiener, whose father was also Ferdinand. We were asked to call our classmate "Fritz," which we did. "Ferdy" was out, way out. Fritz had at least one older sister, whose name was Margaret. She impressed me as very learned and was, I believe, a college graduate when we were in high school. Fritz's family subscribed to the *Saturday Evening Post*, the popular weekly magazine of Philadelphia's Curtis Publishing Company. The *Post* offered a varied plate of fiction and nonfiction, and its cover paintings by Norman Rockwell and others were wonderful. Fritz was very fond of a series of stories about Alexander Botts and the Earthworm Tractor Company by William Hazlett Upson. He loved to tell us about them, at length. I tried to read one he selected for me, but it couldn't sustain my interest at the time. I have just received from Amazon a 1977 collection of these stories. I was amazed at how many new and used copies of these obscure Botts stories are available on the Internet today as well as erudite articles about them.

Classmate William "Billy" F. Hewitt lived nearby on Liebig Avenue with his widowed mother. She was a very lovely woman but seemed to make demands upon Billy that limited his playtime. She and my mother became good friends. Billy did not go to "our" high school and, consequently, sort of drifted away from the core group. He did get an engineering degree at Manhattan College in 1954 and moved to Ohio shortly thereafter. When I next heard from him, he had become a lawyer and married a local girl. Unfortunately, the marriage did not last.

Some classmates never joined with us, even though they lived nearby. Among this group were Billy Whelan and Jackie Loftus. They seemed older and more worldly wise. Billy's father was a NYC detective, and we got to see his gun on one occasion. Jackie Loftus lived near the school, but we rarely saw him out of the classroom where he was treated as a black sheep. He lived on a corner lot near the school. His father took a notion to build a masonry wall around their lot, probably to keep schoolkids from cutting through. The wall was so crooked, it became a laughing stock. Mr. Loftus tried to straighten it by adding concrete along the top, making it even more of an eyesore.

Mike Stanton lived some distance away and became popular when his father, an army general, became commander of the Kingsbridge Armory.

The armory had one of the largest enclosed spaces in the world; in fact, so large that automobile races were held in it. The space had a huge surrounding balcony that was used for long-distance running events. Best of all, to Mike's classmates, was a lovely garden outside of his father's office. Some strictly supervised class parties were held there.

In later years, I was to become somewhat close to two of the girls, Mary "Chris" Finn and Julia Cahillane, who were good friends. Chris lived nearby, and we had the taxicab business in common. Her father and uncles ran cabs out of the IOTA West Fifty-sixth Street garage. I dated Julia a few times and recently learned that she had been a widow and had died. Chris was the oldest of six sisters who were known collectively as "the fluff of Riverdale." I guess this was because they had a lot of money and flaunted it. Some of her sisters were good friends of my sister Patricia. Chris married Gene Fox, an attorney, and moved to Yonkers. Gene died in 2005. Chris left their house to her son and moved to the Finn family compound at Point Lookout, Long Island.

While I was in college, after mass one day, I recall meeting Chris and Julia who told me about a wonderful book they had read. They insisted that I read it. I did and it has always been a favorite. The book was Salinger's *The Catcher in the Rye*.

Moving On in School

In the seventh grade, we had Sr. Bernadette, whom I think of as a boy hater. I was getting used to the ways of the school. Our class seating was always arranged with the "smartest" boys and girls front and center and the rest proceeding downward and outward in descending class rank. Bob Foran and I were usually in front of the boys, followed by Fritz Wiener. Betty Hughes, Ursula Mahoney, and Mary Jeanne McMahon usually led the girls on the right facing front. It was difficult to get to know the people sitting behind me, especially those at the back of a large classroom. Despite Sr. Bernadette, I enjoyed seventh grade because we covered New York State history in detail. I loved learning about how the early Dutch settlement gave way to the British, yet Dutch influences persisted and were of benefit to the New York colony in the war with Britain. Of particular interest to me were the battles that took place in and around New York, especially along Lakes Champlain and George. I very much wanted to visit those sites and eventually did so on many occasions.

When the teacher left the room, she would usually assign some work and appoint a "monitor," whose duty it was to report those who misbehaved in her absence. This was a most unhappy duty. When Sister left, one of the goals of the more forward boys was to entice girls into the clothes closet at the back of the room. They were often successful. After a particularly long and boisterous absence one day, Sr. Bernadette returned to find the room in chaos. As she looked around to find a culprit to punish, she noticed that one of the girls had spilled ink all over her white school shirt. Sister went up to her and slapped her face, calling her a "bold article," one of a nun's favorite expressions of disdain. I've always remembered this occasion because of the injustice of it, for several reasons. One, it wasn't the girl's fault her fountain pen had leaked. Two, this particular girl was one of the more backward students and thus easily picked on. Three, I don't remember that the girl played any part in the chaos, though she was out of her seat when Sister arrived. I guess you can tell that Sr. Bernadette was not one of my favorites.

Fr. Kilcoyne, in his capacity as youth minister, decided that there would be dances for the upper grades in the school's basement clubroom. With so few boys in the class, dates for these school dances became a problem for the girls. The boys were generally content to wander over to the dance and hang around until it was over. The girls' mothers seemed to prefer that a boy "go with" their daughters. My mother would complain that some girl's mother had stopped by to chat her up, apparently to find out if I had a date for the dance and, if appropriate, to make a suggestion in favor of her daughter. I hadn't mentioned the dance to Mother, so she wouldn't know what the woman was getting at. I always thought it was nutty as I had little or no interest in going to the dance at all.

We had a music teacher, Eleanor Lanning, who came once a week. She taught us music fundamentals such as octaves, scales, and notes and directed us in musical productions. Her 1945 project was to have the seventh and eighth grades put on the new and popular musical *Oklahoma!* My classmate, handsome Billy Whelan, was chosen for the male lead. All went well until they started to apply his makeup on show day. There was no way they would put lipstick and rouge on him, said Billy. After a big pow-wow, some kind of settlement was reached, and the show went on. After my acting skills had been evaluated, I was put in the chorus and mostly served as a stagehand. But it was my first taste of theater.

We had some good-looking, as well as smart, girls in my class. One such was Mary Jeanne McMahon who was wonderful as the female lead in our *Oklahoma!* production. I had several dates with her while I was in college. I guess that by then, my service and college experience had given me sufficient self-confidence to ask her out. When I left Riverdale for New Jersey, she left for Chicago, events that were totally unrelated. I know that I've been told what she was doing in the mid-West, but I can't remember.

At St. Margaret's, we also had weekly gym and art teachers. I don't recall anything at all about gym. In art class, I remember learning about primary colors and perspective. We were also introduced to some of the famous "classical" painters such as Hals, Gainsborough, and Rembrandt. There was nothing about impressionist or postimpressionist art. In one session, we were asked to do an original work with our crayons on special paper provided to us. I did a colorful composition using a series of geometric shapes. It went over very well and was hung in the school hall for all to admire. Was I an embryonic Picasso or just showing engineering tendencies?

In the spring of 1945, we were stunned by the death of President Roosevelt. Few people knew the declining state of FDR's health through the war years, so his death came as a shock to us. Of course, after-the-fact documentaries show how fragile he was even before his first election campaign in 1932. His health was a carefully guarded secret, and the press aided and abetted the secrecy. It was much like the later cover-up of President Kennedy's health problems and womanizing. So Harry S. Truman, a total unknown to us youngsters, became president and was a surprisingly good one. He was my third president—Hoover first, then Roosevelt, and now Truman.

One of the things I enjoyed most in school was poetry. I liked to read the poetry in class and at home, but hated to memorize poems as demanded by my teachers. Those I remember most vividly were Poe's "The Bells" and "The Raven"; Alfred Noyes's "The Highwayman"; and Longfellow's "Village Blacksmith," "Midnight Ride of Paul Revere," "Evangeline," "Courtship of Miles Standish," and "Song of Hiawatha." Most of these poems had a historic American context and were very accessible to young people. My favorite then was one of Alfred Joyce Kilmer's, not "Trees," rather "The House with Nobody in It." It somehow appealed to my sense of adventure and love of railroads. It opens thusly:

Whenever I walk to Suffern along the Erie track
I go by a poor old farmhouse with its shingles broken and black.
I suppose I've passed it a hundred times, but I always stop for a minute
And look at the house, the tragic house, the house with nobody in it.

Sergeant Kilmer, of New Brunswick, New Jersey, was killed at the Second Battle of the Marne in 1918, at the age of thirty-one. It seems to me that we also had to memorize Lt. Col. John McCrae's short poem "In Flanders Fields," which opens:

> In Flanders fields the poppies blow
> Between the crosses, row on row,
> That mark our place; and in the sky
> The larks, still bravely singing, fly
> Scarce heard amid the guns below.

Oddly enough, John, a Canadian doctor, also died in 1918 while on active duty. Memorization didn't stop with catechism and poetry. We had to memorize and recite Lincoln's Gettysburg Address, among other classics.

Outside of School

Once I had my paper route, my after-school activities became limited. I did find time here and there to play touch football on Delafield Avenue and half-court basketball in the schoolyard. The schoolyard also had a handball court, which kids rarely used, but was used by Fr. Dale and some of the visiting priests. Our street football games were sometimes ended by the police, who had been called by a neighbor unhappy with our noise. During the war, we rarely saw the police because of our remoteness from the Kingsbridge station house and their shortage of cars and our lack of phones with which to call them. One day, there was an accident on our Liebig Avenue corner. One of the vehicles was a truck loaded with scrap newspaper for the war effort, and it had dumped its load in the intersection. Someone called the police, and we kids hung around waiting to see what would happen. Finally, away south on Liebig Avenue, we saw an approaching cloud of smoke. It was a police car making its last trip to Riverdale. Both truck and police car had to be towed away. After the war, there was a sad accident at that same corner. A woman was driving home in her first new car, right from the dealership, and was sideswiped, whereupon she crashed into the front steps of a nearby house. New cars were so rare, even we kids shared her pain.

A very popular activity for a large group of city kids was the game of ring-a-levio. For this game, two teams were created. One was sent out to hide while the other went out to catch them and return them to "jail."

Holding the victim long enough to say "ring-a-levio" three times effected capture. The defending team also had to guard the jail, lest someone from the other team do a jailbreak. The game ended when all of the hiders were jailed. However, the game rarely ended because when it got dark, it was very hard to spot those in hiding. Later we were to play the game on our bicycles. This was to prove quite dangerous, as crashes were frequent. On one memorable occasion, we were seated on our bicycles, all lined up side by side on Delafield Avenue, getting ready to play. Whoever was on the left end fell over to his right, crashing into his neighbor and toppling the whole group. No one was injured, but the wheels of the bike on the right end were bent into a V shape. Suddenly, we had one less player.

In the 1970s, a book was published titled *Ringolevio: A life Played for Keeps*. It was purportedly written by Emmett Grogan, a New York street kid who went to San Francisco to become part of the Merry Pranksters of Kerouac's *On the Road* fame. I was hoping the book would reinforce my memory of the game and dispel the doubts of family and friends that there was such a game. The book did neither and was rather a bore. It has some currency among fans of Kerouac's book, also boring. Some doubt that there really was an Emmett Grogan, but actor Peter Coyote says he is/was a friend, and there is/was too an Emmett Grogan. I've read that manhunt and prisoner's base are variants of ring-a-levio.

A big event in the fall of 1944 was a big, really big, hurricane. It caused many a tree to fall. One in particular, a huge one, fell across Delafield Avenue between our house and the school. We had to climb over it for days until it was removed. On our next trip down to the Hudson River, we found the shoreline had been seriously eroded. The Mount St. Vincent railroad station was many feet closer to the river than it had been before the storm. On one side, the concrete that supported the steps over the tracks was practically in the river. I was to learn that serious erosion had occurred all along the New York Central main line tracks for some distance northward along the river.

The presidential election of 1944—again Roosevelt versus Willkie—gave me my first paid role in politics. Dan O'Neill, our local Republican committeeman, put me to work for the party, stuffing flyers in doors and handing out election material on the way to the polling places. I was to work for them again in the 1948 and 1952 presidential elections. By 1944, the entrepreneurial Riverdale Irish had turned heavily Republican. I later saw this as a betrayal of their roots in the party of Roosevelt and Truman. (There's more on Dan O'Neill in appendix G.)

In retrospect, Riverdale winters seem very severe, with lots of snow and nowhere to put it. After a heavy snowfall, we'd shovel out our driveway and garage, piling all the snow in our small garden area at the back fence. We'd then tunnel into the mountain of snow and make an icehouse deep enough to require a candle to light it. I don't recall what we did with it once we'd built it. Riding our sleds, preferably Flexible Flyers, down hilly streets was a favorite after-snow and after-school activity. There were then no "snow days." Usually a large group of sledders would assemble for the Spencer Avenue slope south from West 261st Street. This was popular because it was a lightly traveled dead-end street. We'd build a mound part way down to effect a kind of ski jump for our sleds. We'd run and jump on our sleds to get maximum "lift" off the jump.

What can go wrong in this bucolic setting? Plenty! In those days, I wore corduroy knickers. While sledding one day, I had a bunch of "barn burner" matches in my pocket. Don't ask me why. These matches would light with the slightest friction and on any surface, not like today's safety matches. When I landed on the sled, there was sufficient friction from my corduroy knickers and sled to ignite the matches in my pocket. Ouch! I never did that again. It was hard to explain the burned pocket of my knickers. While that was the worst accident I had on my sled, my sister Patricia went down a small hill near home on her sled, but went out of control and right into a tree, breaking her arm. Pat did not want to tell Mother because she was afraid she'd be taken to the hospital where they'd remove her arm. We played a game with friends, but Pat was so miserable that I finally told Mother and off they went to the hospital. They had to walk down to Broadway to catch the streetcar up to the hospital in Yonkers. It was not a pleasant trip. Meanwhile, I babysat my sister Maureen.

Speaking of knickers, I can't remember when I finally got to wear long pants, certainly by seventh grade. It seems to me that we were very hard on clothing, or else the clothing we had was cheaply made. I know that each fall I got a new coat, which was called a snow coat or mac. It would be worn out by spring, with holes in the elbows. Footwear was also a problem. I remember getting a new pair of shoes each September and a new pair of sneakers in the summer. The shoes would not last through the winter, so we'd put cardboard inside to cover the holes in the soles. It wasn't a good solution in wet weather. When it was time to get new shoes or sneakers, the whole family would march about two miles up to a shoe store off Getty Square in downtown Yonkers. The store had a fluoroscope machine that you'd put your foot into to see if the shoes fit. I remember the limited summer selection of sneakers—Keds or PF Flyers. That was it. Most Christmas shopping was done at Genung's department store, also on Getty Square.

The topography of Riverdale seemed to make it a trap for snow, ice, and fog. It sloped steeply upward from east to west and, therefore, got little benefit from the low afternoon sun. Remote as we were, snow would lie for days before the City got around, if at all, to plow our streets. In the great snows of 1947 and 1949, there was nowhere to put the plowed snow on the narrow east-west streets. The City hired college students to accompany the plows and use huge shovels to push the snow into catch basins and open manholes, anywhere they could. Low-lying railroad stations along the Hudson River were particularly hard to access and clear. I recall that at Poughkeepsie, the New York Central Railroad hired Vassar College girls to keep the station open.

Riverdale in Spring and Summer

As head gardener by default, I did the springtime planting of flowers and trimming of the hedges around our small front yard. Another task was disposal of the leaves that had been blown into our front and rear yards over the winter. I'd collect them and throw them over the rear fence into the little woods behind our house. One day as I was picking up the leaves alongside the garage door, I felt a sharp sting. I looked down and saw that two snakes had become resident in the leaf pile and were objecting to my intrusion. I was shocked at their totally unexpected appearance and aggression. I subsequently learned that garter snakes are quite aggressive, but their bite is relatively harmless. Thereafter, I was very careful around leaf piles but never again saw any snakes.

One of the rites of summer was replacing the heavy winter carpet in the living room with a lighter summer carpet. The winter carpet was probably wool and susceptible to attack by moths, so it was sent out for cleaning and storage over the summer months. Come fall, the winter carpet was delivered and reinstalled; and in those days, this was done every year. Well, one fall, Mother was notified that her carpet had been destroyed in a fire at the warehouse. She was mighty upset at losing her "good" rug and went on the warpath with the local cleaning shop. She eventually got a settlement, and we got a new rug, but I can't recall if it was as good as the old one. I also don't recall if Mother sent the new one out for summer vacation.

During the long summer days, we'd often go exploring in the woods that were west of Riverdale Avenue and ran down to the Hudson River. There were a number of large estates and institutions scattered through this area. We discovered such prewar wonders as a beautifully executed Depression-era WPA drainage project and an abandoned roadhouse or casino. In the woods around the latter, we found abandoned articles of women's wear, which made us curious about the kind of activities that went on there. A favorite target of our curiosity was the large estate, which we later learned was called Wave Hill. It was occupied at the time by conductor Arturo Toscanini and his daughter, along with her husband, pianist Vladimir Horowitz. One day, we noticed their cherry trees were loaded with beautiful ripe cherries. We went to the front door and ask if they would like us to pick the cherries for them. The woman who came to the door said yes and gave us baskets, saying we could keep half of the cherries and give them the rest. Well, it turned out that picking cherries was a lot of work. So we all picked a basket and ran off with it, leaving no cherries for the householders. Some of the bolder kids would climb the high fence around their swimming pool, mostly after dark, and enjoy a skinny dip. Wave Hill is now a public garden and conference center.

Summer also afforded Tommy O'Rourke and me lots of time for reading. But we also joined our sisters and their friends in games, usually played out of doors. Tommy had a sister, Mary, who was in my sister Patricia's class. Mary became a nun. Their near neighbors, the Deveraux family, had eight children, one of whom became a Jesuit priest. Two Deveraux girls, Nancy and Rita, were near our ages and often joined our games. A favorite card game was Authors, a rummy variant from Parker Brothers, the game people. Tommy and I liked it because it expanded our knowledge of famous writers and their works. I recently bought the game just to trigger my memory of it. The girls played hopscotch (often called potsy in New York) and jacks and jump rope. Tommy and I would sometimes join in. Together we played kick the can, a game not so popular with the girls. They didn't seem to enjoy games where someone was it. Board games such as Sorry!, Clue, and Monopoly were usually played indoors during the winter. We accumulated quite a collection of these popular Parker Brothers board games as well as their quality card games Pit and Rook. Parcheesi was also popular. In later years, my mother and Mrs. Deveraux were neighbors in a nearby high-rise apartment block. Her youngest daughter, Janet, lived with her.

Halloween was new to me when we moved to Riverdale. In the city proper, dressing up and "begging" were done on Thanksgiving Day. In

Riverdale, Halloween was a combination of mischief night and trick-or-treating, with the emphasis on "mischief." The older boys usually went out with stockings full of flour, and these were used to pummel the unwary, usually girls. The more evil-minded filled the stocking with sand or mixed a rock or two in with the flour. You had to be careful who you met up with while out trick-or-treating. As the gang moved along during the night, gates would be removed and tossed over the fence. Another favorite was to ring a doorbell and toss a flaming bag of dog poop on the porch. The fun was in watching the householder rush out and stamp out the flames. Eventually people got wise to this trick, and it became passé.

Living near us was an especially bad dude, big Reggie Ames. Reggie went to public school and usually only associated with us younger kids at Halloween. One night as we patrolled the area, we came to a house with a beautiful jack-o'-lantern sitting behind a big picture window, a rarity in the 1940s. For some reason, Reggie didn't like the setup, so he found a brick or large rock and tossed it through the window. Wow, did we ever run. Reggie did get his just desserts. Wally Herschkel was a recluse who lived by himself in the woods in a house with a corrugated iron roof. Reggie loved to torment Wally by hiding in the woods and throwing the biggest rocks he could onto his roof. The racket would cause Wally to come charging out of the house, but by then, Reggie was gone. Unknown to Reggie, Wally had acquired a gun and next time came out shooting. He hit Reggie in a very sensitive place and put him in the hospital. Reggie came back much chastened, but poor Wally was hauled off never to return.

One of my fondest memories of Riverdale and St. Margaret's was Midnight Mass at Christmas. I particularly enjoyed serving that mass as an altar boy. In those days, we had to fast in order to receive Holy Communion, so we were hungry after mass. On the way home after Midnight Mass, it was customary to stop in and visit with neighbors and share some food and drink. A usual stopping point was the O'Rourke's home, which lay between our house and the church. My sisters and I all had close friends within that family and my parents played cards with Tim and Geraldine, so it was like a second home. I don't recall any holiday experiences that match those wonderful Christmases in Riverdale.

The War Ends, Also My Days at St. Margaret's

The summer of 1945 saw the end of the war with Japan. There was lots of jubilation all around. To us kids, the war seemed to end as quickly as it had begun. We had never been exposed to any of the sciences, so atoms and molecules were all new to us. While the war was on, we'd learn about new "blockbuster" bombs that were being used to destroy whole blocks in German and Japanese cities. So at first, an "atomic" bomb seemed like just another big bomb. Little did we know that it was to usher in the atomic age and the cold war.

But the final war years held some tragedy for a neighbor and for Mother. During WWII, a family hung a special flag in the window with a blue star for each son in the service and a gold star for a son killed in action. Our next-door neighbor became a "gold-star mother" when she was notified that her only son was killed in action. The poor woman screamed and cried for days. We were not friendly with them as their door fronted on Delafield Avenue, and we rarely saw them. However, our houses were so close that we were forced to share her grief. In March of 1944, Mother learned by telegram that her mother, Mary Clarke Quinn, had died. Telegrams always brought bad news in those pretelephone times. Mother was inconsolable for days. It was especially hard for her not having seen or spoken to her mother since 1933. Mother vowed to get over to Ireland to see her father before he died, and she did.

The end of the war gave us kids cause to celebrate and also exhibit our worst possible behavior. Fireworks were in demand and in short supply, particularly in New York City, so we manufactured our own. We'd take a large bolt and thread it part way into a suitable nut, then pack the remaining space with "barn burner" match heads. Another bolt was then screwed tightly into the nut, tightly compressing the match heads. The device was then thrown into the air. If it landed correctly, it would explode and fly into the air. We would then chase it and reload it. It was very dangerous as there was no way to control how high or where the device would go. After the war, illegal standard fireworks started to appear, brought in from out of state. I never thought them worth buying for myself but did join in when others had them.

A sinister favorite firework was a "cherry bomb." This was a sort of kids' weapon of mass destruction. I'm not sure what it was, but it made an awful bang. Because most houses were close and shared a driveway, a cherry bomb lighted and tossed between them made a loud explosion, as they say, enough to wake the dead. Poor Mr. O'Connor lived across the street from the school. I don't think he ever did anything to incur our wrath; he was just a convenient victim. An exploding cherry bomb in his driveway probably made him think his furnace had blown up. He would come charging out of the house as the culprits faded away. He rewarded us with a good show and, thus, was made to repeat his performance every year. His daughter Eileen was a good friend of my sister Maureen, so I would see Eileen at family gatherings. I was always ashamed of how we tortured her father, even though I had very little to do with it other than to watch.

Eighth grade with elderly Sr. Maureen started off well with our first ever class trip. I believe that the entire class fit on one bus. We went to visit warships docked along the Hudson as part of the famous fleet review in the fall of 1945. We visited the USS *Enterprise*, an aircraft carrier that served in the Pacific theater, and the USS *Missouri*, the battleship upon which the Japanese surrendered. It was a thrill to stand on the deck of the *Missouri*, on the embedded plaque that commemorated the surrender spot. On the way back on the bus, the girls started singing, and because of the occasion, a popular choice was "Bell Bottom Trousers." This old sea song had been cleaned up and popularized during the war by the big band of Guy Lombardo, his Royal Canadians. Their chorus went as follows:

Oh, bell bottom trousers, coat of navy blue
She loves her sailor, and he loves her too.

But the version, our girls' sang went as follows:

Oh, bell bottom trousers, buttons made of brass
Loose around the ankles, tight around the . . .

Needless to say, Sister chaperones were not happy with the new version. I'm told that American sailors still wear bell-bottom trousers, but I can't recall when I last saw a sailor in uniform.

Prior to the class trip, my father and I had driven down along the Hudson to observe as President Truman reviewed the fleet from a small boat. There were scores of warships anchored midstream all the way from the battery on the tip of Manhattan and up into Yonkers. As he passed a ship, it fired a salute from its five-inch gun. The ground shook, and the shock wave bounced off the Palisades on the New Jersey side of the river. It was totally awesome. I was glad I never had to experience either end of such gunfire. I recall that my father and I also visited the ships on another occasion. I'm not really sure which ships I saw on the various trips. I may have seen the *Enterprise* with Dad.

As eighth grade got well under way, I began to have problems with Sr. Maureen, rather, she had problems with me. To my way of thinking, it was an institutional problem, that is, the eighth grade was totally unnecessary. In New York State, the Board of Regents monitored student progress by mandating a set of examinations that governed progress through the educational systems. Catholic schools also had a set of exams, but the Regents exams topped them. Historically, the grade-school Regents exams could be taken at the end of the seventh grade or in the eighth grade. I suppose that arrangement went back to the days when people left school at age sixteen. In seventh grade, I took my first Regents exam, which was in geography. I did well, earning a 96. Early in eighth grade, I took the Regents exam in arithmetic earning 100. Although I had yet to take the English I, English II, and history exams, I felt I was high school ready and had no need for the rest of the eighth grade. Consequently, I found class to be a total bore and did little to conceal my feelings. Since I sat up front, Sister could not help but observe my disdain for her and the class. When I started using ink drops from my fountain pen to bomb paper targets on the floor alongside my desk, Sister asked that I leave the room and report to the principal.

I spent much of the remainder of the school year in and out of the principal's office. In retrospect, you wonder why they didn't just send me on to high school or to some other school. Anyway, I worked for the principal, running errands and taking care of the books in the school library. Since few people had phones, one of my tasks was to go on my bike to the homes of absent students to find out why they were not in school. I was not always warmly received, but fulfilled my duties as junior truant officer. Another task was to take the "mite box" money to a bank up in Yonkers. I don't remember how much money we collected each week in the school's mite boxes, but whatever it was, it was collected from the students for the missions in Africa. I had a good relationship with the principal and didn't mind doing the chores she assigned. Anything was better than boring class with Sr. Maureen, who was retired by the time my sister Patricia reached eighth grade.

The highlights of the parish social calendar were a "bazaar" and a minstrel show. The bazaar was our main fund-raiser as the pastor thought raffles were vulgar. The bazaar was held in the school gym-auditorium and consisted of booths selling goodies or booths with games of chance. My favorite game had a big wheel with numbered slots that matched numbers on a game board upon which you placed your bet. As in roulette, the wheel was spun and stopped at a winning number. I played this game for several years and lost all the time. One day I had my money on a number next to the winner. When the operator was distracted, I slid my coin over to the winning number. I was paid off, but the operator looked at me with suspicion. Afterward, I really felt guilty and never played the wheel again.

The other highlight was a minstrel show put on by the men's Holy Name Society. The men dressed in old-time clothes and were blackfaced in classic minstrel show style. I went but never appreciated the show until I was asked to work the coat-checking concession. This was a moneymaker for me and my pals. Interest in the show grew so that it was moved from our gym-auditorium to the auditorium at the College of Mount St. Vincent. Today such a show would probably cause major-league protests in heavily Jewish North Riverdale. I believe that Dan O'Neill was the mainstay of the Holy Name Society and most everything else in the parish. Fortunately for me, I was Dan's "best boy."

The Tyrone Band

As I mentioned earlier, my father was very interested in music. He had a fiddle that he played from time to time and an instrument that I believe was a kind of Autoharp. I used to fool around with it, but I never saw him with it, so I never knew how it worked or why he had it. He also had an accordion that I vaguely remember him playing from time to time. He wasn't very good with it. Dad was a piper in the County Tyrone Pipe Band. He had bought me a small set of bagpipes and tried to get me interested by taking me to Tyrone band practices. To the best of my recollection, they were held in the Seventh Regiment (NY) Armory on Park Avenue at East Sixty-seventh Street. I loved the setting and the thundering music and can hear it my head to this day. The bagpipe tunes that I remember best are "O'Donnell Abu!" and "Garyowen" (see appendix H for the lyrics). However, I think my very favorite pipe tune and song was then and is still "The Minstrel Boy." I've always loved music and regret that I never took the time to learn to play an instrument.

While the Tyrone pipers played and marched around the armory's main assembly hall, I explored the innards of the otherwise empty, spooky building. Downstairs were the battalion and company meeting rooms with their flags and pennants from the many wars of this so-called Silk Stocking Regiment. The Seventh Regiment escorted Lincoln's body when it lay in state in the city on its journey to Springfield. After retirement, General MacArthur lived at the nearby Waldorf-Astoria Hotel. When he died, his body lay in state in the armory.

After practice, the custom was to rendezvous at a local saloon. Among the band members I remember were my godfather Frank O'Donnell who wielded the big bass drum, pipe master Terrence McSwiggan, and piper Hugh Ward, one of the younger bandsmen. He, in particular, seemed to be assigned to get me into the band. Master McSwiggan was rather an aloof and austere man who was probably well suited to his august position.

The Tyrone band played publicly several times during the year, principally in the St. Patrick's Day parade and during the summer encampment of the NY guard regiments at Camp Smith, near Peekskill. This last was probably payback for the use of an armory as a practice hall. Dad took me with him one year to the final day of the encampment. We took a New York Central local to Peekskill, and there were picked up in army vehicles and taken to Camp Smith. As my father and the bandsmen got into uniform and ready to appear for the finale, I explored the encampment. It consisted of a football field of tents, mess tents, and command tents. To my amazement, I came upon my classmate Mike Stanton who was there with his father who was, apparently, the commanding general. As it became time for the final review of the troops, I was invited to ride with Mike in the back of his father's jeep. Thus, I got to see the whole show and hear my father's band in the open field. Dad probably wondered where I'd gone off to and was amazed to see me riding with the general. The train trip back from Peekskill was a horror. As the train limped along, Mike and I sometimes walked beside it.

I think the band was finally ousted from the Seventh Regiment Armory as the band instruments and other equipment ended up in our basement for a time. On St. Patrick's Day, the bandsmen would show up, pick up their instruments, and head downtown by car. I remember that my godfather always put white shoe polish on his drumsticks before he went off to beat the big drum. It was a stirring sight to see the band march up Fifth Avenue and play the music I heard in the armory. Uncle Frank was very handsome and particularly outstanding as he twirled his drumsticks and set the cadence with his big bass drum. It was inscribed with "County Tyrone Association" surrounding the bloody red hand of Ulster.

My sister Patricia, and perhaps Maureen as well, took piano lessons and practiced on a second-hand instrument Mother found at one of her Yonkers shops. They also took Irish step-dancing lessons. These in no way offended my sensibility except when I was dragged to recitals. The girls looked funny in their skirts and capes and were not very exciting dancers. At the time, I had little appreciation for most things Irish and wondered why anyone would want to perpetuate this peculiar dance form. Worst of all was being dragged to the annual Feis, an Irish festival held on the grounds of Fordham University. I believe my sisters and their dance schoolmates performed there, where prizes were given in various categories of Irish performing arts. I'd wander around, looking for friends with whom to hang out, no such luck. Boring! Later Patricia became a member of the County Tyrone Association's Ladies' Auxiliary. Their spiffy uniform was topped of with a tiny cap. I have no idea what the purpose of the auxiliary was, except to look sharp and march in the St. Patrick's Day parade.

In those days, I remember traveling through the city with my father in some kind of vehicle. I have no recollection of where we were going to or coming from when we'd pass a place where a number of streetcar lines intersected. Some lines used overhead trolleys to pick up the electric power and others used a third rail buried in the street. Well, at this busy intersection, there was a pit where a man worked. His job was to reach up and connect or disconnect cars to or from the third rail. My father knew one of the men who worked there, so we would stop on the street and look down to call hello to him. If he wasn't busy, he would climb up and talk.

St. Margaret's Wrap-up

Early in 1946, New York's Archbishop Francis Spellman went to Rome to receive his red hat, that is, become a cardinal of the Catholic Church. Despite the fact that Fr. Kilcoyne referred to the bishop as Fat Frank, this was a cause of much joy in New York's Catholic community. Consequently, our eighth-grade class was asked to write something in honor of the occasion. The best pieces were to be posted in the school hall for all to admire. I submitted the winning poem and John Collier the best prose composition. For many years, I had both of these gems in my archives, but they cannot be found now. I recall that my opening lines went as follows:

> Our dear father is in Rome,
> We hope he'll soon be flying home
> We'll rush to greet him with a cheer.

I felt that John's piece was really well done and that he was a gifted writer. As editor of our high school newspaper, I tried to get him to write for us, without success. John chose to be active in school sports and then worked after school in senior year.

Well, graduation day finally came that June, and I would be well rid of the nuns. As the family's first grade school graduate, I was the occasion of a major celebration. We had relatives and friends from as far away as Long Island and New Jersey. My cousin Barry Quinn was there along with friends Tommy O'Brien and Tommy O'Rourke. We all had a few brews, and no one seemed to mind. By graduation, I had decided to go to Cardinal Hayes High School, a diocesan school, because my closest friends were going there. It would be the beginning of nine years of boys-only schooling. It seemed that the Catholic Church did not approve of mixing boys and girls beyond elementary school. I don't recall having any discussion with my parents about choosing a high school; they just went along with my choice. Besides it was inexpensive, only $5 per month for those who could afford it, and many families claimed they couldn't.

There was some controversy over the awarding of medals and honors during the graduation ceremony. Bob Foran received the boy's "general excellence" medal that was awarded to the top male student. I was awarded a medal for "outstanding scholarship," presumably denoting that I was close on Bob's scholarly heels. To everyone's surprise, Fritz Wiener, a perennial third place in scholarship, was honored as class valedictorian. Fritz was also awarded a scholarship to the St. Regis School, a Jesuit institution, and the city's top classics school. Scholastically, it was a very tough school, and many flunked out.

Many in the Riverdale Irish community felt that I should have gotten the "general excellence" medal and that it was given to Bob only because his parents took good care of the nuns. The nuns were noted for playing favorites and perhaps needed to do so to succeed in their earthly mission. Anyway, it was widely felt that I had been wronged, but I don't think so. My eighth-grade report card was mediocre. I got a 96 in health, my best subject, and a 75 in arithmetic, my worst, with an overall average of 85. I had one lateness and mostly Bs in cooperation and courtesy. However, in separate marking periods, I dropped to a C in courtesy and had a D in cooperation. My eighth grade was not a scholastic success.

Fritz's undeserved honors were attributed to his uncle who happened to be a prominent Jesuit priest. Fritz went to Regis and lasted for two years, whereupon he joined us at Hayes. My dear friend Tommy O'Rourke also had a Jesuit uncle who arranged for him to go to Regis when he graduated from St. Margaret's the following year. Tommy was a good student but very undisciplined and did not survive at Regis. He too, graduated from Hayes. On the other hand, childhood friend Arthur McCreesh of Brooklyn graduated from Regis. The Jesuits did a great job of indoctrinating their students, and I found many of their graduates to be quite insufferable as adults.

Summer trips to Rockaway Beach continued, but in a splendid big car. Dad had acquired a 1939 DeSoto cab that had been repainted a dull green, so let's call it the "green monster." It had "jump" seats and could easily sit seven adults and many more children; there were no seat belts in those days. We'd pack up a crowd of neighbor's kids and head off to the beach. I'm not sure of our route as the Brooklyn-Battery Tunnel hadn't been completed, so I suppose we went over the Bronx-Whitestone Bridge and then onto the Cross-Channel Parkway, which took us across Jamaica Bay.

Off to High School

In September of 1946, I made my first trip to Cardinal Hayes High School at 650 Grand Concourse in the South Bronx, not far from Yankee Stadium. The school was named for a New Yorker of Irish descent, Patrick Cardinal Hayes, who was archbishop of New York from 1919 to 1939. The school was, in fact, not far from our old neighborhood of Mott Haven. The Grand Concourse itself was one of the city's broadest and most impressive streets. Some of the borough's most prominent buildings lined the Concourse north of the school. Among these were the Bronx County Courthouse, the Grand Concourse Hotel, and Poe Cottage, where the writer lived for a time. Yankee Stadium was a few blocks west of the Concourse.

Anyway, it was rather traumatic getting settled in a large school of 2,600 students in one building after the cozy confines of St. Margaret's School. Buying textbooks, getting a locker, and finding the cafeteria (lunchroom) were all challenges. I quickly learned that I had to have what was called a gym bag to haul my books, lunch, and gym stuff to and from school. Hayes first opened its doors in the fall of 1941, so it was bright spanking new when I arrived. Many of the students of the early 1940s never finished, choosing to serve in the armed forces during World War II. In fact, in my senior year, Ben Ptaszenski, a returned serviceman, was elected student president.

Travel to Hayes from Riverdale was, itself, a challenge. It started with a downhill walk of six blocks east to Broadway. There we boarded the C car, which went south on Broadway, then east on Kingsbridge Road, and up a long hill to the Jerome Avenue to the elevated subway line. We then took the train down underground to the East 149th Street station, and hiked up the Grand Concourse to school. The school was located above a corner of the New York Central's Mott Haven rail yards where the tracks of the Central and New Haven lines merged to go downtown to Grand Central Station. It was a cold and windy place in winter and very noisy with constant train activity. One of the hazards of the trip was the possibility of a streetcar derailing on the long hill where Kingsbridge Road ascended from Broadway. A runaway streetcar could wipe out service for hours. We'd then have to wait for buses to come pick us up so we could continue our journey.

Our high school was run by New York archdiocesan priests who also taught many of the liberal arts subjects. Xaverian Brothers generally taught the sciences and some languages, and Marist Brothers taught mathematics. Both priests and brothers pitched in for the mandatory religion classes. There were other teachers for music and business subjects. We soon found out that the brothers were much stricter than the priests. In freshman year, I had social studies (Fr. Joseph McGann), English (Fr. Leo Halpin), Latin (Fr. Victor Pavis, later Msgr.), general science (Br. Norbert), algebra, and religion. There was also the state-mandated health education, which involved classes and gym. In one of the gym runs in the park across from the school, I was recruited for the cross-country track team. Having been brought up seemingly in a cocoon, I had no concept of organized sports and track, in particular. So I declined. I've since regretted that I didn't go out for cross-country. However my knees would probably have failed me sooner, so perhaps, just as well.

I don't remember much of freshman year; it seems to have gone by very quickly. I know that I enjoyed my classes, even Latin. I was quite surprised to finish the year with a 93 in Latin. In English, I had trouble with formal English grammar, something I'd never had in elementary school. Taking Latin seemed to help with English grammar, eventually. I liked Fr. Halpin, and he liked my writing. He was faculty advisor for the school newspaper and suggested that I join the staff, and I did. Social studies were my favorite subject and amplified the geography that I enjoyed in elementary school. I ended the year with an 88 average, sort of a B+. Oddly enough, an 82 in algebra helped pulled me down below an A average, and math was supposedly my strong suit.

There occurred one memorable serious incident in freshman general science class. Br. Norbert was easily rattled and got carried away doing the various demonstrations of scientific phenomena. He also darkened the room to show movies and slide presentations. Alongside the classroom windows were a series of dirt-filled garden boxes that were never used by our class. One day during a film, some of the boys near the window started throwing dirt balls across the room. Their trajectory could be seen on the screen. Of course, the boys on the far side started throwing them back. By this time Br. Norbert, who was sitting on the far side of the room, saw what was happening. As always seems to be the case, he caught a boy throwing dirt back, not one of the boys who started the melee. Brother grabbed the boy and punched him in the face, knocking him down. I think he then kicked the boy before he dragged the bloody lad out of the room. We were all stunned and became very quiet and evermore wary of Br. Norbert. I didn't know the boy, so I never knew what happened afterward, and nothing was ever said to the class. I do know that sometime before my junior year, Br. Norbert disappeared from the faculty list.

In my freshman year, I got my first taste of "jug," as we called after-school detention. Teachers issued jug slips usually for infractions in class. The slips were collected by a monitor and turned in to the office of the dean of discipline, the dreaded Fr. Stanislaus Jablonski, known as Jabbo. The juggees reported after school to a classroom where they were detained for an hour, or longer if there was any bad behavior in jug. Apparently, different faculty members were assigned each day to monitor jug. Priests usually allowed you to read or do homework while seated at your desk. Some of the brothers ran jug like an army boot camp. You were made to stand at attention for fifteen minutes, then sit at attention for fifteen minutes, and so on. If you acted up in their jug, you'd get a smack in the face and maybe a jug slip for the following day. Jug was best avoided, and I did so. I doubt I did more than three jugs in my four years at Hayes.

After we moved to Riverdale, each Christmas, we'd make the pilgrimage to Macy's in downtown Manhattan. They had a wonderful and dazzling display that moved past the many long windows on West 34th Street. They also had a great Santa Claus who, if you paid 25¢, gave you a small gift when you sat on his lap. I was too old for that but had to stand by, impatiently, while my little sisters waited in line to see him. I loved to go to the boy's department and look at the dazzling display of Erector sets, electric trains, and chemistry sets. These items had been in short supply or non-existent during the war years, and I was hoping to get one or more. I think it was for Christmas in 1946 that I got my Lionel train set. It came about because for years I had been pestering Mother to let me buy a .22-caliber rifle like Cousin Barry's. She said there would be no rifle but that I could have a train set, so the deal was made. She knew that I was very interested in trains and railroading in general. The O gauge set was very basic—steam locomotive, tender, boxcar, gondola car, and lighted caboose. Over the years, I added to the set and created a nice setup on a large table in the attic. The children of visiting relatives and friends loved to get to the attic to see the trains; they were quite a novelty at the time. For various Christmases, I acquired my very own Erector set and chemistry set. These were products of Gilbert, in those days a prominent maker of boys stuff, including American Flyer train sets. After doing some of the erector projects and chemical experiments, I put them aside for other more interesting pursuits.

Another well-established New York event was the Christmas show at Radio City Music Hall. In the 1946 season, the entire family went to the show. We saw the very boring film *Anna and the King of Siam*. It had an excellent cast, featuring Irene Dunne and Rex Harrison in the title roles, but it was not a kid's picture. The Rockettes were probably at their very best, but failed to impress me. I was not into dance at that point in my life.

The other big yearly holy day was Easter. The girls and the ladies got new spring outfits to show off—hat, coat, dress, gloves, stockings, and shoes. I don't remember what my dad and I got, but I'm sure I, at least, got something new. We have a picture of my sisters in their new duds, circa 1942. My sister Pat says her dress in the picture was made by Aunt Kathleen Quinn who came and stayed with us while she worked on it. What I remember best about Easter is that it was always cold. We'd look forward to Easter and spring so we could go outside for pictures in our new outfits, and we'd shiver. We always colored our Easter eggs and Mother hid them around the living room so that we could have an egg hunt. I don't think there was any reward for finding the most. As an altar boy while at St. Margaret's School, I had to spend a lot of time in church over the "holy days." I liked the Holy Thursday services but did my best to dodge Friday and Saturday duty. As good Catholic school kids, we'd honor Good Friday by either spending the three hours of Christ's torture and death in church or at home praying in our rooms. There was no tomfoolery on Good Friday.

A major innovation made its appearance shortly after the war, the ballpoint pen. The early offerings were quite large and made to resemble miniature spaceships. In fact, they were promoted as "space age" and were quite expensive. We got early versions for Christmas and found they were quite balky and not quite ready for prime time. However, they were the first steps on the way to eliminating the fountain pens we had to use. These pens were quite beautiful but often leaked, leaving ink stains on your shirt pocket. In primary school, we had ink wells built into our desks, but the ink usually dried before we finished with it. I don't think the nuns were too keen on the ink wells as there were many ink stains on the floor. It was important to have a supply of blotting paper to soak up excess ink. The great fountain pen companies Parker, Waterman, and Esterbrook would soon succumb to the practicality of the ballpoint pen. In the 1960s, while working at RCA in Camden, New Jersey, I witnessed demolition of the Esterbrook plant. Parker continued into ballpoint pens, but I haven't noticed them lately.

Moving Up at Hayes

September 1947 saw me back at Hayes for another year. I was getting used to the travel by streetcar and subway, where seating was often not to be had. The New York City subway system, while mostly underground, was actually 40 percent above ground. When I boarded the subway to school at Kingsbridge Road, I actually had to climb upstairs to an elevated platform. On the way south, the train went underground, so I had to again climb up stairs to the Grand Concourse and on to school. On the homeward trip, the train went above ground in time to see Yankee Stadium off to the west. Not far from the stadium, we'd pass the Ruppert Brewery, home of Knickerbocker beer. Col. Jacob Ruppert built the New York Yankees into the baseball powerhouse it is today. The streetcar and subway cost 5¢, but you'd get a free transfer from one to the other. The 5¢ fare was a bargain that was vigorously defended by various political factions, along with nickel beer. However in 1948, during my sophomore year at Hayes, the fare went up to an astounding 10¢. It was the end of an era of cheap public transportation. As students, we got a monthly discounted pass so that we continued to pay close to the original 5¢. The New York subway fare is now $2, and there is no student discount.

To make life easier for the boys at Hayes, the city provided an express bus for us. It picked us up at Kingsbridge Road and went down the Concourse, nonstop to school. The bus was an old prewar model whose few amenities were soon destroyed by our rowdy schoolmates. Any part of the interior that could be thrown out of the windows, was. That included parts of the very seats we sat on. A special assembly was held for the presumptive riders of the special bus, and we were severely chastised. It was no surprise that the special bus service ended shortly after it began.

While at Hayes, I continued to augment my stamps, particularly stamp collection. Every morning as I walked to school from the subway station, I passed the main Bronx post office. New stamps were issued regularly, so I would often stop in and try to buy whatever plate number blocks were available. Plate number blocks of four were desired by collectors, and the postal workers tended to ration them. I was thinking that I would be accumulating something of value, because older plate number blocks were quite pricey. Unknown to me, the post office was flooding the market with new issues and printing the stamps in large volume. Years later when I sought to sell my collection, first-class postage rates had gone from 3¢ to over 20¢. Consequently, my stamps were virtually worthless; and I had a lot of them. I did sell them but got only 10 percent of the original face value. It was a hard lesson. I think I got about $400 for that portion of my collection that I was able to sell. I kept a small starter collection, in case any of the grandkids are interested. But stamp collection doesn't seem to be of much interest in these days of television, computer games, and iPods.

At school, I became more involved with *Challenger*, the student newspaper. I began to write news stories and became acquainted with the rest of the staff, many of whom were to be close friends while at Hayes and even after graduation. Principally I speak of Tom Gaffney, Jim Rattray, and George Gunning. In my junior year, Bob Lancellotti and Desmond Vella, as sophomores, became important contributors and friends. One of the drawbacks to my involvement with the school paper was my need to get home to deliver my papers. Since we had no phone, I arranged in advance for my sister Patricia to deliver my papers, if I did not get home by a certain time. It got so that she delivered them quite often. I hope I paid her handsomely for her work, but being rather tight with a dollar, I probably didn't. Eventually, I had to give up my *Herald Statesman* paper route, probably by the end of my sophomore year. I was very reluctant to do so as I was earning almost $8 a week, a goodly sum in the late 1940s.

I had a really bad problem with my sophomore class assignments at Hayes. Unknown to me, another James Joseph O'Donnell had been transferred into our school's main building from one of the outlying buildings. Naturally, he got my academic class schedule, and I got his business class schedule. Such mix-ups were common prior to the introduction of Social Security numbers as positive identifiers. By the time I got things straightened out, I ended up in one of the lower-achieving homerooms and its related set of core classes. While I didn't like being in that particular group, it probably helped my grade point average.

My second year courses were world history, geometry, Latin 2, Spanish 1, English 2, and Religion 2. Of my teachers, I only remember Fr. Blust in world history. He had a difficult time controlling a very unruly class. I loved history and still do and got a 98 in his class. With that 98, I won the runner-up prize in history that year. Oddly, my *Challenger* colleague and friend Tom Gaffney got first prize. After history, my next best mark was a 93 in geometry. I ended the year with a 90.1 grade point average, which I suppose is an A. I had one lateness and one absence. As sophomore year came to an end, I turned sixteen, old enough to get working papers, so I could take advantage of possible new employment opportunities.

Adventures In and Out of Town

As we aged, so to speak, we began to go farther afield for fun and profit. It started with bike trips across Yonkers to the Bronx River. There we found a wonderful bike path along the river that we were able to follow all the way into downtown White Plains, a trip of about sixteen miles. This was a trip we made when we had nothing better to do, often with Tommy O'Rourke, but sometimes with others. I had gotten a Kodak Vigilant 620 camera for Christmas or birthday and took it along on our various journeys. It was a folding bellows type of camera that, with flash attached, looked pretty spiffy. I sold it recently on eBay. I recall taking some pictures as we explored the New York subway yards in the Bronx. I then started developing my own black and white photos, tying up our main bathroom for my darkroom. This was so unpopular that I had to give it up as one of my bad ideas.

To supplement his newspaper income, Bob Foran had been caddying upstate with his older brother, Arthur. When Arthur graduated from college, Bob was left without a companion for the long journeys to the wealthy golf venues in northern Westchester County. Bob invited me to join him and I did so, not knowing what I was getting into. I don't remember the full details, but I know we left home in the dark and started hitching car rides through Yonkers, to the Saw Mill River Parkway, and then onto the Taconic State Parkway to where it then ended at Chappaqua. We then had to get one or more rides to the Mount Kisco Country Club, where Bob and Arthur had been working. There we went into the caddy yard and waited for a chance to tote heavy golf bags till dark. If we were lucky, we got to make two eighteen-hole rounds, carrying two bags each time. Each bag was $1.50, so we could make $6.00 a day at most, with an occasional tip. I knew nothing about golf, so I was not a very good caddy and, therefore, rarely got tips. After the long day at the golf course, we'd have to start the long journey home, usually arriving well after dark. One thing that I enjoyed at Mount Kisco was the proximity of their "summer stock" theater. I never got to see a performance, but enjoyed seeing the performers, some of whom were quite famous, rehearsing or working on scenery. I soon found there were better ways to earn money than making the all-day, thirty-mile trip to Mount Kisco in order to carry heavy golf bags for "peanuts."

Bowling was another out-of-school activity. Initially, we took the streetcar down to the lanes near West 231st Street in Marble Hill, a section of Manhattan that's physically in the Bronx. It was only possible to bowl if a pinsetter was available. I believe we paid the pinsetter 10¢ a game, plus a nickel tip. We'd roll the coins down the alley to the setters who were often injured by flying pins. Near the bowling alley was the RKO Marble Hill movie theater that we also patronized. It wasn't one of the real palaces like those in Yonkers that we later patronized. However, nearby was Luhr's ice cream parlor, truly a wondrous emporium with nice booths and sexy lighting. It was an after-the-movies date place, but I don't recall ever taking a date there. I know many of our local girls had been there, probably on dates, possibly not.

In later years, we moved our activities up into Yonkers. A bowling alley on South Broadway had the new-fangled AMF pinsetting machines and nicer alleys. They also served beer, a real attraction for us in a time when there was no formal age-checking. After bowling, we'd usually stop at a pizza parlor near the city line. They had about the best pizza I've ever had and served beer. We also started patronizing the Yonkers movie theaters, the RKO Proctor's and the Loew's Yonkers, both on South Broadway near Getty Square. These were splendid movie palaces, but have not endured. The Proctor's is now used for office and retail space. The Loew's became Brand's when the film studios were forced to divest their theater chains and has since been demolished.

It was in one of the Yonkers movie palaces that I saw the first 3-D film, 1952's *Bwana Devil*. Going in, we were issued the cardboard glasses with red and green plastic lenses that were needed to get the three dimensional effect. The glasses and the movie itself were terrible and people started walking out long before it mercifully ended. We felt we had been ripped off. Tommy O'Rourke often went with the group to the movies. He had an easily triggered laugh that would burst out and be heard above everything else in the theater. If the movie was a comedy, we'd usually make him sit well away from us. I imagine that he got a few laughs out of *Bwana Devil*.

Classmate Bob Foran and his family belonged to the Riverdale Neighborhood House. This was a private facility that offered tennis, swimming, and a lending library. Because there was no nearby city library, we had access to the Riverdale library for a small fee. Only Bob of our group had access to the full facilities. Apparently he had been playing tennis there and invited me to play as his guest. At that point, I hadn't played tennis in years. Much to Bob's disgust, I beat him. I don't recall that we ever played again.

Another summertime activity was softball. It was played weekday mornings in the schoolyard at nearby PS81. It had a short left field so the better players could hit the ball off as well as over the fence, the latter being a home run. I was a scatter hitter and could never power the ball out of the yard. It was fun though. The Neighborhood House hosted a tough senior softball league that played weekday evenings. As kids, we often wandered over to watch the games. While in college, I had a chance to substitute, and I actually had a few hits and walks against some of the league's best pitchers. These were guys that had dazzled me over the years. In the field, I played wherever I was needed, often as the short fielder. In college, I played on the 1953 intramural championship softball team, along with my future brother-in-law. As the poorest player, I was safely installed in right field. I never had much of a chance to play ball after that year.

Mother's New Housing Arrangements and Cousin Packey

There was a severe housing shortage after the war, and Mother decided that we would take in a boarder, namely an elderly woman who needed a place to stay. She gave the woman my small hall bedroom next to the bathroom. We then had to totally rearrange our sleeping arrangements. Mother and the girls took the front bedroom and Dad and I the back bedroom. Both rooms were quite large and easily accommodated us. I don't recall a great deal of this period of our home life, but I do recall my father and me listening to the radio in our shared room. We particularly enjoyed the Henry Morgan comedy show that came on every weekday evening for fifteen minutes. Morgan and his sidekick Gerard, played by Arnold Stang, mocked everyone and everything, including their sponsors. It was a lot of fun while it lasted. I was a lifelong fan of Henry's, but we parted ways after an altercation some years later. There will be more about this later.

Mother's boarder did not work out, so she decided to create a separate rental apartment downstairs. The apartment would consist of our kitchen, dining room, half bath, and pantry. The renters would use the side entrance that also gave access to the laundry equipment in the basement and to the "apartment" at the top of the stairs. To accomplish this, we would create a kitchen and dining room upstairs in what had been the back bedroom used by Dad and me. I don't remember in detail how it all got done, but I do recall helping my father with some of the electrical work, including hanging the light fixtures in the new kitchen and dining room. The sleeping arrangements had to change again. Mother, Dad, and my sister Maureen went to the front bedroom; my sister Patricia took the small hall bedroom vacated by the boarder; and I went to the room on the third floor. I was to remain in that cheerful room for the rest of my stay in New York. The stairs were no problem and, eventually, there was plenty of room for my friends to stay overnight.

One of the drawbacks to the downstairs apartment, as originally conceived, was the lack of a shower or bathtub. This problem was solved by the fortuitous arrival on our shores of Cousin Patrick Quinn, oldest son of mother's older brother Joseph who had inherited the family farm. "Packey," as he was called, arrived, as did most greenhorns in those early postwar years, in a long overcoat that almost touched his shoes. Uncle Mike Quinn sponsored Packey, met him at the airport, and dropped him off at our house where he was to stay until he got a job. Mother immediately put him to work installing a steel shower in the near end of the pantry and clothes rods in the far end. With what amounted to a full bath and closet, the apartment was soon rented to a young couple. I don't recall ever seeing or speaking to them during the years they lived in the back of our house. However, my sister Maureen remembers playing with their son, Ricky, who was her age. She and Ricky would play Monopoly for days and just leave it set up on a table. They would go back and forth through the door between our living room and their apartment.

I guided cousin Packey on trips around the city as he looked up friends and relatives from the Old Country. However, he soon found construction work in New Jersey and moved into Uncle Mike's home in Clifton until he found his own place. He was drafted into the army and gained U.S. citizenship upon discharge. Back in New Jersey on leave from Westover Field, Massachusetts, Packey married Frances McHugh of County Mayo, Ireland. I missed their 1951 wedding as it occurred while I was away in the air force. Packey went on to a very successful career as a construction foreman and superintendent. He and Frances had a very strong marriage and had five children, four boys and, lastly, a girl. Sadly, Packey died of colon cancer in 1999 at age seventy.

Eventually the postwar housing boom caught up to demand, and our tenants moved out. My sisters and I then demanded that Mother restore the house to the way it had been originally. As we got older, we didn't like our friends to think that we had to take in boarders or renters to make ends meet, when we didn't need to. Again, I have no recollection of how or when it was accomplished, but I think it was done by the time I graduated from high school in 1950.

Another facet of living with my mother was her clandestine shopping trips to the used furniture stores on Riverdale Avenue in Yonkers. She had no transportation other than her strong legs to carry her up and back. But, magically, strange new (to us) objects would appear from time to time. Among these were a beautiful set of wicker furniture for our sun parlor, a china closet that we used as a bookcase, and an *Encyclopedia Americana* set, vintage 1926. These last were beautiful books but sadly out of date in the atomic age. We had wanted a record player, so there appeared one day a windup Victrola console, along with a number of 78-rpm one-sided records. These recordings were really ancient then and were by such favorites as John McCormick, Enrico Caruso, and Bing Crosby, very early Bing Crosby. We were not crazy about the records but enjoyed winding up the old machine and getting music out of it. It was not electrically powered, so much winding was needed to keep a record session going.

In the postwar period, new recordings were appearing on the small, RCA-developed, 45-rpm disks. My sisters and I begged Mother for a new electrically powered machine that would take that speed as well as the earlier and larger 78- and 33-rpm platters. The age of electronic music came to us in the form of a big Philco record player, so no more windups. We did have to buy special needles so as to minimize wear on the record surfaces. On Riverdale Avenue in Yonkers, there was a firm that installed and maintained the new Seeburg 45-rpm jukeboxes. Records taken out of the machines were then put on sale in their shop. There I was able to buy for pennies the popular recordings of the day—by Bing Crosby,

Frank Sinatra, Nat King Cole, and others. Novelty songs were big in the 1940s and had a short life in the jukeboxes, so they quickly ended up in the bins of my favorite (used) record store.

I suppose one of the most important "magical" acquisitions was a prewar Maytag washing machine that had been upgraded for the new milieu. It consisted of a large copper tub on wheels with a hand wringer on the side. The washing action was performed by a set of three copper suction cups on a frame that turned while plunging up and down in the tub. The cups were removable to give access to the clothes, which were then dragged over the edge of the tub and through the wringer. It sounds really clumsy, but it was a great improvement over hand washing. After a year or so, the machine was upgraded again, this time by addition of a spin ringer unit. It wasn't fully automatic as the washers are today; I'd call it a semiautomatic washer. One problem with the machine was that it "walked" around the basement, the plunging and rotating action was so potent. We were afraid that the machine would knock out the nearby boiler for our steam heat or crash through the garage wall. I don't recall when the Maytag monster was replaced with a more up-to-date machine. It was a truly beautiful piece of machinery.

I Become a Junior

Back at Hayes in the fall of 1948, I had no newspaper route to distract me from my duties as News Editor of the high school newspaper. This was a demanding job that involved my own writing as well as editing that of others. In addition, I had to write headlines, do page makeup, crop pictures, and take copy and pictures to the printer. John McNamee was our editor in chief, but I have no memory of him. Fr. Halpin was moved to a new post, and Fr. Loughman took over as faculty advisor. While Fr. Loughman took his duties more seriously, we never found him difficult to work with. In time, he too pretty much left us to our own devices. He remained with us until we graduated.

As a senior editor of the paper, I was given a pass to the press box for our football games. When at home, our team, the Cardinals, played at Triborough Stadium on Randall's Island in New York's East River. The stadium was quite large, but we had a good view from the press box. Having no responsibility for sports writing, I was pressed into service as staff statistician, a job I held through senior year. This duty involved keeping the game charts so that we knew about passes, runs, yardage, tackles, and so on. Our sports staff, ably headed by Tom Gaffney, used the data to write their game stories. Our team, in red and gold, played some tough schools, including Brooklyn Prep, Chaminade, and on Thanksgiving Day, Mount St. Michael's Prep. In the previous season, we had a 4-4-1 record, where the tie was 0-0 with All Hallows Prep. In that year, we even beat the Mount 24-0, so the season was considered a success. I don't know the results of the 1948 season but suspect it was about the same as the prior year and in 1949 when we went 3-4-1 and tied the Mount 14-14. Win or lose, our Cardinals always played hard, and I got so that I enjoyed the games immensely.

I have only the vaguest recollection of how I got to the games. I know that the school provided a bus for those who wanted to go to the away games, as I remember the long bus trip out to Chaminade Prep on Long Island. We always seemed to lose to them. For the home games, I believe we were on our own. For me, this meant the same trip I took to school, but now I stayed on the train to 125th Street in Manhattan, East Harlem to be exact. Then I had a two-mile walk across East 125th Street, over the Triborough Bridge, and down the seemingly endless steps and ramps to the island. Our Thanksgiving Day games with the Mount were all played on the island, as the Mount did not then have a suitable field. Later, the stadium was called Downing Stadium and music events were held there. In 2002, the stadium was torn down, rebuilt using lights salvaged from Ebbets Field in Brooklyn, and renamed Icahn Stadium.

In those days, our family Thanksgiving dinner was held at my godmother Aunt Mary's house in Mount Vernon. Christmas was then held at our house in Riverdale. Aunt Mary, my mother's younger sister, had married a Swede around 1936. His birth name was something like Johannes Alexanderson, which he had legally changed to the handier John Alex Sander. John was good-humored and an excellent host, often plying us with Swedish caviar and strong drink, some of which he brought back from trips to the Old Country. My sister Patricia remembers that "orange blossoms" were a favorite cocktail at the Sanders home. In 1938, Mary and John had a son, John Patrick, who was always called Jackie within the family. Age-wise, he sort of fit in between my two sisters. In later years, he looked upon me as an older brother. As an only child, Jackie was badly spoiled by his mother, much to the disgust of his father. I suppose it was to be expected, as Mother often spoke of her sister as spoiled and pampered.

From Riverdale, my parents and sisters went by car to Aunt Mary's. For me it was a challenge as to how I would get to Thanksgiving dinner in

Mount Vernon from an island in the East River. Well, I retraced my steps westward across the Triborough Bridge and across East 125th Street to the railroad station on Park Avenue, instead of the subway station at Lexington Avenue. Local trains of the New York Central and New Haven Railroads, to and from Grand Central Terminal, made a stop there, perhaps some of the long distance trains did as well. There, I was able to take a New Haven train to Mount Vernon and walk from the station to Aunt Mary's. Being a train aficionado, I enjoyed the train ride, short as it was.

My junior year classes were American history (Fr. Bernard Corrigan), Latin 3, Spanish 2, Algebra 2 and 3 (Br. John Bosco), English, religion, and the ever-annoying health education. Of these, history was far and away my favorite. Father Corrigan was an excellent teacher who recommended and encouraged outside reading. He got me started reading the wonderful Kenneth Roberts novels: *Rabble in Arms, Arundel, Northwest Passage*, et al. These were typically set around the American Revolution, and in many of them, Benedict Arnold was the hero. As a result, I wanted very badly to retrace Arnold's campaign down Lake Champlain in 1776-7 and visit his heroic battlefield sites at Saratoga and elsewhere. Even as I write, I fondly remember those books as among the best I've ever read.

I recall my introduction to calculus in Algebra 3 with Br. John Bosco, a Marist brother and fine math teacher. At that point in my life, I had no idea that I was to spend many more years in calculus classes. In Spanish class, I sat next to our star basketball player, Lou Gigante. Lou was also in my senior homeroom and was to become, without a doubt, the most famous member of my class. As a priest of the New York Archdiocese, Lou became known as the Mafia Priest because his brother was Vincent "the Chin" Gigante, known mobster and hit man. Vincent was able to elude conviction for major crimes through various subterfuges but did go to prison on lesser charges and died there in 2005. Fr. Lou unwisely became involved with Vincent's defense, hence the soubriquet Mafia Priest. Vincent's family is suing the federal government for mistreating his many illnesses while he was in prison in Springfield, Missouri.

Lou's basketball skills earned him a scholarship to Georgetown. There, he was a standout for several years and became cocaptain in 1954. After Georgetown, Lou went to Dunwoody, the New York Archdiocesan Seminary in Yonkers. As a priest, Lou became famous for his leadership in various causes for relief of poverty in the South Bronx and for new housing in blighted areas. He even served on the city council and was a candidate for congress in 1970. Lou left a strange message in my 1950 yearbook:

"Lots of luck to a guy who always mixed up the class meetings." I guess I was continuing to practice subversion of authority in the classroom, a skill I had honed in the eighth grade.

Junior year ended with a falloff in my grades—religion (93), English (89), American history (91), algebra (84), Latin (81), and Spanish (80)—for an average of 86, or a B. My excuse for this reduced academic performance was that I was spread too thin. In addition to my duties on the school newspaper, I was working after school and during major holidays. These jobs were somehow arranged through the school, but I don't remember how I was chosen and how they came about. In any case, both were memorable.

After-School Jobs

My first "real" job was with the Triborough cold storage plant on West 125th Street in Manhattan. It was a huge facility, some six to eight stories high and occupied most of a city block on the far west side. The work there was seasonal and involved storing and retrieving fur coats. For some four weeks in the late spring, we unloaded trucks full of coats and placed the coats in the proper place in the plant. In early fall, we did the reverse. Some weeks I worked for Triborough almost forty hours after school and weekends. The headmen at Triborough were Irishmen, so that's probably why boys from Hayes got the temporary jobs.

This seemingly simple job was actually rather complex. The springtime receiving task was relatively easy. Since each truck picked up coats from specific furriers or department stores, the furs could be taken, en masse, to their proper place in the plant. The more difficult task was to retrieve the coats in the fall when each boy was given a list of coats to pick from many locations in the plant. Top-line furriers and shops had their own locked cages within the plant. So before you set off to pick your list, you needed to know the floor and location on the floor and have the key for any locked cages involved. You grabbed a rolling rack, if one was available, and took the elevator to one of your floors. When you got off, you would be in near-total darkness as we were supposed to turn off the lights when we left a floor. As you felt around for the light switch, you might hear a blood-curdling scream from out of the gloom. You might then decide to go to a different floor and try your luck there. If not, you'd find the switch, turn the lights on, and warily search for your coats, alert to any potential tricksters. You'd also see the famous injunction of Jesus, slightly paraphrased and posted over the elevator door: "For many are cold, but few are frozen."

On some floors, furs were on three open tiers of racks. So you'd first check to see if any feet were showing under the lower coats. The floors were so huge, that it was hard to spot the feet of any potential trickster(s). Moreover, someone could be hiding among the upper coats, to play a trick on you. A typical trick was to get to the light switch and turn it off, leaving you on an upper tier of coats in total darkness. Another trick was to wait until you came down from an upper rack and were standing on the floor. Then a heavy wooden coat hanger would come skidding across the floor aimed at your ankles. They hurt if they found the target. The office had sound monitors on each floor and could sometimes detect our hanger fights, if they became too boisterous. Then we'd get a nasty announcement to cease and desist.

In climbing to get a coat from the upper racks, we usually had to step on the coats below and drag the coat out from its place if they were densely packed. This sometimes meant torn linings. An even worse situation occurred in the cages of the top-line stores. The cages were also tiered, two or three high. The latter were not high enough for you to stand, so you'd have to crawl in and out on your knees and drag the coat behind, be it mink, sable, or ermine. Since the cages were lined with metal mesh, it was easy to snag and tear the coat linings. We'd sometimes have meetings where the Irish bosses would rant and rave about the sorry condition of the coats and that we would be losing accounts because of our carelessness.

I enjoyed working at Triborough and especially the fine coffee shop across the street. They had the best apple dumplings. As best I can recall, I worked there during my junior and senior years. Some weeks, I put in almost forty hours and didn't get home until nearly midnight. We Hayesmen all became well-known to the bosses and them to us. Well, in senior year, the employees got all upset by some management action. It was decided that we should probably strike. So at break time, we all trooped down the stairs to go to the coffee shop and discuss our plans. The stairs led to the front office, and as we walked past the bosses, I happened to be first. I guess they'd overheard our talk because the big boss said something like "O'Donnell, are you leading a strike? If so, you will be fired." I convinced him we were just talking and would be back after break. When we went back to work, the problem, whatever it was, seemed to have evaporated. It was my first labor action and my first semistrike. There would be other opportunities in the years ahead. I'm guessing that the rise of air conditioning and the decline of the fur business combined to doom Triborough as I can find no listings for it anywhere.

My other job was with the House of Flowers just off Fifth Avenue on West 53rd Street. There I delivered flowers during Christmas and Easter holidays. But delivering their flowers was to be more than a job, rather an adventure. This was a very big and well-known flower shop whose logo on a plant or flowers was prized. There were many delivery boys and vans and a delivery manager who made daily assignments. Only a lucky few ever got to go with the vans, never me. I always seemed to draw the most difficult deliveries. The manager stressed that we must always come back to the shop at the end of the day and turn in our time cards, or we would not be paid.

My first assignment was to deliver two large plants to the Upper East Side. I was given some change for the Madison Avenue bus and pushed out the front door with a "Godspeed." Well, the plants were so big I couldn't see where I was walking and almost fell off the curb as I crossed Fifth Avenue, heading east to Madison. I had to struggle onto the bus and find a seat for my "friends" and me. My fellow passengers either laughed at my struggles or looked very displeased with my presence on the bus. The first delivery was to the famous Metropolitan Opera soprano, Lily Pons. Somehow, I was able to get the doorman to let me deliver the plant in person, probably because it looked too big for him to take up. I had become a big opera fan and looked forward to meeting the singer. When I arrived at her apartment, Lily opened the door, grabbed the plant from me, and closed the door. No tip, no thanks for Jimmy. The other plant went to an elderly woman farther uptown, in a much less classy setting. She gave me a tip of $1, which was a lot since my pay was only 25¢ an hour.

Allow me a brief digression here. I had begun to listen to the Saturday-afternoon opera broadcasts from New York's Metropolitan Opera. I found that I enjoyed the between the acts (entr'acte) sessions with Milton Cross and other opera fans and critics more than the opera itself. These sessions appealed to my love of history and biography as opera plots, historical context, and past and present performers were discussed. As a result of the broadcasts, I became an opera fan even though I was not to experience a live opera performance for almost fifty years. In the same time frame, I happened to hear a radio dramatization of William Faulkner's novella *The Wild Palms*. It was very powerful and probably very much like the earlier productions of Orson Welles's *Mercury Theater of the Air*. I remember it so well, even now, but can't find any record of it on the Internet.

Well, back to work. One of my most famous flower deliveries was
to Centre Island off the north shore of Long Island. Again, I had to lug
a very large plant through the subways and onto a Long Island Railroad
(LIRR) train at Penn Station. The train took me to Oyster Bay, a seaport
town, where I'd been told I could get a ferry to the island. When I got
there, I found that the ferry only ran during the summer. Rummaging
around town, I found that there was a road to the island, but it was a
long way around, and the cab fare would have been prohibitive as I only
had a few dollars for the mythical ferry. I called the store and asked for
new instructions. They must have called the customer because when the
delivery manager called me back, he told me to leave the plant at a specific
place, and the customer would pick it up when next in town. So back on
the LIRR and onto the subway I went, with no tip for my pains.

My most infamous delivery was to Norwalk, Connecticut. I carried the
plant via bus or subway, I can't remember which, to Grand Central Terminal
on East 42nd Street. There I bought my ticket to Norwalk and took a New
Haven train to Stamford, Connecticut, where I would transfer onto the
branch line north to Norwalk. While waiting for the train at Stamford, a
family noticed that I had a House of Flowers plant and wondered if it was
for them. It turned out that it was for them, and they offered to take it and
let me head on back to New York. I gave up the plant, and they gave me a
tip whose amount escapes me. Now here's where I blundered.

Instead of getting on a train back to New York, I decided that since
I already had a ticket to Norwalk, I would go on and do some exploring.
John Collier, my classmate and fellow House of Flowers delivery boy, was
ostensibly dating a girl from Redding, Connecticut. I had noticed that
Redding was a train stop on the way to Norwalk. So I cleverly decided
to ride on to Redding, get off there, walk around town, and tell John all
about it. When the train stopped at Redding, there was no station, only
an overturned wooden soft drink box to step onto. There didn't seem to
be any town, either, just a wide spot on a very narrow road. I had to start
walking to find a road where I could hitchhike back to New York. I was
able to catch a series of rides that actually took me past my house, but I
couldn't get off there. I had to go on downtown to turn in my time card.
I arrived at the closed store after 10:00 p.m., twelve hours after I left, and
pushed my card through the slot in the door. I arrived home via subway
and bus around midnight. It was all in a day's work. Nothing was ever
said about my strange odyssey, and I was too ashamed to talk about it.
In spite of my misadventures, I enjoyed working in the heart of the city
during the holidays. It was exciting.

The Family Goes to Montreal

In the summer of 1949, after my junior year, I finally got to make my journey to the major historic sites along the Hudson and Lakes George and Champlain. My father decided that we would all go on a trip to Montreal, and I'd be able to visit my sites along the way. We had recently acquired a used 1948 Plymouth four-door sedan and would put it to the test. We set off along the Saw Mill River Parkway and onto the Taconic State Parkway as far as it went. Then we took secondary roads on our way to Saratoga Springs. Along the way, we visited the so-called Saratoga Battlefield sites along the upper Hudson River between Schuylerville and Bemis Heights. General Benedict Arnold was the on-field hero of the American victory in the two major battles fought there in 1777. Saratoga is considered to be one of the most decisive in world history as it brought France into the Revolutionary War as an ally against the British.

We stayed overnight in a rooming house in the village of Saratoga Springs. Next day, Mother went to the famous New York State Spa to soak her bad back in the mineral waters. Meanwhile, the rest of us walked around the beautiful grounds. After the bath, Mother said it was good and helped her back. We all had to take a therapeutic drink of the vile-tasting stuff before leaving the spa. From Saratoga Springs, we went up to Lake George by way of Glens Falls. We somehow missed the falls and got to the Village of Lake George. There, we were disappointed as historic Fort William Henry at the foot of the lake had been allowed to fall into ruin, and there was nothing to see but the lake itself.

We journeyed up the west side of Lake George, taking advantage of any scenic overlooks. The lake, originally called Lac du Saint-Sacrement by the French discoverers, was then considered to be one of the most beautiful in the world, and perhaps it is still. At the head of the lake, we visited Fort Ticonderoga, then being reconstructed after years of neglect. Ethan Allen and the Green Mountain boys captured the fort early in the war. In 1777, Burgoyne's force from Canada drove the Americans out of the fort by setting cannons on an overlooking hill. The fort was recovered when Burgoyne surrendered at Saratoga later that year.

From Ticonderoga, we drove up the west shore of Lake Champlain with a stop at Ausable Chasm, advertised as the Grand Canyon of the east. We took the raft ride through the canyon on the white water Ausable River. This was our first significant trip out of New York City, so the canyon and river greatly impressed all of us. We crossed the border at Rouses Point and followed the Richelieu River to Montreal. Along the way, we passed through some really poverty-stricken Native American/Canadian villages. Even we survivors of the poverty-stricken South Bronx were appalled.

As we crossed the St. Lawrence River, my first impression of Montreal was not good. It seemed to be a dirty and gloomy industrial city, a sort of Canadian version of bleak 1940s Pittsburgh. We stayed in a downtown hotel and did a few tourist spots around town. It became apparent that Dad had an ulterior motive in visiting Montreal. He had immigrated to Canada, and his first job was on the Marshall family farm near the city. We set out in the car for an area southwest of the city where Dad thought he had lived and worked. We drove around but never found the place. I don't know if we looked for the Marshalls in the phonebook or not. We were not used to phones or phonebooks in those days, as we had gotten our phone only a short while before. In later years, we would reconnect with the Marshalls. Their daughter and a friend made several shopping trips to New York and each time stayed with us for a few days. I rarely saw them as they spent most of their time downtown, presumably shopping. They found prices in New York much lower than in Canada and the selection better.

On the trip back from Montreal, we went west from Rouses Point to Malone, New York, so we could drive through the heart of the Adirondacks. It was a wonderful trip, seemingly through the forest primeval. My mother and sisters were a little fearful as the trees grew thickly from the side of the road so that you could see nothing but the dark forbidding forest. We stopped at Saranac Lake and Lake Placid for walks around town and visited souvenir shops. We motored on, and nightfall found us alongside the Catskill Mountains. We decided to spend the night in one of the tourist cabins that passed for motel rooms before the advent of Holiday Inns and interstate highways. These primitive places had outdoor plumbing and virtually no lights. I think you were expected to provide your own. It was scary going out in the dark to a potty; there were all kinds of noises in the woods. I guess Mother took the girls and bravely challenged the dark and whatever lay out there.

The summer of 1949 is vivid in the memory of Yankee and Red Sox fans. It was the year Joe DiMaggio was out of action most of the year with a foot injury. He returned to the fray for a vital Red Sox-Yankee series in late summer and was absolutely sensational. He single-handedly knocked the Sox out of first place and put the Yankees back in the pennant race. Because we were traveling in New York State, I was usually able to listen to the Yankee games and follow the amazing American League pennant race. The race went down to the last series in September, and was won dramatically by the Yankees who swept all three final games with the Sox. I was at a Hayes football game on Randall's Island when I heard that the Yankees had finally taken the pennant from the Red Sox on the last day of the season. *Summer of '49*, David Halberstam's excellent book about this exciting baseball season, was published in 1989. David has written a number of excellent books on a variety of subjects. He died at age seventy-three in a 2007 California car accident.

That same summer, the family went south in our "new" car to visit cousins in Philadelphia. To get there before turnpikes and interstates, we had to go to New Jersey, then over the then-famous Pulaski Skyway, and onto U.S. 1. Almost all New York-Philadelphia traffic was on this four-lane road, which ran right through all of New Jersey's major cities. Needless to say, travel was stop-start in a choking midsummer heat—there were no auto pollution controls and no auto air conditioning. We were physically and emotionally drained by the time we got to Philadelphia. We first visited the downtown historic sites and were, I believe, able to go into Independence Hall. Then we went out to West Philadelphia to look for Mother's cousins—the Cartins and the Magees. I don't think we ever found them, as many Irish American families had moved out of the city by then. We were often victimized by the infamous "Chinese Wall" that divided the city. It carried commuter trains downtown and always seemed to block our way as we searched for the addresses we had. The "wall" was later replaced by the current commuter tunnel, a major civic improvement. I do remember going out to Ardmore to try to find Gertrude Sinnott, a Devine cousin of Dad's. I don't think we found them either. We stayed overnight at the city's best hotel, the Bellevue-Stratford, a real step up from the tourist cabins of New York State. In terms of seeing the cousins, the trip was a failure. However, we did enjoy seeing the historic sites; and the overall experience was, obviously, memorable.

Senior Year

After a wonderful summer, I looked forward to senior year while my sister Patricia looked forward to her first year at Elizabeth Seton Academy where many of her classmates were going. This school was part of the local Sisters of Charity's "empire" and was located on the grounds of the Mount, as we called the College of Mount St. Vincent. As a private school, Seton was relatively expensive, but was only a short walk from home.

Unfortunately, Patricia was not to finish high school at Seton. The college needed the space, and the school relocated up to Hastings-on-Hudson after her second year. This meant some form of transportation—car, bus, or streetcar—I'm not sure which of these was available. Anyway, Mother decided that it would be less complicated for Patricia to go to the Cathedral School for girls. This was a less expensive diocesan school, but it was located way down town on the east side of Manhattan. Patricia was to spend many hours on bus and multiple subways going to and from Cathedral. She was very unhappy over this disruption of her school plans, but she soldiered on and made it work for her.

By now, my little sister Maureen was also well along in school. She liked to play card games, and I taught her to play cribbage and bridge. At an early age, she was able to join my parents and me in partnership contract bridge, and she played well. There's a family photo of her at the bridge table wearing a coat and hat because she had the mumps. Later I taught her to play chess, and I taught her too well. One day we played, and she beat me in four moves, the so-called "fool's mate." I don't think I ever played chess with her after that. Maureen recently reminded me that we used to play "store" and that I always had to be the shopkeeper and her the customer. I had a little cash register that made play more realistic, but as the customer, she didn't get to use it.

Our aunt Nell O'Donnell, estranged wife of my dad's brother, Frank, was a frequent Sunday visitor. She was very fond of Maureen and brought gifts for her birthday and other holidays. She gave me my first gift books—*Autobiography of Benjamin Franklin* and *Trail of the Lonesome Pine*. Undoubtedly, they were used books that someone had given to her. They were definitely not boy's books, but when I finally got around to reading the latter, I found that it was quite good. The best thing about Nell was her stories. She worked in one of Manhattan's upscale "beauty" saloons and told hilarious tales about her clients in a very chichi voice. She lived alone in an apartment block in the Inwood section of Manhattan's far north. Her death in 1973, at age sixty-six, was mysterious. She was found lying dead at the foot of her building's stairs, and her purse was missing. Was she mugged, or did she slip and fall? She is buried near my parents in Valhalla, NY.

Other frequent visitors were the O'Briens, our neighbors from Washington Heights. Generally, it was Charlotte and the two boys, Tommy and Kenny, and they would come by subway and streetcar. Kenny was younger and played with my sisters. Tommy and I hung out together and, weather permitting, we'd go exploring on Jew Hill and on down to the Hudson River. Sometimes Tommy would come by himself. We were both altar boys and very familiar with the mass liturgy, so we'd often play "priest." Other frequent visitors were Nick and Anne Browne and family. Anne was a Devine and a second cousin of Dad's. Patricia, their oldest daughter, and I were very good chums. In later years, Patricia and Tommy O'Brien dated, seriously, at least on Tommy's part. But Pat opted for a life of adventure, and joined the Peace Corps, ending up in Pakistan and marrying there. Kitty and I and our boys had Christmas dinner in 1990 with her at her home in Santa Clara, California. She was divorced from her husband and had three boys about the same age as ours, so it was a nice family gathering.

Back to high school. Late in my junior year, I had come to realize that I needed another "hard" science course for college admission. For some reason, I'd been focusing on languages—Latin and Spanish—even though I didn't do all that well. I had been reading some of the great Spanish-language writers, in translation, and enjoyed them very much. But class was much less satisfying. Anyway, after much deliberation, I chose physics over chemistry or biology. As it turned out, it was a fortuitous choice, but it put me at the tender mercy of the fierce Xaverian, Br. Lucas. Brother ran a very tight ship from his desk that was on a high platform in the lab. His favorite trick was to ask a miscreant to come up and stand alongside his desk while he continued with his lecture or experiment. When least expected, he would reach out and smack the offender in the face and send him back to his seat. Sometimes he would have several miscreants lined up for the "treatment." I was a senior in a class of juniors, as most students took physics in their third year. I should have been wiser, but alas, I did suffer the "treatment" of Br. Lucas at least once.

My senior class schedule consisted of Religion 4, English 4, one semester of PAD (Problems of American Democracy), one semester of health education, trigonometry, Spanish 3, and the aforementioned physics. For English 4, I elected the creative writing option, taught—very loosely—by Fr. Tom Dunn. There will be more about him and the course later. I ended the year with an 83 average that was dragged down by a 75 in health education. Allow me to digress on that subject.

Health education was mandated by the state and consisted of actual physical workouts, indoor and outdoors, and, in senior year, a semester in the classroom. In order to get a passing mark each year, one had to submit a clean bill of health from a doctor. Because we didn't have a "family doctor" in those days, this was always a nuisance. Until your doctor's OK was submitted, you were given a 58 in health; afterward, you usually went to 90 so long as you attended gym and class. In senior class, I had a weird Marist, Br. Martinian. Our classrooms had front and rear doors. One day, during class, a group of our star football players were going down the hall, probably returning from a team meeting. For some reason, they thought it would be fun to cut through our class by coming in the front door and going out the back. We were all appalled by this display of arrogance, most of all Br. Martinian. He got over his initial shock and grabbed the last one coming through and threw him out the rear door, across the hall and up against the far wall. Who was lying on a heap in the hall but our star, all-city halfback, Fred Rossetti. I'm sure he, as well as the class, was impressed by the strength of Br. Martinian. While I was very careful not to offend Brother, I must not have taken health very seriously as shown by my poor grade.

The most interesting class by far was creative writing with Fr. Tom Dunn, one of the very popular Dunn brothers. The other was Fr. Joe Dunn who had headed the English Department but had been transferred out of Hayes prior to my senior year. Fr. Leo Halpin, my teacher of English 1, had succeeded Fr. Joe as head of the department. Fr. Dunn's creative writing class was populated by a lot of my friends, some of whom were with me on the school paper. Class opened with delivery of the day's newspaper to your desk, either the *New York Times* or the *New York Herald-Tribune*. You were free to read the paper for quality of writing and possible degree of bias or other infractions of the Dunn sensibility. While reading the paper, you had to be careful not to block Fr. Dunn's view of the room or vice versa. Thus, we had to learn New York's famous "subway fold," wherein you folded the very large newspaper vertically into quarter sections so that it could be read in a very cramped area.

Fr. Dunn had his hand in many things beyond the classroom. He was New York Archdiocesan Director of Television and moderator of the school's drama club, the Patricians. Hence, he was greatly interested in theatrical matters, particularly scripts for stage and the new small screen (television). Some classmates were assigned proposed scripts for review and comment. During class, Fr. Dunn would comment on the day's news, with focus on what interested him, reviews of new films, plays and television shows. You could give your own contrary opinion if you dared. While he did give us useful insights into the performing arts, it was from an elitist perspective. That is, he and we his special students were morally and intellectually superior to the benighted heathens of the world. We would readily condemn any works of the day that did not meet our lofty Catholic standards. I remember him particularly abusing failed Catholic John O'Hara upon the 1949 publication of his novel *A Rage to Live*. My own contribution to his class was usually to ask to be excused so that I could go to the *Challenger* office and, presumably, work on the school paper. Fr. Dunn would always excuse me but with some pithy remark to display his displeasure at my deserting his class. He signed my yearbook with a parting shot: "And how are you? Aren't you in the *Challenger* office?"

Fr. Dunn was a notorious matchmaker. He wanted his boys to meet proper Catholic young ladies and know how to treat them. Accordingly, he arranged for dance classes after school, and you had better be there, or you would be denounced as a heretic. He arranged for some of us to go to the home of the dance instructor where we would meet some suitable Catholic girls and put our dance lessons to practical purpose. Unfortunately, the dance instructor's home had a billiard table and that distracted us from our duties vis-à-vis the girls. It got back to Fr. Dunn that we were not up to snuff in the dance department and that we needed more work and his resolute assistance. After a few further weeks of effort, the project was abandoned, and he was off on some other scheme of which I had no part, thankfully.

My homeroom of 4I was considered to be among the elite of the nineteen senior homerooms—*A* through *R*. We had standout athletes in Joe Barba, track and field, and Lou Gigante, basketball. Kevin Duffy became a federal judge for the southern district of New York. In 1985, he presided over the racketeering case of Mafia chieftain Paul Castellano, who was gunned down outside Sparks Steakhouse midway through the proceedings. As a judge, he is reputed to be tough and cantankerous. Classmates Bob Foran and Tom Gaffney were in my wedding party. The most annoying classmate was John Patrick O'Neill who sat next to me. JP, as we called him, seemed dedicated to improving what he considered to be my poor behavior in class, and he would pray for me. With difficulty, I avoided him anywhere and everywhere, in school and out.

The classmate who impressed me most was Robert Emil Kesting. He and his younger brother William commuted to Hayes in the South Bronx from Nyack, New York. I don't know how Bob got to school before he got his motorcycle, but thereafter, he and Bill rode that machine to school every day. Our yearbook had home addresses for all seniors, so I checked

Yahoo! Maps for Bob and Bill's trip. It came up as thirty-six miles but over roads and river crossings that did not exist in the 1940s. To cross the Hudson, they had the choice of the Yonkers-Alpine ferry or the George Washington Bridge, neither very attractive while on a motorcycle. Bob and Bill went on to earn engineering degrees at Manhattan College.

Catholic Institute of the Press

While juniors, the *Challenger* editors were invited to participate in the journalism program of the Catholic Institute of the Press (CIP), and we gladly did so. The program continued on through our senior year. During the CIP term, we met once a week, downtown, for seminars with leading reporters and other staff members of the many New York-based newspapers and magazines; and in those days, there were many of them. There were participants from many of the city's Catholic schools. I suppose the purpose was to ensure a steady supply of motivated and well-trained Catholic journalists. We were to meet and be taught or entertained by some very famous people, including Fulton Oursler (writer), Meyer Berger (*New York Times* columnist), and Eddie Dowling (Broadway actor). Each CIP participant was assigned a mentor, and for at least one year, mine was Sylvester Pointkowski, a truly unforgettable name. I don't recall what his daytime position was, but he was director of journalism for CIP. My mentor gave me assignments and reviewed and critiqued my submissions. I recall that one of my assignments was to interview the stationmaster at Grand Central Terminal. As a railroad buff, this was right up my alley, or perhaps I should say tracks. So one night, I went over to the terminal and inquired as to where I would find the headman. I finally tracked him down to an office high up near the ceiling of the main hall, which is considered to be an architectural gem. It turned out that he was the assistant stationmaster and assigned to night duty. I don't have the text of my story, and I don't recall how my mentor evaluated it. The stationmaster was most cordial, however. For both years, I received a certificate of merit and have newspaper clippings citing my accomplishments. It would seem that I was headed for a career in journalism.

As a by-product of a CIP session, we were to have a small part in a historic cold war drama. One night in 1949, as the Hayes contingent left class and were on our way to the Lexington Avenue subway, we came upon a disturbance at the Waldorf-Astoria Hotel on Park Avenue. We were drawn into the protest that involved chanting anti-Communist slogans as we marched around the perimeter of the mammoth hotel. There was a huge police presence, with ranks of cops drawn up in the hotel's garage and visible to us as we passed the entrance to the garage. We really had no clear idea what it was all about, quickly got bored, and left after a few turns around the hotel. Later we learned that Communist sympathizers were holding a "Scientific and Cultural Conference for World Peace." At this conference, the United States was denounced and Soviet Russia hailed as a citadel of peace; and civil disobedience against the United States was recommended. Richard Boyer, a writer for *The New Yorker*, allegedly spoke openly for the Communist Party: "It is the duty of Americans to defy an American government intent on imperialist war." Needless to say, this "peace rally" drew an army of protesters from veterans and religious groups strongly opposed to the Communist message, and we were among them.

The *New York Daily Mirror*, a popular tabloid newspaper, sponsored a forum for aspiring journalists at a downtown hotel. A number of Hayesmen were invited, mostly from the *Challenger* staff. I don't remember the forum agenda, but we were served a fine lunch, with Frank Sinatra singing live and in person. Jim Rattray, our sports editor, became a *Mirror* reporter after serving in Korea with the Marine Corps. When the *Mirror* folded in 1963, he went to work for the Central Intelligence Agency. He has since retired to Hilton Head, South Carolina.

Other Activities

My involvement with CIP and my *Challenger* duties did not prevent me from participating in other school activities and from working my jobs at the fur storage plant and the flower shop. At school, we had a group called the student council or SCs as they were known. I don't recall how they were chosen, but these seniors enforced the traffic rules in the halls and on the stairways during class turnovers. An SC could take the name of a student offender and report the miscreant to the student court. The court comprised six members who were elected by the student body from among the eligible senior class candidates. I ran, was elected, and served for the year alongside my St. Margaret's classmate Fritz Wiener. The court had power to issue various punishments, including "jug" slips for serious offenses. I'm hazy as to how it actually worked and what was actually accomplished. It was kind of strange that as an SC, I could issue a summon to a student to appear in court, where I would then hear his case. There's a picture in the yearbook of the entire group of SCs. Seated together in the last row are Lou Gigante and me.

Another student officer and friend was Mike Ward who had also lost his scholarship elsewhere and come to Hayes for his junior year. He joined the *Challenger* staff and served as business manager and became involved in many other student activities. A relative unknown, Mike was surprisingly elected student comptroller. Although he was both smart and talented, Mike seemed to have trouble focusing on the concrete stuff of life. Inscribed in my yearbook by Mike are the ambiguous words "To the 'big exec.'" I was to see him from time to time in the 1970s when he took a job somewhere in the Philadelphia area and had a house in Cherry Hill. When he moved away, I lost track of him.

During football season, I continued with my scorer duties, tracking all the plays and players for the school paper's sports staff. From a cold start and total lack of football knowledge, I became a fairly competent observer. Our school used the T-formation that was the standard in those days. It was also called a "full-house" backfield as it employed four backs—quarter, left half, right half, and full. We occasionally shifted into the single—or double-wing formations. Our lily white teams usually played well but rarely had winning seasons.

I even got to see pro football on one occasion. Wellington Mara, son of the Giants' owner, lived in Riverdale, or perhaps Fieldston, and passed some free tickets down to a few of us through one of our parish power brokers. We went to the Polo Grounds for a game with the Los Angeles Rams. Our team was then called the New York Football Giants to distinguish it from the baseball Giants who also played at the Polo Grounds. The quarterback for the Giants was Charley Conerly and for the Rams, Bob Waterfield. Bob was famous for recently marrying the sexy film star Jane Russell. Bob was a multiple-threat—he played offense, defense, punted, and place kicked. The Rams won the game on a beautiful bootleg run by Waterfield. The thing about the occasion that sticks in my memory is the emptiness of the stadium. There couldn't have been many more than five thousand people in the stands. No wonder we had free tickets.

As Christmas break approached, I got a job with the post office for the holiday mail crush; I have no idea how it came about. I was assigned to the Bronx Central annex, which was close to the school. There I sorted for the mail carriers, an entirely manual operation in those days. I must have impressed my superiors because they pulled me off sorting, gave me a mailbag, and sent me out on the street as a carrier. It was cold work, as that high area of the Bronx is open to the cold westerly winter winds. I had to hike to my group of apartment blocks, carrying my heavy bag of mail. But, if it had been other than a heavily Jewish area, I suspect it would have been worse. Entering a lobby, I would use my master key to open access to the mailboxes, and the noise would alert the tenants to my arrival. Some would come down to get their mail directly from me. Others would invite me in for a cup of coffee or tea or whatever. I wasn't sure what the regular carrier did with these invitations, but I was very selective. If I was very cold and the offer came from a respectable looking and fully dressed woman, I would occasionally accept. I was very suspicious of the men who invited me. I soon found it was easier to get in for a cup than it was to get out and decided thereafter to buy my own coffee. After a few days as a carrier, I was put back to sorting for the remainder of the season. It was very boring after my adventures while carrying the mail.

Late in senior year, I was suspended from school; it came about in a very strange way. After class one day, a group of *Challenger* staff were smoking while working in the newspaper office. Because we routinely smoked in the office, we attracted visitors who came by to have a smoke. The office was located at the south end of the second floor and had a mail slot in the door that gave a view of the long hall. For some reason, someone looked through the slot and saw Msgr. Waterson, the principal, strolling down the hall. Now none of us could recall that Monsignor had ever set foot in our office, so we assumed that he would be going into one of the classrooms. However, it soon became obvious that, purely out of curiosity, he would come into our office. Since our smoking privilege was perhaps unknown to Monsignor, we decided to hide our cigarettes and open the windows to let out the smoke. Some of our guests were worried that since they didn't belong in the office, they'd best hide in the closet. Monsignor came in, and we greeted him as he looked around without saying anything. Unfortunately, when he opened the closet door and several of our guests tumbled out at his feet, he realized that something wasn't right, but he wasn't sure what it was. He asked all of us on the scene to follow him down to the office of Jabbo, dean of discipline, and delivered us to his tender loving care. Jabbo wasn't sure what the problem was either but decided that a week's suspension would do all of us some good. Since the "infraction" occurred early in the week, our suspension was only for the balance of the week, and we were back in class the following Monday. While suspended, many of the group congregated at a pool hall where we spent considerable time. I do recall that Mother had to go to school with the other mothers to be briefed on the reason for our suspension. Since we were never charged with illegal smoking on school property, no one could fathom what the charge was, except possibly shocking the principal in the performance of his duties.

As senior year came to a close, I had to address the question of college. I assumed that I would go to college as all of my friends were. I never had much of a discussion with my parents about it; I just went ahead and applied to a few schools. I was very unsure about a major. Many of my friends and classmates—Bob Foran, Fritz Wiener, Roy Capaldo, and Billy Hewitt—were committed to engineering at Manhattan College. I was inclined toward journalism and so decided to take a liberal arts major, also at Manhattan. But before graduation, I had to undergo a rite of passage. I had turned eighteen that spring and had to register for the military draft that had been in effect since 1940. I went to the designated draft board in the Bronx, was treated shabbily, and got registered. My classification was 1A, meaning first to go.

The graduation ceremonies were held at St. Patrick's Cathedral on Fifth Avenue in Manhattan. As plagued me throughout my Hayes years, the program of the day listed the graduates, but failed to list both James J. O'Donnells. There may have been some confusion, but I did receive a diploma from the hands of Francis Cardinal Spellman, archbishop of New York, who was known to some of his disgruntled clergy as Fat Frank. As we graduates entered the cathedral, we learned that war had broken out in Korea. Some wags suggested that army trucks would pick us up after the ceremony and take us directly to Fort Dix for infantry training, as the military draft was still in effect. While that didn't happen, we were concerned that military service lay in our immediate future.

Back home in Riverdale, we had a fine graduation party, with all the usual suspects. Tommy O'Rourke was there, along with Tommy O'Brien and his brother Kenneth. There were undoubtedly lots of relative, the Quinns from New Jersey, the Sanders from Mount Vernon, and the Brownes from the Bronx. Naturally, my classmates were attending their own parties, although I think some of them stopped by to sample our supply of beer. I'm sure I received lots of cash from relatives and friends. Best of all, my parents gave a Philco portable radio. It was a big and beautiful beast whose battery was as big as a paving block and about as heavy. But it was a great radio, and it served me well for many years. I wish I still had it. Now on to college and new adventures!

Riverdale's beautiful Church of St. Margaret of Cortona as it was in our time.

St. Margaret School, 2005. In my time, nine nuns lived on the top floor.

Our house at 437 West 261st Street, Riverdale, 2004. The lower roof covered what was called the "sun parlor."

Riverdale station on the Hudson River. It was originally a New York Central commuter stop. Now it's used by New York's Metro North Service.

My sister Maureen, 1941.

My sister Patricia's Confirmation Day, 1944.

Looks like an Easter Sunday to me, mid-1940s.

Quinn trio: Mother, Aunt Alice, and Uncle Mike.

Dad's first new taxicab, a 1947 DeSoto seven-passenger car, shown sitting in our cramped backyard. By 1955, he and his fellow owners were using standard Plymouth sedans with automatic transmissions.

Riverdale's College of Mount St. Vincent, also a Sisters of Charity motherhouse. We called it the Mount.

The County Tyrone Association pipe band of the 1940s. My father was a piper and my Uncle Frank the bass drummer. The man in the center is probably Terence McSwiggan, the pipe master. Frank is on the far right, and my father is just off McSwiggan's right shoulder.

New York's Seventh Regiment Armory on Park Avenue at Sixty-seventh Street. The Tyrone pipe band practiced here in the 1940s. It was quite a thrill to be in the main hall while they played. My favorites were "O'Donnell Abu!" and "Garryowen" (see appendix H). On his death, Gen. Douglas MacArthur was laid in state here.

The Kingsbridge Armory was an enormous Bronx landmark. Classmate Mike Stanton's dad commanded this unit of the New York National Guard.

A favorite hangout was the Yonkers Public Library. This building is no longer in use as a library and may have been torn down. Sad!

The Riverdale Neighborhood House was a private facility with library, swimming pool, and tennis courts. I had a library card for which I paid a small annual fee. Tommy O'Rourke and I spent many happy hours browsing the new books shelf. It continues under the same name, but is now a "settlement" house that provides a helping hand to families, adults, and children.

The great estate of Wave Hill, overlooking the Hudson River, was occupied by many famous Riverdalians, including maestro Arturo Toscannini. It is now an arboretum open to the public.

The main line of the New York Central along the Hudson River. The view is from the bridge at Mount St. Vincent station, West 261st Street. We spent many hours watching trains go by at this spot.

I can't stay away from trains. This was one of the first pictures I developed and printed in my own darkroom.

My St. Margaret's graduation picture, June 1946.

Graduation partygoers. *Left to right:* Me, Tommy O'Brien, Kenneth O'Brien, and cousin Barry Quinn. We were all so skinny then.

Altmore Chapel graveyard, 1946. *Standing left to right:* Minnie Quinn (wife of Uncle Joe), Aunt Cissie (Quinn) McVeigh, Ms. McComisky (Mother's teacher), Cousin Mike Quinn (brother of Patrick and Sean), and Grandfather Quinn. Below left is cousin Eileen Quinn on her first trip to Ireland. Seated next to Eileen are cousins Sean and Brendan Quinn.

Riverdale during the record blizzard of Christmas, 1947. That's Mother trying to move some snow.

The bucolic Bronx River. We biked along the river to White Plains and back on balloon tires.

More of the blizzard of 1947. I had to deliver my papers on my bike in the snow.

The very busy scene at Broadway and West 242nd Street. Both NYC and Yonkers streetcars delivered passengers to the IRT elevated subway station.

On its way to the Hudson River, the Sawmill River ran under the buildings in downtown Yonkers.

This is the type of electric locomotive that pulled the New York Central's long-distance passenger trains past Riverdale and on into Grand Central Terminal.

The beautiful double-decked Henry Hudson Bridge that connected Manhattan and the Bronx at Spuyten Duyvil Creek. Dad went over this bridge every day on his way to work. It was only 10¢ then.

The 1947 graduating class of Our Lady of Good Council School, Moorestown, New Jersey. Catharine "Kitty" Carlin is second from left in the front row.

In the winter of 1949, Riverdale was hit with another blizzard.

The Carlin sisters, Kitty (left) and Pat, probably late 1940s.

Cardinal Hayes High School on the Grand Concourse of the Bronx. The school opened in the fall of 1942 and was named for a former archbishop of New York. I graduated in June of 1950.

The Hayes home football games were played here at what came to be called Downing Stadium on Randall's Island. I spent many hours in the press box (top center). The stadium was torn down and rebuilt and given a new name and the lights from Brooklyn's famed Ebbets Field.

Hayes student officers 1949-50. I'm standing at the left and my pal Fritz Wiener is standing at the right. We were on the student court. Seated left is the student comptroller Mike Ward who was very talented but undirected. Seated center is our president, a returning WWII veteran.

Hayes newspaper staffers attended a forum given by the *New York Daily Mirror*. From the left are Tom Gaffney, *Challenger* co-editor in chief; George Gunning, news editor; and me. At the far right is Desmond Vella who succeeded us as chief editor. Des became a priest of the archdiocese.

Every year, the Hayes senior class marched in the St. Patrick's Day parade, one of the rights of passage. Here our staff photographer caught *Challenger* staffers Tom Gaffney, me, and Mike McParlane on the march. The funky school hats were required, and the long coats were needed for the traditional cold weather.

PART 3

College Years (1950-1956)

Summer of 1950

As the family looked toward summer, there were some important choices to be made, principally by me. My mother and my sisters were going to Ireland, and I was invited. My father was staying in New York to manage Pete Toal's taxicab fleet as well as his own small but growing fleet of cabs. Dad said that if I stayed behind, I could work with him at the West Fifty-sixth Street garage. I said that I would do so if he would give me automobile driving lessons and help me to get a driver's license. I was eighteen years old and ready to hit the open road on my own. Furthermore, I'd be paid while learning to drive. Dad said OK, so the deal was made. Mother and the girls would go abroad without me.

On the appointed day, we loaded up the car and drove to New York International Airport at Idlewild in Queens. It was a very primitive place in 1950. In fact, we had all been at Idlewild for the dedication ceremony in 1948. We had been sitting on the beach at Rockaway when we noticed a lot of air activity to the northeast. Dad remembered reading about the dedication and asked if we'd like to go, as we were nearby. Off we went, and we were actually able to park and make our way onto the field. President Truman and Governor Dewey were the guests of honor. In addition to dedicating the new airport, the occasion was honoring the fiftieth anniversary of Greater New York, that is, the incorporation of Brooklyn and Queens into the city. There was a wonderful air show that featured flyovers by jet fighters, a type of aircraft unknown to most of us. We were able to climb on a grounded F-84 Thunderjet fighter plane. All in all it was an unplanned day that turned out to be absolutely wonderful.

Well, the airport in 1950 sported a single hastily put together terminal building with an open observation deck. Passengers walked out on the tarmac and climbed a set of stairs into the plane. We have pictures of Mother and my sisters at the airport and getting on the plane. They are wearing their Sunday-best tailored suits and 1940s-1950s style hats. The airline was American Overseas Airlines (AOA), which was later incorporated into the, now defunct, Pan American World Airways (PAA). The plane was a then state-of-the-art, four-engined, propeller-driven Douglas DC-4. Similar planes had been widely used during World War II. From the observation deck, Dad and I saw the plane off and then headed back to Riverdale. We received postcards from Newfoundland and Iceland and perhaps other places, where they stopped for refueling. The Atlantic crossing in those days was done in a series of hops between fueling stations.

I know Mother and the girls enjoyed their first visit to Ireland, but Mother complained bitterly about the cold and damp weather. They had to stay in the old Quinn house and barn; Mother was probably in the house and the girls in the barn. Mother always talked about the problem of "pulling" flax, the stuff that was used to make linen in the mills of Belfast. Flax was harvested in late summer and was laid on the ground to dry before it could be sent to market. It was hard to dry flax in the damp Irish weather, so dealing with flax was an unpleasant task. Mother tried to help out while visiting but hated every minute of it. The Irish summer was unusually damp, and Mother was glad to get back to Riverdale after two months of rain and cold.

I asked my sister Patricia if she had much time with Grandfather Quinn. She said she didn't as he spent most of his time talking with Mother. Also, at ninety-five, he did not get around the farm much, although he did go to town—Pomeroy—with them one day and had a drink or two. Mother's first cousin Fr. John Cartin, of Philadelphia, also visited while they were there. He wanted to walk in his mother's footsteps, so they all walked from Cornamaddy to Altmore Chapel with him. Grandfather talked to him about Philadelphia and where he had worked there as a young man. It was fortuitous that Mother saw her father in 1950 because he died the following year. They did bring back a fine picture of Mother and the girls with Grandfather Quinn.

Cousin Eileen Quinn Dodd recently sent pictures that were taken when she and her mother, Aunt Kathleen, visited Ireland in 1946. They probably spent most of their time with her mother's family in the Cookstown area. However, they did visit with the Quinns (and Clarkes) around Cornamaddy. Some fine pictures were taken then as well. My favorite was taken in the Altmore Chapel graveyard. It has Grandfather Quinn standing near the headstone honoring his deceased wife, Mary. He is very well dressed with a neatly creased fedora on his head.

My Working Summer

I believe that my father worked six days a week in 1950, and I went along each day. We drove downtown via the Henry Hudson Parkway and the West Side Highway and arrived at the garage around 8:00 a.m. In those days, the group were still running the large seven-passenger DeSoto. They made great taxicabs but were expensive to operate. They had manual gear shifters and, given the city's stop-and-go traffic, burned out clutches were a major problem. A city commission, known colloquially as the Hack

Bureau, regulated taxicabs. Hack inspectors were stationed at key points throughout the city to enforce the bureau's rules and regulations.

Mornings, while Dad checked to see that all the cabs were already on the street or in the shop, I started on the day's receipts. The arrival of a coffee wagon made for a nice midmorning break. At the end of each shift, the drivers deposited their daily trip sheet(s) and money down a chute that led into a large steel safe. The safe was opened only in the morning, so a full day's receipts were on hand. If all cabs had been out for both shifts the previous day, there would be several hundred individual submissions. In addition to counting the money, I did a rough check that the taximeter and automobile mileage readings submitted by the driver were more or less in order. If the mileage was high and the meter low, we'd know that the driver cheated by not dropping the "flag"—starting the meter—when he picked up a fare. The driver would be canned if this developed into a regular pattern. Sometimes it was necessary for us to physically spot-check the readings within the car to build a case against a driver.

Since my father ran several fleets, including his own, I had to separate the proceeds and the trip sheets by fleet. The money was then counted, and a deposit ticket prepared for each owner. Of necessity, trip sheets were saved and filed by date to permit easy access when requested by the police. On the way to lunch, we made the day's deposits at a local bank. Lunch was almost always in a diner on West End (Twelfth) Avenue. I had to get used to diner lingo and diner food. There were usually a few of the other cab operators lunching at the same place. I'm not sure if it was because of this early diner experience or not, but I've always been fond of diners.

Afternoons were kind of slow at the garage, so I'd wander around both inside and out in the surrounding neighborhood. I became friendly with the master mechanic who ran the repair concession. He was a German who had been trained by Mercedes Benz. I don't remember how or why he came to the United States. Anyway, he had no children to take over from him and asked if I would like to be his heir. I would work with him and eventually take over the operation. I thanked him but let him know that I was going off to college and would not be interested. All I could think of was the dirt and grease around his shop. It just wasn't for me. The other people in the garage were all from Ireland and tended to talk together about business, politics, and the Old Country, nothing of much interest to me. The surrounding neighborhood was industrial rather than commercial and not very interesting. Finding nothing much of interest, I started to bring books with me to read in the office.

After 3:00 p.m., things started to pick up as the early morning drivers came in for the shift change. My father and the other fleet managers checked fluids and tires, while the drivers reported on any problems with the car. I don't think the cars were refueled at that time, probably before the second shift driver took the car away. If the car was serviceable, the driver took it out of the garage and parked it on the street if the night driver wasn't there to take the car away. The day driver then deposited his trip sheet and cash before leaving for the day. The surrounding streets were all "no parking," but we had many more cars that the two-story garage could house. Therefore, a lot of police payola went toward insuring ticket-free street parking. If an incoming car was not fit for the night shift, it was driven or pushed to the service area. A supply of recapped tires was always on hand for quick tire changes. It was important to keep the cabs on the street bringing in cash. I believe that the taxi fare at the time was 20¢ when the flag was dropped and 5¢ per quarter mile thereafter. The meter ran while the cab was stopped in traffic, but at a slower rate. So many taxi trips around Manhattan cost a dollar or less. It doesn't seem like much today, but everybody in the business was making money.

We headed home after rush hour usually with two extra passengers, taxi drivers from the Old Country with whom my father was friendly. Instead of taking the West Side Highway on up to Riverdale, Dad got off at Dyckman Street, in the heart of the Irish ghetto known as Inwood in northern Manhattan. We'd go to a local saloon and have a few drinks before heading up Broadway to Riverdale. One of the passengers left the car in Inwood and the other stayed with us to pick up the Yonkers streetcar at Broadway and West 242nd Street. I don't recall what we did for an evening meal while Mother and the girls were away. Perhaps I walked to the Riverdale Steakhouse for one of their delicious hamburgers that I washed down with nickel beers. After all, I was eighteen, the drinking age in New York State.

During that summer, as promised, Dad gave me driving lessons and set it up for me to take my driver's test. These were given across the Bronx near the courthouse and not far from my high school. On test day, Dad took me over and we discussed whether or not we should slip the examiner a few "pounds." I don't believe that we did. I was told to go up a nearby hill and parallel park on the hill, which I thought I'd done very well. He then had me continue up the hill to the traffic light and make a left turn. This was a problem as I found that I needed an extra hand—one for the gearshift, one to steer, and one to put out the window to signal the turn. I guess that was my downfall, as I did not pass. On the retest, I think we did slip the examiner a few pounds and, voila, I passed. Did you remember that at the time the British pound sterling, whose symbol was £, was worth $5. The Irish I knew often spoke of pounds, not dollars, as a sort of code. Well, by the end of the summer, I had my driver's license and started to borrow Dad's car. I think he was a reluctant lender. I believe that one of my first jaunts was to Playland.

In 1928, Playland, a brand-new amusement park, was constructed in Rye, New York. Located on Long Island Sound, it covered 273 acres of land with beautiful gardens and elegant art deco architecture. The park cost five million dollars to develop, and two hundred thousand dollars were spent constructing three roller coasters. They were the kiddie coaster, the Dragon Coaster, a gentle ride that still operates today, and the spectacular Airplane Coaster. The park had a swimming beach and rental rowboats. Rye's Playland was dedicated as a National Historic Landmark in 1987. It is the only government owned and operated amusement park in the country and was featured in the 1988 Tom Hanks movie, *Big*.

Many high schools had outings to Playland, usually by day excursion boats that ran to Rye from the Bronx and elsewhere. However, all of my visits were by bus and car; and I can't recall when I started making the trip—while in high school or college. Once you've been to Rye, every other amusement park seems second rate. I loved to take dates through the wonderful and dark Tunnel of Love and go boating on the lake. There was so much to do and so little time to do it. The first times I went, I would only venture on the Dragon Coaster, as all feared the Airplane Coaster. Finally, Tommy O'Rourke and I decided to take the plunge and it was white-knuckle all the way. Both of my legs cramped as I held on for dear life, so I could barely stagger off when we finally came to earth. Tommy lost his wallet during the ride, and I don't think we were able to find it. Sadly the Airplane Coaster was torn down, probably for safety's sake. I've been on other and bigger coasters but not matched that one for sheer terror.

On to College

I don't recall thinking a whole lot about college. I had a vague interest in journalism as a career, but my high school aptitude test did not jibe with that notion. Rather, from a hindsight perspective, I realize they probably were pointing me toward accounting. However, at that point in my life, I knew nothing about accounting as a career. Relatively nearby, Columbia University had a respected journalism major and I thought about going there. But I don't recall doing much of anything toward that end. Many of my schoolmates were going to Manhattan College, so I just went along with the herd. Most of them went into the School of Engineering, whereas Tom Gaffney, my *Challenger* coeditor, and I went into the School of Arts and Sciences.

The Brothers of the Christian Schools ran Manhattan College. The brothers were the bearers of a long educational tradition, going back to seventeenth-century France. The Catholic Church honored their founder, John Baptist De La Salle, as the patron saint of teachers. When I arrived, the college's arts program was organized by time periods and first year was devoted to ancient history, arts, and literature. This meant, for example, that we read Herodotus for history and Homer for literature. I also had advanced algebra and Spanish and the obligatory religion class. The latter took us through ancient religions through the rise of monotheism. I got a poor mark on a paper for touting the virtues of polytheism. I guess I was already starting to veer off the Christian path.

The war in Korea drew at lot of our attention because we were of draft age. The war was fought on the ground and, very dramatically, in the Security Council of the United Nations. When the Russians boycotted the meetings, the council authorized the United States to organize a force to push back the North Korean invaders. The United States rushed troops in from Japan to help the South Korea forces, known as ROKs. These American troops were not up to the task, and the allied forces, initially just ROKs and Americans, were pushed into a small perimeter at the southeastern tip of the Korean peninsula. This dire situation dictated that a much larger American force would be needed and, consequently, more men drafted into the armed forces.

By start of school in September, it seemed judicious to do something to keep from being drafted into the army. Tom Gaffney, Tom O'Brien, and I decided to apply to the air force reserve so that if called to duty, it would be air force, not army. The air force was in a chaotic state, as it had only become a separate branch of the service in 1947, a few years before. In the "skills" section of the application, I had inserted typing and shorthand because I had taken summer courses in those subjects as preparation for college. I was accepted into the reserve on the basis of those skills, and the others were turned down. I was given an enlistment date of January 5, 1951.

That winter, I was ordered to report to Floyd Bennett Naval Air Station (FBNAS) for my first weekend training session. I have no recollection of how I got there; my father might have driven me or I might have taken a subway train. FBNAS was in the far reaches of Brooklyn and where Howard Hughes landed at the conclusion of his record-setting flight around the world in 1938. There Katharine Hepburn and an army of photographers met him. When I got to FBNAS, it seemed to be almost deserted. I found my way to the building were my group was to meet. It was a small group that seemed essentially leaderless. We just sat around and told stories on Saturday, got our "tickets" punched, and went home. That was it for my first weekend in the military. Most of the men present were still in army uniforms, as the new air force uniforms hadn't gotten down to the reservists.

It may have been about this point that I first heard of the *Catholic Worker* from my cousin Patricia Quinn. This was a publication put out by Dorothy Day from St. Joseph's House of Hospitality at 223 Christie Street in lower Manhattan, just off the Bowery. I enlisted John Collier to go there with me for one of the Friday night speaker meetings. Dorothy Day and Peter Maurin had founded the house and promoted something they called the Green Revolution, or back to the farm. They had a farm all ready in Upstate New York. They also offered beds to the indigent at Christie Street. The group was very much into spirituality and social justice, and their speakers usually focused on those themes. Afterward there was coffee and black bread and lots of discussion. Oddly enough, it was a good place to meet interesting young ladies. John and I attended several sessions and afterward stopped in at some of the old time saloons on the Bowery or Third Avenue. The old el (elevated railway) kept the Bowery in appropriate semidarkness day and night. Now the el is gone, and the area known as SoHo has become gentrified. I still see the *Catholic Worker* newspaper from time to time and note that they've had to move north of Houston Street, an area perhaps called NoHo.

We Get Television

Up to now I haven't written about the advent of television in our home, mostly because I don't know when we first got a television set. As mentioned earlier, television manufacture and broadcasting were shelved during the war in favor of radar development. Radar had turned out to be a critical necessity during the war. After the war, small television sets began to appear and were bought despite the utter lack of programming. My sisters and I don't remember much about early home television, but we do remember watching the ubiquitous test patterns and hoping for a show to come on, any kind of show.

My father enjoyed listening to the popular Sunday-night radio program of Fr. Fulton Sheen, later to be Bishop Sheen. It was rumored that Sheen would get a thirty-minute television show, and he did, starting in 1951. My sister Patricia remembers shopping for televisions all over the Bronx, but I don't. I believe it was sometime in 1950 that Dad decided that we would get a television set. He and I went to a shop in Yonkers and looked at all the sets on display. I chose the one I thought would be best. Unknown to me, my father had already decided on an Admiral, probably because they sponsored the good bishop on radio and would do so when he moved to television. So our first set was an Admiral, a popular brand in those days, along with Dumont, Philco, and other long gone brands. It was, of course, a black-and-white set; there would be no color for some years. I forget the screen's size, probably around twelve inches. But it was in a beautiful big wooden console as was common in those early days of television. I believe that it was delivered and set up by people from the store. Basically, they plugged in the rabbit ears and turned them for the best reception. I think we got a roof antenna in later years.

Well, disappointment was to ensue for my father. He loved comedy shows on radio, but they were slow to move to television. However, there was the Milton Berle show on Tuesday nights and Dad enjoyed it very much. When Bishop Sheen's program started on television, it was placed opposite Berle's show on Tuesday night. My sisters and I didn't care much for either show, so it did not affect us, and I don't recall that my mother watched much television. So on Tuesday nights, my father had to make a difficult choice between slapstick comedy and the Holy Word. I don't recall what he chose, but I seem to recall that both put in an appearance on the small screen.

The show my sister Patricia and I enjoyed most in those days was *The Lucky Pup Show* that featured The Great Foodini and his sidekick Pinhead. Later, when the show moved from CBS to ABC, it was called either *Foodini, the Great* or *Pinhead and Foodini*; I'm not sure which. This was a puppet show that was on weekday evenings for only fifteen minutes. Foodini was a magician whose plots always came to naught in spite of the best efforts of Pinhead to save him. The show was very funny and had adult humor, including political satire. That's probably why it went off the air by the end of 1951. Sponsors seem to fear sharp comedy.

Television programming was so bad in the early 1950s that we turned to wrestling for something to watch. My father and I enjoyed such stars of the day as Gorgeous George and Antonino "Argentine" Rocca. Wrestling was televised live and had not yet become purely show business. Rocca had a lot to do with the wrestling revival after the war. Later in the 1950s, we had *The Jackie Gleason Show*, in my humble opinion, the best show ever on television. His show, an hour long and done live, featured comedy skits and the June Taylor dance troupe. It was a true variety show, but Jackie played all of the characters, including the Poor Soul, the incompetent Rudy the Repairman, playboy Reginald Van Gleason, and bus driver Ralph Kramden. The college and drinking crowd particularly enjoyed the antics of Reginald who never saw a drink he didn't like. In a famous skit, he walked into a shoe store whose walls were lined with shoeboxes. Looking around for a moment, he walked up and picked a box; lo and behold, it contained a bottle of booze. During the show, Jackie always had one of the lovely dancers bring him "coffee." Later, for reasons of economy, his show was shrunk down to the half-hour *The Honeymooners* and became a star vehicle for Gleason's pal Art Carney.

As the 1950s advanced, so did the quality of television programming. The late 1950s are today considered to be television's golden age. Indeed there was a plethora of fine variety and dramatic shows, all done live. Eventually, a process was developed so that shows could be filmed for rebroadcast to the West Coast. Film's drawbacks led to the development of magnetic tape, a superior way to record shows, initially for rebroadcast, then to allow for multiple "takes," editing, and the addition of "canned" laughter.

Reliability of television sets remained poor up through the 1970s, when transistors and integrated circuits started to appear in new sets. The vacuum tubes that were the heart of the early sets were considered to be inherently unreliable and also used a lot of power. If you had a problem with your set, you'd take out all the tubes and run them through a tube tester. These were conveniently located in drugstores and supermarkets, as well as television repair shops. If the tube tested no good, you'd buy a new one. Unfortunately, the testers were not always up-to-date, that is, didn't cover some of the tubes you hoped to test. In my experience, in spite of conventional wisdom, I almost never found that the tubes were a problem. Failures within the high-voltage power supply or with the picture tube were the more common problems. No matter, the tube sets were very unreliable by today's standards.

Henry Morgan and I

I mentioned earlier that my father and I were great fans of Henry Morgan's 1940s radio program. It was on daily for fifteen minutes and featured Arnold Stang, who played Morgan's sidekick Gerard. Morgan was considered the bad boy of radio and was often cancelled by his sponsors, and so he would disappear off the air for a time. But he would resurface, sometimes in the movies and then on television. In the 1950s, he became a panelist on the long-running *I've Got a Secret*, a show where he was able to put his sharp wit to use on the hapless guests. He also frequently appeared with Morey Amsterdam on the original *Tonight* show that was broadcast live from New York. Morgan was briefly blacklisted after his name appeared in the infamous anti-Communist pamphlet "Red Channels." That he was any kind of Communist sympathizer was unlikely, but his former wife had known leftist affiliations.

For a while, I didn't see or hear much of Morgan, except when I occasionally saw him on *I've Got a Secret*. One day I read that Morgan would be appearing at Hutton's, a mid-Manhattan nightclub and restaurant and would broadcast on radio from there. I tried to catch his show as often as I could, but the press of college gave me few opportunities. At Hutton's, Morgan would interact with the audience and make them the butts of his wit. Anyway, he did or said something that offended me, but I have no idea what it was. So whenever the gang and I were downtown, we'd stop at Hutton's and heckle Morgan while he was on the air. Well, one night we, and perhaps he, had too much to drink, and it got rather nasty. I also believe that he recognized us as troublemakers from an earlier visit. The upshot was that Morgan had my pals and I ejected from Hutton's and told never to return. Morgan's radio and television career was not adversely affected by our altercation as many successes followed for him. He died of lung cancer in 1994 at age seventy-nine.

Another popular radio program in the 1950s was the morning talk show of *Breakfast with Dick and Dorothy*. Dorothy was Dorothy Kilgallen, a newspaper columnist for one of New York's "rags," the *Journal-American*. Dick was Richard Kollmar, a Broadway producer. Their program was broadcast on WOR and always seemed to be playing on the kitchen radio, so we listened as we had breakfast and prepared for work or school. Dorothy and Dick seemed to know all the "top" people in New York and recounted tales of their doings the previous night. Remember, this was the age of nightclubs, and they seemed to make the rounds every night. Their talk was always entertaining, and you had to admire their stamina, out all night and up early for their radio program; and Dorothy also had her newspaper column to write. Dorothy served for fifteen years as a panelist on the popular TV program *What's My Line?* The cause of her death in 1965 at age fifty-two was undetermined and considered by many to be suspicious because of her attacks on the Mafia and government misdeeds.

She is interred at Gate of Heaven Cemetery, the resting place of most of New York's Irish Catholic community.

Back to College

As freshman year wore on, I became less certain that I was on the right track in so far as my major was concerned. I was also concerned about the cost to my family, although it was never mentioned. I decided to apply to some of the newspapers for a job, citing my high school newspaper experience and my involvement with the Catholic Institute of the Press. I was sure I could count on recommendations from my mentors in that program.

For the Christmas season, I got a job in the shipping department of Brooks Brothers, a really upscale men's store on Fifth Avenue. My job was to strip the price tag off the merchandise, pack it in a gift box, and finally pack it in a shipping box. The boxes were then stacked for pickup by UPS or the postal service. Of course, UPS then was nothing like the monster it is today. We, lucky Manhattan men, who got to work in the shipping department were staggered by the prices of the merchandise. A tie was $12.50 at a time when my mother was buying ties for me at Alexander's, a Bronx institution, at three for a dollar. Of course, I never wore them. The Brooks Brothers job was in the basement of the store and dreary. It was a very boring job by comparison to the ones I'd held during my high school years.

For some time, a group of St. Margaret's classmates had been getting together for bridge and, sometimes, poker. The core group was John Collier, Bob Foran, Fritz Wiener, Tommy O'Rourke, Jimmy Conway, and I. We would take turns hosting play. Jimmy Conway didn't care much for bridge, so at his house we played poker for money. He was a steady loser, and somehow his mother found out that he was gambling and losing, so she put a stop to his poker playing. As the group became restricted to playing contract bridge, controversies arose, particularly due to Jimmy's lack of bridge etiquette; he liked to quit in the middle of a rubber when he was losing. We couldn't drop Jimmy from the group as we were having more and more difficulty getting up a foursome, particularly when John Collier started to date regularly. So in order to get our games under control, I drew up a "constitution" for the group. I still have the original document that was signed by all six members early in 1951. We didn't know what to call the group, so we simply called it the Undersigners. We continued to play bridge until I left for the service in the spring of

1951. I recall one horrible experience while playing bridge, either at Fritz's or John's home. Someone brought cigars, and we proceeded to light up and smoke. I don't think I'd ever had a cigar before. In any case, I became ill and had to rush home. I didn't know what to do, so I took a cold shower, whereupon I threw up, and then went to bed. Thereafter, I was very wary of cigars.

As the end of winter neared, I received notice that my reserve unit was being called to active duty. I was to report to FBNAS on May 1. I don't recall that I was told where I would be going from FBNAS. I was also told that there would be no more meetings before call-up. I was rather amazed at the whole idea of activation of my "unit" as it seemed to be rather a ragtag bunch to me. I made arrangements with Manhattan College to finish my courses early, and they were most accommodating. Consequently, I received full credit for all of my courses for the year. My school buddies held a farewell party for me in a saloon over near Fordham University. One of the things I regretted leaving was my longtime service as an altar boy. For years Bob Foran and I served the main Sunday mass at 11:00 a.m. I had served so long, most of the parishioners assumed that I'd go off to the priesthood, not the air force.

When the day came, I packed my stuff and my trusty Philco portable radio. I wasn't sure if I should wear the old army uniform I had been given in January, but whom could I ask? Anyway, on activation day, I put on the uniform and, accompanied by the entire family, was driven to FBNAS. I guess we were all apprehensive about my going off to war, but I looked upon it as an adventure and an escape from my career indecision. Unbelievably, when my parents got home from dropping me off, there was a letter from the *New York Times* offering me a job interview. Obviously, I was not destined for a career in journalism.

Off to War

When our unit was assembled at FBNAS, I learned that we were going to MacDill Air Force Base near Tampa, Florida. We boarded a DC-4, the same kind of plane Mother and the girls had flown to Ireland the previous summer. There was a lot of gambling on the plane that I avoided, but my first flight was rather boring. We arrived at MacDill in the dark and were taken to the barracks of the headquarters squadron of the 305th Bomb Wing, Eighth Air Division, and a unit of the Strategic Air Command (SAC). I learned later that SAC was commanded by the flamboyant Curtis Lemay of WWII bombing fame. His idea was to bomb everyone else back to the Stone Age.

Because I came in as a clerk-typist, I was initially assigned to the headquarters' typing pool. However, I was given an interview with the captain who headed personnel. Because I had a year of college, he set it up for me to interview at various sites around the base, including the base newspaper. I quickly determined that I wasn't interested in any of the suggested jobs but that I'd like to be a truck driver. I explained that I needed a break from pushing paper. He went along with this and had me transferred to the motor vehicle squadron. On my way out of the captain's office, the master sergeant that ran the headquarters told me that if I knew what was good for me, I'd get back to my desk in the office. I didn't listen and went off to be a truck driver.

On reporting to the motor vehicle squadron, I found that I couldn't drive anything as I didn't have a military driver's license. I was assigned to the maintenance unit where I worked under a wonderful boss, a master sergeant leftover from WWII. My job was to check that vehicles on return to the motor pool had been properly serviced and gassed. I also scheduled vehicles for their periodic maintenance. The job was easy but boring, and I told my boss that I really wanted to become a truck driver—the lure of the road. My boss had taken a real liking to me and wanted me to stay with him. He said the billet would be mine when he retired, and I'd make sergeant quickly. When I persisted, he arranged for me to go to driver's school. Now I'd offended two master sergeants.

Driving school was a typical military overkill, wherein a seemingly simple process is made complex by dense protocols. There were morning classroom sessions followed by afternoons taking turns driving some old army trucks in the field. Since it was now summer in Florida, it would seem sensible to do the fieldwork under the hot sun in the morning with classroom sessions in the afternoon. It probably didn't matter much as the classrooms were not air-conditioned. After lunch, we usually sat in the shade up against the building and sometimes dozed off. One day, I dozed off to the extent that I didn't get out to the trucks on time. The nasty master sergeant that ran the driving school flunked me out, and I was sent to become a truck tire changer. I later learned that the sergeant had been a colonel during WWII. Changing tires was not a job anyone wanted, especially the half-dozen blacks that had been flunked out with me.

In talking with my fellow flunked driving-school classmates, I felt that we had a case against the sergeant who ran the school. I made an appointment for us to see the captain who commanded the entire motor vehicle squadron, school included. Beforehand I briefed my fellow flunkies as to what to say. We impressed the captain, and he reinstated us so that we all quickly received our military driver's licenses. I finally got what I wanted but had alienated yet another master sergeant. Where would it all end? While in the motor vehicle unit, I lived in a segregated barracks, blacks down and whites up. Upstairs was filled, so I bunked downstairs with the blacks. We got along just fine, but they objected to my choice of music. They hated hillbilly music, now called country and Western, as it was the music of white oppressors; but I loved it. I also loved what was then called Dixieland and Nat King Cole. I had my trusty Philco rewired to accept a 45-rpm record player and started to build a nice record collection. I was able to make peace by toning down the hillbilly and playing up Nat's. Eventually I had to move upstairs and found it very uncomfortable living with a bunch of real drunken rednecks. I just didn't fit.

I enjoyed many months of driving trucks and cars on and off base. I got a plum assignment to deliver batteries to the various rescue vessels that patrolled the waters that surrounded the base—Tampa Bay and Hillsborough Bay. I would drop off newly charged marine batteries and pick up the spent batteries. It took all day to do this, as I'd have a few beers with the crews on their boats. Each day I'd go to a different boat, and I could stop along the way to check out the scenery or purchase a Baby Ruth bar, still my favorite. I was also asked or ordered to be the guard on the base's mail truck. A .45 Colt automatic was strapped around my waist, and I was pushed onto the mail truck driven by one of my redneck sergeant bunkmates. It turned out to be a fun task. We'd take the mail into Tampa's main post office and pick up the mail destined for the base. We'd kind of swagger around town like John Wayne, armed and dangerous. The local girls were impressed. However, I decided that I really didn't like the job when, as I drove along the beautiful bayside road back to the base, my boss fired his weapon out through the back window of the vehicle.

Our bomb wing consisted of three bomber squadrons: 305th, 306th, and 307th—with two of them serving in Korea. Our aircraft were old B-29s, the famed Superfortresses that bombed Japan during WWII. The planes were barely airworthy, and there were many accidents. Sometimes, on landing, flaming engines dropped off, hitting houses near the base. The base had a reputation for bad accidents on the ground and in the air, with alcohol a contributing factor. It seemed that the last of our

B-29s and support personnel were going to be shipped to Korea, so the men were ordered to get a full complement of inoculations. Naturally, the day of the shots was very hot, and many men passed out from the deadly combination of heat and needles. The planes went but not the ground support personnel, so I stayed at MacDill. It was typical of the military, hurry up and wait. To keep the wing operational, a squadron of B-47s was assigned to us. This was a beautiful, newly developed, swept-wing medium jet bomber that, unfortunately, proved to be a death trap—at least at first. A sad assignment was to join a convoy to go into south central Florida to recover bodies where one of our B-47s had gone down. In addition to driving back and forth for several days, I was part of the search line that found a few pieces of human remains. What little we found of the three crewmen was buried in a single grave. Some years ago, I read that the families had been asked if they would like the grave exhumed and the remains DNA-tested to permit separate burial. The families wisely declined.

My most bizarre driving experience involved taking three officers to the Avon Park Bombing Range, also in the wilds of south central Florida, not far from Sebring. We were to stay for the duration of SAC's annual weeklong bombing accuracy competition. There were no accommodations on the base, so the officers stayed at Avon Park's finest hotel, the Jacaranda. I stayed in a boarding house where I had a room and three meals for $1.50 per day. Each morning, I drove the officers in my carryall to the base for the day's events. They were then taken by jeep out to the triangulation towers in the boondocks. I'd hang out at the so-called base headquarters until it was time to go back to town. On a field radio there, I was able to listen to the aircrews as they made their bomb runs, as well as the scorers in the towers. Some days, cloud cover caused cancellation of the day's runs. The family that ran my boarding house had two charming daughters a year or two younger than me. On off days, I would take them to the old swimming hole, and we'd hang out there. If I was needed, the officers would have the police find me and tell me to get back to work.

To get to work meant driving out of town onto a dirt road that ended at a chain. There I'd get out of the vehicle, climb over the chain, and crank a field telephone that was connected to the base's fire station. Someone would come down and drop the chain so that we could drive up to the base's headquarters and get to work. Sharing the base with us was the air force's tropical survival school. Among the things they learned was how to catch, cook, and eat snakes; and there were plenty of them all around the base, mostly big diamondback rattlesnakes. The officer who ran the school had a big pet blue indigo snake that he usually kept with him so that it wouldn't be caught and eaten. While my family packed a box lunch for me every morning, the officers had nothing. When our group arrived one morning, it was decided that we would raid the store of rations kept by the survival school as they were up early and out in the field. I was ordered to help carry out the stolen goods. Naturally, we chose only the best stuff. After a couple of raids, one morning when we arrived, a big sign was posted on their barracks: "We're waiting for you!" We decided to do without raiding that day.

Many days I was taken by jeep to explore the backcountry and visit with men in radio vans out in the wilds. They lived in the vans for a week and then had a week off. Everything was kept off the ground to avoid unwelcome guests. We spotted some really big ones that the guys tried to pick off with the survival kit's .22-caliber rifles. The fast-moving snakes were hard to hit, not like in the movies. It was all very strange to me, like Africa. As we were driving back to MacDill after the competition ended, I asked one of the majors how it had gone. He said it was terrible as many of the bombardiers were hung over and puking on their bomb runs. This report was not building up my impression of the United States Air Force. It seemed more like a college fraternity than a fighting force.

I had a visit from the family that summer. Mother, Dad, and the girls drove down in the family's trusty Plymouth sedan. It was their first trip to the Southland, and they found it sweltering. Neither the car nor their hotel in town was air-conditioned. I took them around the base and to the few tourist sites I knew of; not having a car myself, I wasn't really into touring. I remember taking them to one of the Gulf Coast beaches, so different from the Atlantic beaches—the gulf water is so warm, so shallow, and so calm. My sister Patricia dove in and instantly came up with a nose bleed, something that had happened to me at my first swim in gulf waters. The warm salt water apparently allows blood to seep through the nasal membrane until the body adjusts. One day, a terrific Florida monsoon erupted while we were on the causeway between Tampa and the beaches. The rain was so hard that water started to come into the car around the windshield. We had to pull off the causeway onto one of the small islands that dot the bay and seek shelter under the trees until the rain abated. In spite of their discomfort, the family enjoyed their stay in Florida. Mother, in particular, took every opportunity to visit in Florida. My Dad and my sisters never found Florida much to their taste.

First Leave and Off to Biloxi

My first leave was a fiasco from start to finish. I figured if I'm in the air force, I should be able to get a free plane ride to New York. I hooked up a plane ride to Mitchell Field on Long Island, but they wouldn't take me unless I had a parachute. I checked out a chute and got on board an old WWII C-46 "Commando" transport with several other passengers, including a servicewoman—a rarity in those days. The plane flew at a low altitude right through the clouds, so it was rock and roll the whole time. We lay on the floor and did everything else we could think of to keep from vomiting—it was bad. To make matters worse, we had to make a stop at Fort Bragg, North Carolina, right in the middle of their airborne maneuvers. When I got to Mitchell Field, they would not allow me to return my chute for credit, I had to take it back to MacDill. That meant I had to carry my chute and duffle bag onto the Long Island Railroad and then on the New York subways and buses. Of course, people stared and some may have thought I was Howard Hughes just returned from an around-the-world flight. When I got home, I slung my chute under a table in the living room where it caused much comment during my stay.

The folks on the home front criticized my choice of duty assignments. The general opinion was that I should be learning something useful, not driving around having fun. Ever mindful of the opinions of others, I went back determined to make something of myself. This time I took a train from New York to Tampa, no more el cheapo air force flights for me. Back at MacDill, I returned my parachute and set about finding a more upscale job. I don't recall how it came about, but I got myself assigned to an office duty in the orderly room (headquarters) of the Motor Vehicle Squadron. There, while serving as finance clerk, I was permitted to review all the opportunities for the air force schools. Because I could read and write at an advanced level, I soon became indispensable to the office crew. I did many of the reports that they found difficult to do and processed new people into the unit. They loved me, but I wanted to move on. Before I could do anything further, I was assigned to a month of basic training, right at MacDill in the middle of summer.

It was a grueling experience, made so by a really demonic captain from West Point. Among the many follies was parachute jump training. Without proper facilities, we were taught to jump and roll from twenty-foot towers. We made a series of long hikes along the back roads around the base. My favorite was a hike through the town of Rattlesnake to the docks at Port Tampa. As we went through town, people on their front porches waved handkerchiefs and small flags. Port Tampa had been the jumping-off place for troops going to Cuba in 1898. I guess the locals thought we were again going off to war. Since we returned to base by a different route, we didn't disabuse them of that notion. But the worst part of the whole experience was the extended calisthenics we did late afternoons and out of doors on concrete that had been heated all day by the Florida sun. What I remember most about my spring and summer at MacDill was the heat. At the end of the day, my fatigue uniform would be white from the salt in my perspiration. After I took them off, my salt-impregnated fatigues were almost stiff enough to stand by themselves.

Once back at the squadron from basic training, the office staff did everything they could think of to keep me interested but to no avail. One day my boss came to me and said, "We have a school for you. First, you are going to take a math test that you will surely pass, and then you are going to radar school." MacDill was obligated to send six qualified men to school and was hard-pressed to find them. I was one of the few possible, so I was selected to uphold the honor of the command. Failure was not an option. I took the test and got a perfect score and my wondrous "accomplishment" was reported in the Riverdale and Bronx newspapers. I still have the clippings I received from family and friends. I have rarely taken so simple a test.

In October, I prepared to leave MacDill for the Air Force Electronics School at Keesler Air Force Base, Biloxi, Mississippi. One day another airman destined for Keesler found me and suggested that we make the trip together in his car. Unknown to me, this was the beginning of a long association with Norman Thierer, a fellow New Yorker and a Jew. No one was more inept at anything military than Norman—he couldn't march, he couldn't dress, he couldn't salute. But he was smart enough to get into radar school and thus, became my classmate for the next nine months. We set out in his vintage 1930s car with all of our gear, as we were not sure where we'd be assigned after school. In the Florida Panhandle, we ran into a huge forest fire and the main route, U.S. 90, was closed. We were forced to detour onto back roads and wander through the smoke and dodge flames as we sought a way west. Not an auspicious start on our journey to Biloxi.

At Keesler, Norm and I were assigned to a student squadron, but had to pull KP (kitchen duty) while awaiting the start of class. Between duty assignments was the worst possible position in the service as you got all the dirty work assignments. Once classes started, you were OK as you were busy with classes and homework. Norm and I were assigned to the early school shift, 6:00 a.m. to 12:00 noon. We were called out for breakfast at 4:00 a.m. and queued up around 5:00 a.m. to march to school. Forming up right behind us was a squadron of women, so a lot of badinage went on between the men in the back of our formation and the women in the front of theirs. As winter wore on, mornings became very cold and we had to wear our battle jackets over our summer uniforms on the way to school and carry them under our arms on the way back. Marching back from school at noontime, we would sing slightly ribald ditties. Our unit commander was Lieutenant Bruno, who may have been of Russian descent. We'd often march along, chanting, "Bruno's chain gang. Ugh!" to the tune of "Song of the Volga Boatmen." Bruno loved it.

My first six months at Keesler were spent in a course called electronic fundamentals. There, we learned all the basics and built a small oscilloscope from a kit. Because the pace was geared to the slowest man, some sessions were boring, and I found it hard to keep awake. If you were caught dozing, however, you were ordered to stand in the back of the room while holding a wastebasket. Since electronics was entirely new to me, I did learn a lot during the course and had no trouble keeping up. I hated the early morning turnouts and hoped for a better schedule the next time out. Otherwise, school at Keesler was very pleasant duty, much like college.

Off hours activities included swimming either in the base's pool or in the nearby waters of the Gulf of Mexico. Keesler kept a fleet of rowboats that you could take out and cruise the bay, north of the base, and do some fishing if so inclined. The sight of pelicans diving into the bay after fish was an interesting new experience for me. On one occasion, we rowed all the way across the bay and landed on the far shore. Meantime, a storm came up, and we couldn't launch the boat against the waves crashing on that lee shore. We had to wait out the storm and got back late to dinner, almost too late. Another favorite thing to do was to go into Biloxi, see a movie, and pick up a box of southern fried chicken at a special place known only to the real veterans among us. This was a new treat for me, and I still think it was the best I've ever had. Many nights we simply stayed on the base and went to the PX (post exchange) for a beer or two. Sunday Mass was held in the base's movie theater, and a regular collection was taken. Comedian Bob Hope and his troupe put on a show for us in the base's football stadium.

Travel Adventures and Misadventures

By the time my group moved on to the advance electronics course, we had started to travel farther afield during our leisure hours. Along the beachfront between Biloxi and Gulfport, there were numerous restaurants. Because our mess halls basically shut down after breakfast on Sundays, we'd eat dinner at one of beachfront restaurants that offered a Sunday special, usually all you can eat for a dollar or so. We'd each order, eat, and swap meals on the refills. It was unbelievably good and inexpensive. Other off-duty hangouts were the honky-tonks that lined the back roads around Biloxi. They usually had a band and dance floor, and even ladies to dance with on occasion. The band usually had a "kitty" that lit up to remind you to drop a little something in the basket. Honky-tonks were definitely "country," and it was easy to get into fights with the locals.

If Norm, or someone else with a car, volunteered to drive us, we'd go to New Orleans for the weekend. At first, we'd sleep in the car. Later we found a hotel that gave us a room for a dollar a night. Next-door was a Morrison's Cafeteria that offered good food at a fair price. We all enjoyed exploring the Crescent City's parks, bars, and jazz joints. We were not welcomed in the better restaurants because we looked like what we were, badly dressed and semisober soldiers on leave. We were able to see some of the hot shows, especially those featuring Lilli Christine—the Cat Girl—and Kalantan, two of the biggest names in 1950s burlesque.

A fascinating place was the Old Absinthe House, which was built in 1806 and then housed the Old Absinthe House bar and restaurant. The drink for which the building and bar were named was outlawed in this country until quite recently—it allegedly caused blindness and madness. However, we did enjoy Pernod, which we were told was similar to absinthe. As you sit at the bar, you feel at one with the famous people who came before you. Listed on a plaque outside are William Thackeray, Oscar Wilde, Sarah Bernhardt, and Walt Whitman. Andrew Jackson and the Lafitte brothers plotted the defense of New Orleans here in 1815.

A favorite hangout was the public beach on Lake Pontchartrain. The beach and a nearby dancehall were very popular with the younger set. There always seemed to be something happening in New Orleans, and I was there for Mardi Gras in 1952. On Bourbon Street during carnival, I found a sailor from Riverdale drunkenly trying to direct traffic. Another nice feature of New Orleans was the regular Saturday-night moonlight cruises on the Mississippi. The boat, an old-fashioned stern-wheeler, offered a band for dancing and there was, of course, plenty to drink. Norm and I met some girls on the boat and took them one Sunday on the paddleboats in City Park. One trip to New Orleans in Norm's car almost had serious consequences. The main road, U.S. 90, ran along the beach and was under construction. One night as we approached the serious right-hand bend in the road near Bay St. Louis, the car's lights went off. In the dark, we shot off the road onto the beach, barely passing between standing concrete storm drainpipes. Norm's battery cable had come off, very nearly sending all of us to the hospital.

For 1951 Christmas leave, a group of us decided to drive from Biloxi to New York with Norm Thierer in his jalopy. As we got into Alabama on a two-lane road, we became enveloped in thick fog and were sideswiped. We got out to survey the damage and saw that the left rear fender was rubbing on the tire. Norm grabbed the fender to pull it away and got his fingers burned. The group decided that Norm was an unsafe driver and put him in the backseat. We then took turns driving. I remember driving through Birmingham and marveling at the beautiful homes and the famous tall statue of Vulcan that overlooks the city. As we entered the hill country beyond Birmingham, the car's transmission and clutch started to act up. We limped into Raleigh, North Carolina, and pulled up at a traffic light. We could not get the car into gear and got out to discuss the situation. A policeman harassed us as we had traffic tied up right in the center of the city. Norm said he would stay with the car, and the rest of us could do whatever we wanted. I elected to hitchhike. I got lucky and a trucker on his way from Miami to Philadelphia picked me up on the promise that I'd help keep him awake. Unfortunately, near Danville, Virginia, we both dozed off and ran off the road. We woke up when we heard the truck tires hit the gravel verge, so no damage was done, but it was scary. When we got to Philadelphia, I decided to take a bus to New York. I was finished with hitchhiking.

Over the holidays, I was able to get in touch with some of my fellow northbound travelers to arrange for the return trip. A driver had been found, and we were to rendezvous at the Greyhound bus station in Scranton, Pennsylvania. I decided that I would be better off in another vehicle, so I didn't try to contact Norm while in New York. My parents drove me to Scranton where I met the group for the trip to Biloxi. While it was New Year's Day, the weather was beautiful. When we got to the Bluegrass Country on U.S. 11, we put the convertible top down. I really enjoyed seeing the flood control works of the wonderful Tennessee Valley Authority (TVA), a

1930s project of the Roosevelt Administration. As darkness came, we were in Tennessee, then a dry state. In a few towns, we asked for the local bootlegger and, in one case, were sent down a back road to meet him. The rendezvous didn't come off, so we kept going through the night and eventually ended up in a famous roadhouse in, presumably dry, Meridian, Mississippi. This joint I'll call "Al's" was well-known throughout the South, and we enjoyed a few libations before pushing on. As a result, we were late getting back to Keesler. While making our excuses to the first sergeant, he asked how long we'd spent at Al's. We were astounded at his prescience, but he pointed to an Al's matchbook sticking out of one of our watch pockets. This caper did not enhance my standing with the first sergeant. Unknown to us, Norm Thierer had gotten his car repaired and pushed on to New York. On his return trip, he had a bad accident in the mountains. His car was totaled, and he was hospitalized for over a week. Norm was a good-hearted guy, but without a doubt, the world's worst driver.

Final Days at Keesler and Back to MacDill

Near the end of my time in the basic course, I had a pleasant surprise. Upon entering the barracks after lunch, I found my buddy Tommy O'Rourke stashing his gear. After graduating from Hayes a year after me, he had signed up for a four-year hitch in the air force. We were in the same unit, at least for a while. He arrived in the dead of winter. It was so cold in the unheated barracks that water froze overnight. The cheap World War II blankets we had been issued were small and didn't come up to our chins. So at night, we had to pile on our coats and anything we could find to keep warm. Once the sun came up, there was a rapid warm-up and all was well. We were all going to write to our congressmen to complain about the wretched conditions at Keesler. I don't know if anyone did so. Some years before, the U.S. Senate had investigated drinking and prostitution in and around Biloxi and presumably, the corruption of innocent servicemen. The resulting cleanup put the local bars off-limits to military personnel. However, there was plenty of beer available on base at the PX (post exchange) and liquor at the clubs for officers and NCOs (noncommissioned officers).

Once I'd completed the basic electronics course I was on "casual" status while awaiting start of my advanced course. This meant that I was available for kitchen duties, known as KP. There had been no KP at MacDill, so I didn't know what to expect. I quickly learned that I wanted no part of it when I "pulled" several days of KP. However, shortly after Tommy arrived, I was transferred to a different squadron for the advanced course, which turned out to be on airborne-radar bombing systems. I was also assigned to the afternoon shift in the only air-conditioned building on the base. Tommy and I rarely got together thereafter, but we do remember a golf outing to a hotel course in Gulfport. I had warned Tom to be careful around the water and not to reach for a ball with his hand. At one point, I saw that he had hit a ball, down near a water hazard and was headed that way. I heard a blood-curdling scream coming from his direction and went over to him. Contrary to my direction, he was about to reach for his ball when he spotted a snake. I asked what kind was it, a water moccasin? He said he didn't care what kind it was; he wasn't going back for his ball.

Life in the new unit was much better as it was an honor squadron and had special privileges whose details I've forgotten. The course was absolutely marvelous, and I learned a great deal about radar bombing and navigation. Our focus was on the APQ-24 system that was used on the later B-29s and the replacement B-50s. The latter was a B-29 outfitted with Pratt & Whitney engines instead of the unreliable Curtiss-Wright engines used in World War II aircraft. At that time, the Q-24 was the latest and greatest system deployed on the bombers of the Strategic Air Command. However, it was superseded when the new jet bombers went into service. I became totally fascinated with electronics and resolved to go to engineering school when I got out of the air force.

As graduation neared, our squadron held a wild party at a hall in nearby Pascagoula, Mississippi. Our lead NCO was Master Sergeant Billy Penn, of Tennessee. Over time—in my opinion, over small things—Penn had developed an antipathy for me. He probably didn't care for Yankees, in general. I happened to be talking to Billy when the band struck up Dixie. I stood and started singing along with him. When he noticed that I was singing with him, he said he was impressed and that, maybe, I wasn't such a bad guy after all. With that, I clapped him on the back and, as he was drunk and unbalanced, pushed him face down into the table at which we had been sitting. Unfortunately, he remembered that episode and became further resolved to make life difficult for me. As an honor student, he couldn't do much, but after graduation, I became "casual" as I awaited orders to go back to MacDill. He was then able to stick me with every dirty job that came along. The worst was an assignment to go with a team to pull up tree stumps on the base's radio range. To do this job, we were issued pickaxe, spade, and block and tackle. It was in late June, so it was hot and dirty work. After a week or so of this duty, I thankfully received my orders back to Tampa. So Norm Thierer and I set out for our home base in his badly used car.

We arrived back at MacDill in early July of 1952, after an uneventful trip. I don't recall where my buddy Norm was assigned, but I reported for duty with the base's A & E (Armament & Electronics) squadron. Since I now held a radar mechanic's rating, I expected to be busy fixing aircraft radars. This was not to be. We had no aircraft left on base with the systems for which I'd been trained. Any transient aircraft with a Q-24 system was preempted by those with higher ranks than I had, meaning just about everyone in the unit. The headquarters people were paying me back by squelching the promotion to corporal that I expected. Other than occasional guard duty and truck driving assignments, I had no duties. On one occasion, I was driving an old WWII weapons' carrier on the flight line and decided to let it rip. My speed attracted the attention of the APs (air police) who claimed they had chased me for quite a while with siren going. Of course, I didn't hear them speeding as I was in an open vehicle. I got a speeding ticket and a reprimand from the squadron commanding officer. I probably got a bad mark in my permanent file.

The new B-47 jet bombers that I mentioned before had a state-of-the-art radar/optical bombing system. The new K system was classified Secret, so my Confidential clearance precluded my even seeing that equipment. However, I did pick some casual information about the system and its programming. It sounded great and I really wanted to get my hands on it, but that was not to be. I only remember one other casual assignment during this period, painting the electronics shack on the flight line. This task was assigned to a group of us on a hot summer day. The shack sat near one end of our ten-thousand-foot main runway, one of the longest in the world. As we were working, admittedly very slowly, we saw a thunderstorm approaching from the far end of the runway. We went up on the roof of the shack and lay there as the rain fell upon us. It was wonderful.

With no steady job assignment, I was hard-pressed to keep out of sight so as not to get put on any more nasty casual assignments. I couldn't stay in the barracks during duty hours, so I had to take advantage of other hiding places on base. I could go to the base swimming pool and to the base golf course, maybe once a week or so for each, but not every day. The officers' wives, who frequented those places, might wonder why if I were there regularly during the week. I found the one place that was safe and that was the base library. I could read there while hidden behind the bookshelves. I suspect that the officers and their wives were not regular patrons of the library as it was not air-conditioned. I had a near miss on the golf course one day. My partner and I, a twosome, were playing faster than the foursome ahead of us, so we were gaining on them. As we approached the ninth green, I realized that my squadron commander and his wife were in the group that was about to enter the refreshment shack. We picked up our balls and skipped over to the eleventh hole to avoid them. An oddity of the course was that it was more sand than grass; the only real grass was on the greens. So in the hot and humid Florida summer weather you'd finish covered in sand, a sure tip-off as to where you had spent the day. The course had a memorable hole of 620 yards in length, the longest I've ever seen or played.

The base and the city of Tampa did offer some fine social amenities. There were frequent dances on base that were open to young ladies from off base. Best of all was the Young Catholic Club run by the Jesuits of Sacred Heart Parish, one of the city's largest and oldest parishes. The club met for socials during the week and sponsored weekend trips around Florida. The priest who ran the club was an avid ballroom dancer and arranged dance lessons for us with a young woman, also a club member, who was an Arthur Murray dance instructor. I especially enjoyed learning to tango with her. Among the memorable weekend trips was a visit to Silver Springs, where glass-bottomed boats gave a fabulous view of underwater life. We often went to Lido Beach near Sarasota, an empty place then but all hotels and condos now. I well remember how hot the sand was, it squeaked as you walked on it, like glass. The only facility on the beach then was a hot dog stand. My favorite though was St. Petersburg Beach because you could pitch your blanket under the trees that grew right alongside the beach. The trip from Tampa to St. Petersburg was over water via Gandy Causeway. Often, as you drove the long causeway, a porpoise would jump out of the water and seemingly look into the car. It was eerie but fascinating, especially to a city boy. I would have liked to date girls in the group, but without a car or a motorcycle, it wasn't feasible. So I was confined to group activities, where transportation was provided.

Since my arrival in the air force, I had been disappointed in its seeming lack of purpose. Now, in the absence of any meaningful work, I was even more determined to quickly serve out my term, say good riddance, and go on to an engineering school. Consequently, I turned down an opportunity to go to OCS (Officer's Candidate School) because I would have to thereafter serve two more years. One of the good things about the service was that my meager pay of $89 per month went a long way on base. I was able to send about half of my pay home to be put in my savings account. While I was hiding out on base, the Korean War was winding down. A rumor was circulating that men would be released early if their skills were no longer needed. I felt that I was surely in that category, so I wrote away for applications to various engineering schools. Among them were the very best—MIT (Massachusetts Institute of Technology) and RPI (Rensselaer Polytechnic Institute). MIT was a natural as it had been the foremost radar laboratory in the country, going back to World War II. On the bulletin board one morning, I saw that my specialty was listed as no longer needed. I quickly applied for release and was told it would be granted promptly. Before I knew it, I was headed back to New York and hadn't yet received any of the requested college applications. After release from active duty, I was supposed to fill out the time remaining in my active reserve enlistment of three years. However, I was never called to meetings and received my honorable discharge in January of 1954. While this suited me just fine, it further diminished my respect for all things military.

Leaving MacDill, I chose to hitch a ride to New York so as to have one last adventure in the Southland. This time I decided to go through Asheville, North Carolina, as I was a real fan of the writer Thomas Wolfe and that was his hometown. I knew it was risky as I'd be off the main roads and into the really rural south. Magically, I got a great truck ride from just north of Tampa all the way to Asheville. I spent the night in a small hotel the trucker, who lived near Asheville, had suggested. Next day, I couldn't find any of the Wolfe sites and decided to move on out of town. I quickly found that decent rides would hard to come by. Being in uniform helped some, but rides were usually of short duration, and the drivers seemed to be either Bible salesmen or Bible belters. I finally got picked up near nightfall by some young guys, who were headed east but who wanted to stop at an illegal honky-tonk for some liquid reinforcement. I feared for my life with this group, but there was no plan *B*, so I stayed with them. Next morning, when we got to a main road in the eastern part of the state, probably U.S. 1, I left them to try to get a ride north. My memory fails me as to how I managed to get home from there

Back to College

When I got back to Riverdale in late September and took stock of my situation, it became apparent that I would have to go back to Manhattan College for my engineering degree or miss a semester or two while I applied elsewhere. Since I left in good standing, they welcomed me back, and I was in class the first Monday after I'd left Florida. I believe my Dad paid my tuition that first semester, but I was soon receiving my monthly stipend under the GI Bill. As a war veteran, I would receive about $1,100 per year while I was in school. That easily covered my tuition, books, and fees; the tuition was something like $275 per semester. I had a very limited wardrobe and no time to buy anything new, so I wore various parts of my uniform to class. This got me in hot water with the ROTC (Reserve Officer's Training Corps) people on campus, as they were not allowed to wear any parts of their uniforms. After I explained my veteran status, they stopped bothering me. I would meet other veterans on campus, and we would bond to form a small bloc within the campus community.

The engineering school was far more rigorous than the liberal arts school of my first college year. Along with a few noncore courses, including religion, I started off with calculus, physics, chemistry, and mechanical drawing. For the latter, I had to purchase an expensive set of K&E (Keuffel & Esser) drawing tools—pencil, compass, dividers, scale, French curve, and accessories. These were rarely put to their intended uses after I completed the course. Most engineering students rushed out and bought an expensive full-sized K&E slide rule to show off to friends and family. K&E was started in NYC by immigrants from Germany and became a major manufacturer of land surveying equipment. I don't think K&E has survived the electronic revolution of calculators and laser instruments.

The big slide rule was a sort of clan badge for what are now called nerds. I myself was reluctantly forced to buy a slide rule, but I opted for a Pickett pocket model that I could carry in the pocket of my shirt or jacket. I even wore it on dates as it was a good conversation opener. I don't know what's happened to it. Although I was an engineer, I adopted the prevailing campus preppy dress code of the School of Arts and Sciences. For years, I wore a tweed jacket with leather patches at the elbows. It went well with my regimental-look ties and oxford cloth shirts with button-down collar. However, I didn't go for the popular white bucks, just slip-on loafers.

As for my freshman classes, I had a very fine teacher for calculus and enjoyed that course. For physics, I had a real oddball who enjoyed needling the students. On one occasion, he suggested that the members of the class were so retarded that, after graduation, they would retire to the Pinewood, a local bar, for a grilled cheese sandwich; and it wouldn't even be Friday. I suppose he was suggesting that our thinking was rutted. I was two years older than most of my classmates and younger than the other veterans I came to know well. I don't recall a whole lot about that first year. However, I did make the dean's honor list with second honors both terms. Scholastically, I typically started fast and finished (not) last, but back in the pack.

The 1952 presidential election campaign between Adlai Stevenson (D) and Dwight Eisenhower (R) was very interesting. Adlai was very well-spoken and my man, but "I Like Ike" buttons were everywhere. Dan O'Neill, our local Republican Party boss, asked me to again work for the party on election day. Fortunately, Dan did not have any way to test my true party allegiance. Consequently, I became chairman of the board of elections for the North Riverdale precinct. My role was to challenge anyone whose voting privilege was suspect and to count and certify the vote. In addition to the expected votes for Adlai and Ike, we got many write-in votes for Douglas MacArthur who had recently been fired by President Truman, as well as for Mickey Mouse and Donald Duck. Ike won handily and went on to be a good president, but we Adlai fans didn't think so at the time. I can remember being at Uncle John Sander's house for dinner during one of Ike's campaigns. Victorious Ike used to hold his arms up in a V as Richard Nixon was later to do. To me, Ike, with his round bald head and his arms up, looked like a monkey hanging from a tree. Uncle John disagreed; he thought Ike looked like an angel. Because of my longtime Republican Party connection, some disgruntled Riverdalians believed that strings had been pulled to get me out of the service in time to work the 1952 election for Ike.

Some good things happened in the spring of 1953. Dad helped me to find a used car that I could afford. I bought a 1949 Chevrolet Fleetwood for $800. The Fleetwood was a four-door sedan with a steeply sloped, presumably aerodynamic, back deck. At school, I played on the intramural softball team of the engineering school, and we won the championship. I still have the key that we each received at an awards ceremony. I played right field, rather poorly, but we won anyway, primarily because we had an outstanding pitcher—Jim Stangroom—also Hayes, class of 1950. I was asked to pledge for one of the social fraternities and decided to join. However, the hazing seemed so dumb I soon tired of the notion and opted out. On North Broadway, right near the college was an excellent record shop that had a large selection of jazz 45-EPs. One of the best music purchases that I ever made was an album called *Dixie by Dorsey*. Jimmy Dorsey's group really swung on the Dixieland and blues classics. I have since transferred the album to CD so that Dorsey's group still swings on special occasions.

After regular classes ended that spring, I went off to surveying camp. At the time, all engineering students were required to learn the rudiments of land surveying in a two-week course. I packed a bunch of classmates into my car; and we set off for the camp that was located on Croton Point, just south of Peekskill, New York. In the car, we all wore the straw "boater" hats that we'd worn on a recent college boat ride to Bear Mountain State Park. As I pulled out of New York's famous Hawthorne Circle onto the Taconic State Parkway, I noticed a state trooper vehicle hidden alongside the road. I was driving very carefully, but the trooper pulled out, gave chase, and pulled me over. There were two troopers in the car, and they only seemed interested in our hats and what we were up to. When I explained, they lost interest and sent us on our way.

At the camp, we stayed in the usual unheated, screened cabins with central toilet and shower facilities. Washing and shaving was done out of doors. It got cold at night, but I had brought along a bottle of Bourbon whiskey to keep my cabin mates and me warm. During the day, it was hidden under a loose floorboard, just like in the movies. After a few days of classroom instruction, we spent the remaining time in the field, doing closed traverses (polygons). So each morning, teams would go off in different directions, each team member carrying a piece of surveying equipment—transit, level, chain, and rod. Our team wore our distinctive hats and looked quite spiffy on the road. The downside of surveying was arithmetic—calculation of angles and distances by hand to five decimal place accuracy. Remember, this was before calculators and computers, and our slide rules were only accurate to three decimal places. We'd spend many nights doing the extensive calculations needed to turn in our survey results. Free nights were spent at local taprooms or visiting nearby girls' camps. One silly final test was your ability to hold a plumb bob steady over a dime. After hitting the Bourbon, I had a problem with that one. While at camp, atom spies Julius and Ethel Rosenberg were electrocuted at nearby Sing Sing prison. We always believed that we saw our lights dim on that occasion. In a time of fierce anti-red passion, their execution seemed so right and justified. Now it just seems sad.

A Hospital Stay

The rest of the summer of 1953 passed in a blur. I had been bothered by a pilonidal cyst that was constantly draining into my shorts. At that time, my mother was working at Columbia-Presbyterian Hospital, as was Tom Gaffney's mother. I've forgotten how they ended up working there and what they did. As an employee, Mother arranged for me to see one of the doctors on staff. He recommended that I have a pilonidal cystectomy, surgical removal of the entire cyst. Such cysts are at the base of the spine and become very annoying and prone to infection. Surgery was performed by one of the residents, the very blond and handsome Dr. Malm. It turned out that my cyst was very deep, and since it had to heal from the inside out, I had to remain in the hospital for several weeks. I was given morphine for pain and had to lie on my stomach. I played a lot of scrabble with my bedmates in the surgical ward, most of whom were only in the hospital for a few days, even those with heart and lung operations. My visitors were often put off by my drug-induced lack of attention to their chatter. Knowing that I was a fan of T. S. Eliot's poetry, Tom Gaffney brought me a gift of his collected works, a volume I still have in my book collection.

While I was suffering in the hospital, my sister Patricia, age eighteen, was off to Ireland again. She and her friend Anna Ward sailed to Southampton, England, on the USS *United States*. It was the newest and fastest liner on transatlantic service. From Southampton, they went to London for four days, staying with our aunt Kitty, Mother's sister. From there, they flew to Belfast, Northern Ireland, where they split up. Anna went to visit relatives in County Down, and Patricia went on to our relatives in County Tyrone. I believe she stayed at the O'Donnell house in Gort. It must have been a wonderful visit because she stayed for four months, when at Mother's insistence, she flew back to New York.

I did get to play some golf that summer. I had been playing off and on since high school with close neighbor Bob Foran. Bob was also into tennis and was dating a girl at a club he aspired to after graduation the following spring. I found another golf partner in Eugene Donovan, who had been a teammate in surveying camp. Gene and I were very compatible and got in quite a few rounds that summer. Sadly for me, the more we played, Gene got better, and I got worse. I had the car and picked him up and dropped him off at home, so I was in good standing with his mother. I remember sitting in his house with a beer and watching baseball on television. Because Gene was a Dodger fan, I had to watch National League games. I particularly remember seeing a lot of Cincinnati's big Ted Kluszewski, who was then at the top of his game. His biceps were so big that he had to cut off his uniform sleeves, much to the consternation of baseball's officialdom. Ted went on to a long Hall of Fame career.

That fall, in my second year of engineering school, new subjects were introduced. The one I remember best was hydraulics with Prof. Nikolas Feodoroff. It was sometimes difficult to understand his thick Russian accent, and he had a sly sense of humor. I also had him for a strength of materials lab the following year. In that lab, we tested the resistance of various materials, including steel bars and concrete objects—cylinders and spheres. My class of electrical engineering majors didn't care much about materials, so we tended to fool around and irk Feodoroff. A classmate, Jim Daley, asked if we should apply enough force to break the concrete specimen under test. Feodoroff replied, "Don't bust balls." The professor wore the most outrageous big neckties in a time of outrageous big neckties. While sitting in the back of the class, a nearby student leaned over and said he thought the czar of Russia, that's how old it was, had given Feodoroff's tie to him. I should add that we were sensitive to ties as we students had to wear shirts and ties to class. I also remember wearing jackets to class, but I think they were optional.

During my second year in E school, I became a staffer on the college's engineering magazine. Appropriately enough, it was called the *Manhattan Engineer*. Other staffers with whom I became close were Bob Tomasulo and the aforementioned Jim Daley. Our editor in chief was Tom Quirk, a junior. Since seniors were expected to spend their free time job-hunting, the chief was always a junior. I also became active in the student chapter of the Institute of Radio Engineers (IRE). This was the closest thing we had to an organization for electronic, rather than electrical (power), engineers. I started to become noticed by the powers that be at Manhattan, specifically Prof. Robert Weil, electrical engineering department head, and Br. Amandus Leo, dean of the School of Engineering. I again made the dean's honor list—second honors—for the first term, but not the second. I would not show up on that list again. Bob Tomasulo and Jim Daley earned first honors all four years. In reviewing my saved lists, I noticed that my high school nemesis John Patrick O'Neill also got second honors but in the School of Arts and Sciences. By year-end, despite my academic shortcomings, I was inducted into Eta Kappa Nu, the national electrical engineering honor society.

Manhattan College had discontinued football during WWII and never brought it back afterward. So the big campus's sport was basketball, and it was very popular among the schools in the New York metropolitan area, many of whom had also discontinued football. Just before my arrival back on campus, Junius Kellogg, Manhattan's star player, had blown the whistle on a point-shaving scandal that rocked the local sports scene. A campus organization called Follow the Five promoted travel with the team to away games. Bob Foran and I joined the team on a 1953 Christmas holiday swing through New York State. We drove in my Chevy to Syracuse for a game at LeMoyne. Driving from Syracuse to Albany on U.S. 20, in perfectly clear weather, we went over a hill and down into Cherry Valley and into a blinding snowstorm. Cherry Valley sits in a bowl and is a notorious snow catcher. A car pulled out of a side road, stalled in the middle of U.S. 20, and unable to stop, I hit it broadside. The doors flew open, and kids spilled out on the road—no seatbelts in those days. Fortunately, the snow cushioned the impact, and no one was seriously injured. When a state trooper arrived, he looked at me and asked if I wore glasses. I said no and he asked why I had eyeglass marks on my head. When I explained that I had been wearing sunglasses back on sunny U.S. 20, he was satisfied. I was able to get my car going and limped into Albany for the game with Siena College. There I got a radiator fix that enabled me to get home for more durable repairs to the car. It was my first and last Follow the Five trip.

In the spring of 1954, probably during Easter recess, Manhattan's senior class was encouraged to make a "retreat" in order to reinvigorate their spiritual side and prepare for graduation. The college arranged for it to be held at St. Paul's Abbey, Newton, New Jersey. I was not a senior, but chose to go along as many of my graduating friends were going. A bunch of us set out in my car for the beautiful drive across New Jersey. At St. Paul's, we were welcomed by the Benedictine clergy and assigned to our rooms—actually cells. Though free to roam about the abbey in contemplation and to attend talks on spiritual themes, we were urged to join in the community prayers and devotions at the canonical hours—Matins, Lauds, Vespers, Compline, and so on. These hours didn't suit us, so we tended to skip most of the religious observances and hung out with the girls we'd discovered were living at a nearby dormitory. We had a lot of fun serving each other at coed mealtime. In this religious setting, William Blake's poem "The Tyger" came to mind, and we started calling the servers "tigers." I don't know if the Benedictines thought this was humorous or blasphemous, as Blake's "tyger" symbolized Christ. On the other hand, I suppose the Benedictines, like a good watch, had experienced all kinds of tomfoolery and kept on "ticking."

The summer of 1954 was devoted to partying and to my weekday job at the Cantrell & Cochrane (C & C) canning plant in New Jersey, an outpost of a big Anglo-Irish firm. The job was an abomination for a person of my artistic sensibilities. Among the duties were unloading railroad cars full of boxed empty cans and "spotting" the cars at the doors. To spot them, we moved the railcars by hand using huge jacks. Thankfully, most of our empties arrived by truck. When not working on the railcars, we sealed and stacked cases of cans. Worst of all, every afternoon we had to man the orange juice line. Gallon cans of orange concentrate rolled down to us, and we packed them six to a carton, by hand. Having just come out of the soldering (of the top) operation, the cans were very hot, and we were given no protective gloves. OSHA came along too late for us. I was glad to see the end of that summer. Sadly, I spent all of the money I earned and had nothing to show for my summer job.

One of the ways I spent my time and money that summer was to make weekend excursions to Greenwood Lake, New York. This was a town located at the north end of a large lake that straddled the state line and was partly in New Jersey. The town had many bars and night clubs, popular hangouts for the younger set. Since New York's drinking age was eighteen, the town drew many people from the Garden State where drinking wasn't legal until age twenty-one. On several occasions, I met Cousin Barry Quinn who made the trip up from Clifton. One hot summer evening, some of the group, myself included, decided to swim across the lake to some interesting-looking places on the far side. Unfortunately, when we got across the lake, we were met on a dock by a man who appeared to have a shotgun. He advised us that we were on private property on that side of the lake and had best swim on back from whence we came. So without a chance to rest, we had to swim back. In a time before the physical ravages of smoking and alcohol took hold, the long swim back was difficult but not life-threatening.

Junior Year

Back to school that fall of 1954, I faced the challenges of thermodynamics (with lab), DC machinery (with lab), and differential equations (thankfully, no lab). Again we "EEs" saw little point in studying some of these arcane subjects—bring on the computers. The labs were a serious problem for us in that exhaustive reports had to be written and submitted after each session. Many times the observed results did not coincide with theory, and the report had to be fudged. The fraternity guys that resided on campus had access to generations of reports and could do some expert fudging. Unfortunately, I was on my own. Here's an example of what can go wrong. In a thermo lab experiment, we had to measure the capacity of a pump by transferring six hundred pounds of water from one tank to another, where each tank sat on a large scale. We started the pump and, as expected, the water drained from the first tank. However, it never made it to the second tank; it just disappeared in the lab's piping system. Of course, this caused quite a stir at Manhattan, and my lab team came under suspicion of sabotage. As far as I know, they are still looking for the water.

I was elected or appointed editor in chief of the *Manhattan Engineer*, the college's publication by and for the School of Engineering. Jim Daley became my second in command. The publication was partially supported by advertising that came to us through the Engineering College Magazine Association (ECMA). It was very difficult to get technical articles written by our undergraduate students, so we relied heavily on professional articles on current topics that came to us via ECMA. We, ourselves, filled in with nontechnical articles about staff and course offerings. Each year ECMA held a convention at its headquarters at the University of Minnesota. Jim Daley and I flew out from Newark on a Northwest Orient Airlines Boeing Stratocruiser, a state-of-the-art four-engine prop plane. The plane then went on to Tokyo with a stop at Anchorage, Alaska. Jim and I had a really enjoyable stay in and around Minneapolis and particularly enjoyed the "strong" local brews that could only be sold well away from schools. We also got to see a Minnesota-Northwestern football game from seats alongside the university's president, right on the fifty yard line.

My time at Manhattan was a time of exciting baseball, particularly the Yankees-Dodgers string of World Series in the 1950s. I can remember escaping to the college cafeteria to catch the latest scores. The Yankees eked out victories in 1952 and 1953. A monumental event occurred in 1955 when the lowly Dodgers pulled out a victory in the seventh game, for their first World Series win. Joy in Brooklyn turned to ashes a few years later when Dodger chieftain Walter O'Malley announced that the team would move to Los Angeles. Though I was not formally a Dodger fan, I did love the team and thought the move was a terrible blow to Brooklyn and to the entire city. I don't think the Mets will ever engender the same affection among the city's baseball fans.

Meanwhile, social life was very active during the school year. We went to "tea dances" held at Manhattan and to those held at girls' colleges. However, we much preferred the evening dances at the many Catholic girls' colleges in the Bronx and adjacent counties, but never in Brooklyn,

Queens, or Long Island. Some of the girls' colleges in Westchester County were very snooty, and Yale men coming over from New Haven outclassed us. Our favorite was the girls' College of New Rochelle (CNR). It was fairly close by, and the nuns and students welcomed the boys from Manhattan for the weekly Friday night dance. Quite often, we sneaked off with the girls and went to a famous hangout on the coast called the Barge. There we could drink and dance without the supervision of the nuns. We did have to get the girls back before the "witching" hour. During sophomore and junior years, I dated a CNR student from Downingtown, Pennsylvania. Weekends, Adrian and I went on excursions and to football games at West Point. She told me that she had a "serious" boyfriend back in Downingtown, so I was disappointed but not surprised when she didn't come back to CNR for her junior year.

While I was away in the air force, my classmates kept up many of our old pursuits, except for bridge, but made an important discovery. They found a German deli in Yonkers that made the best roast beef sandwiches I've ever had. So the deli became one of our regular stops after bowling, movies, and tea dances, whatever. I dated some of the girls from St. Margaret's and especially remember Estelle Halley, a classmate. She lived in a fantastic house in very upscale Fieldston. I think they had a maid who came to the door when I rang. Estelle was nice, but we didn't seem to hit it off. Meanwhile, the guys in my St. Margaret's class had all graduated from Manhattan in June of 1954 and gone into the military as ninety-day wonders or on to jobs, many out of the area. John Collier was still around for parties and such, but his family had moved up to Bronxville, and he worked full-time in downtown Manhattan, limiting his availability. By this time Tom Gaffney, my *Challenger* coeditor, had returned from army service in Korea and was back at Manhattan. We began to double date and were occasionally joined by other *Challenger* staffers. We liked to take dates to the various jazz joints in the city. Most memorable was the Metropole that featured hot Dixieland and where the band marched around the hall while playing "Saints." Eddie Condon's in Greenwich Village was also popular. We didn't neglect Birdland, the Mecca of "cool jazz." I never saw Charlie "Bird" Parker there but did see Stan Getz, Lee Konitz, and others.

By this time in my life, I made the Riverdale Steakhouse my home away from home. In addition to stopping by with my buddies for some cold nickel draft beers, I often ordered up one of their delicious cheeseburgers on real Jewish rye bread. When working late on lab reports or other homework, I'd sometimes take a break by going to the steakhouse for a few brews and a burger. That would restore my strength for more hours of work in my attic room. After coming back from the service, I'd learned that while I was away, my room had been occupied by visiting Irish priests and their bishop from Africa. One of the priests, Fr. James Redmond, invited me to go with him on a tour of the Passionist Fathers' missions in the West Virginia coal region. On the way, we made a breakfast stop at Luray, Virginia, where I had my first taste of scrumptious patty sausage. I became a lifelong fan. I think we also took a tour of the beautiful Luray Caverns. It was an eye-opening trip for me into a dismally poor area, whose progress had stopped at about 1920. I don't think there were many Catholics around. In Powhatan, I served an early mass for Fr. Redmond, and he whispered that I should take up the collection. I turned around to find an empty church. We had a hearty breakfast afterward and moved on to the next stop, Bluefield, Virginia. It was a very interesting trip, and I wonder if Fr. Redmond was recruiting me.

My class hours often allowed me a leisurely breakfast, but a solitary one. Both of my sisters were up and out early for school, and Mother and Dad were at work. Mother usually left a pot of hot porridge on the stove for us, but it was often lumpy by the time I got to it. While enjoying my morning coffee, I usually listened to Dorothy and Dick to get the latest doings in New York society. Mother had been working at Columbia-Presbyterian hospital in Washington Heights. For some reason, perhaps the long commute, she left and took a job at the Hebrew Home for the Aged on Riverdale's "Jew Hill." Mother had spent many years working for Jewish families and got on well with them. She later worked at the monastery of the Sisters of the Visitation, now a home for retired archdiocesan priests. Though it was a fairly long walk to either place, Mother usually got a ride with a coworker, one of the other Irish working women. I imagine work was fun for her, a good way to get out of the house, and have some of her own money. We children were not too happy with Mother's employment, but she would not be dissuaded. She was a very determined woman.

As junior year came to a close, I received some splendid news. I was offered a summer engineering position with General Electric Company at Syracuse, New York. While I was not in the top rank scholastically, I was known for my leadership in college activities. In addition to my position as departing editor in chief of the engineering magazine, I had been elected president of our chapters of Eta Kappa Nu and IRE. Moreover, I had taken an examination for a New York Regents War Service Scholarship, and it was granted in the summer of 1955. It added some welcome dollars to my college fund for senior year as my GI Bill funding had run out.

As a parting gift to our fellow students, we put out a very arty last issue of the *Manhattan Engineer*. We had the usual brief biographies and pictures of the editors but associated each with quotes from T. S. Eliot poems. My fellow engineer-staffers were not crazy about this idea but went along on my assurance that it would be great. I've saved all my back issues of the magazine and can now say with confidence that it was a bad idea. It was too affected; my liberal arts background had sent me round the bend. I don't think any of us suffered serious damage to our reputations. I remained friends with my staffers and one, Bob Tomasulo, was in my wedding party. Bob was one of the top-ranked students in the class and played a great game of chess. We'd play between classes, and he'd win every time. After a time, I just wouldn't play and Bob couldn't find another victim.

Summer in Syracuse

In early June, I left for Syracuse and my job at General Electric Company where I was assigned to the Court Street plant, part of the Defense Electronics business segment. GE set me up with a nice room in a boarding house near Syracuse University. Edwin Jones Jr., a fellow intern who was from the University of West Virginia (UWV), occupied the next room. His father was head of the department of electrical engineering at UWV, and Edwin went on to become a professor and department head at Iowa State University. Unfortunately, Edwin didn't drink and kept pretty much to his room, so I saw very little of him. A short time after I started work at GE, I was invited to attend a meeting of the Engineer's Club of Syracuse whose meetings were held at the Congress brewery, which offered free beer. It seemed that the first order of business was to elect a president and a social chairman for the upcoming term. When the time came for nominations for social chairman, someone I had considered a friend nominated me, saying I had been in town longer than most of GE's new hires and knew my way around. I was elected unanimously despite my protests. I soon learned that the principal duties of the social chairman were to arrange dances and get local girls to attend.

The working environment at Court Street was very pleasant and low-key. The section to which I was assigned was headed by Jim Chapman, a truly charismatic leader. The section was involved in a radar project for the U.S. Navy. I was assigned various small tasks, usually involving design and construction of a piece of test equipment. On my first assignment, I needed a small chassis upon which to place the components I was using. In the lab, I found a piece of sheet metal about the right size and put it in a device called a brake to bend it into the proper shape. When I was done, I found that I couldn't get my chassis out of the brake; it was very embarrassing. So while no one was looking, I took the brake apart and extracted my piece. I put the brake back together as best I could and sneaked out of the lab. Sometime later, while at my desk I heard a loud voice coming from the lab, saying something like: "What idiot screwed up this brake?" My face was red.

Meantime, I had to work on my most important job that of social chairman. I decided to visit the women's dorms at the various educational facilities in and around the university. I hit pay dirt on my first stop, which turned out to be a nursing school not directly part of the university. I became a favorite of the elderly housemother by reading for her the fine print of the stock market listings in the local newspaper. In turn, she helped turn out the girls for a dance I had arranged at the Salt Cellar in downtown Syracuse. This was one of the oldest and best-known places in Syracuse, which was then known as the Salt City. Locally, salt was extracted, rather than mined, by the famous Solvay process. Quite a few nursing students came, all women at the time, so the dance was a success. I met a girl whose name was Carol, and we dated for the rest of my time at GE. I don't remember where Carol came from, but it was some distance away so that she was in town weekends and holidays.

There was plenty to do in and around Syracuse, at least in summertime. Carol and I enjoyed driving up to the Erie Canal and hanging out with the barge men when they tied up for the night. They usually had campfires going and someone making music with a harmonica or an accordion. Another favorite spot was Jamesville Reservoir south of the city. The reservoir stored water for the Erie Canal, not for drinking, so swimming was permitted though there were no lifeguards. This was a popular evening hangout for young people who swam and sat around campfires smoking and toasting marshmallows and such. Most people smoked in those days, but Carol and I only smoked cigarettes. I remember taking Carol to see the popular movie at the time, *Love Is a Many-Splendored Thing*, starring Jennifer Jones and William Holden. It wasn't a first-rate film, but had a good musical score and a great song of the same name. We also went to a play at a nearby "summer stock" theater. Gloria Vanderbilt played the female lead, otherwise I don't remember anything about it.

While dating Carol in Syracuse, I drove back to NYC every other weekend to make a home visit and to see Sheila. She and I had met at a dance at her downtown college where she was what we called a day hop student. She attended Marymount Manhattan College, a four year liberal arts school for girls. It was somehow associated with Marymount College in Tarrytown, New York. We had been dating for several months, even though she was also seeing another Manhattan College student who was in the School of Business. Sheila was very well-read and very bright, so we hit it off on the intellectual level. I believe we visited the Catholic Worker's House of Hospitality together, and I remember that we went to Rockaway Beach together. As always, college boys' visits to the beach involved lots of cases of beer and somewhat rowdy behavior. Sheila was not a very physical person and was not happy with our group's behavior. She became very apprehensive about my drinking, mostly because her father was an alcoholic. Though I often visited at their apartment in Manhattan, I never met either of her parents. While I liked Sheila, I realized we were not going anywhere together because of my drinking and because I planned to leave NYC upon graduation the following spring. When I returned to school in September, I began dating other girls. In those days, we called young women "girls" until they married.

Back at work, Jim Chapman's section held a Friday late-afternoon outing every month at a state park outside the city. Staff and families were invited. The office girls saw to it that all the single men went to the party. Naturally I went when I learned there would be free beer and food. These parties were the best company parties I ever attended. There were lots of pretty girls and women for dancing, there were hot dogs and burgers for eating, and a variety of activities: swimming, softball, horseshoe pitching, and more. I decided that I would come back to Syracuse and Jim Chapman, if GE offered me a permanent job after graduation.

The summer season came to an end, and the city emptied out when the summer students left. As the weather cooled, you got the feeling that winter would arrive any day. It was eerie. I said my farewells to my GE colleagues and to Carol and headed back to Riverdale on a Friday afternoon in September. My Chevy had become somewhat balky, particularly in hot weather and while driving at high speed on the newly-opened New York State Thruway. My usual route home was via the Thruway to just south of Albany where I crossed the Hudson River to pick up the Taconic State Parkway. I preferred the latter route because it was toll-free and more scenic. On my final trip home, I made good time on the Thruway, but once on the parkway, the engine kept conking out every few miles. I'd have to stop and let the engine cool down. When it restarted, I'd drive as fast as I could to the top of the next hill and coast down the other side, hoping the engine would not fail for my assault on the next hill. This tactic worked, but the trip took many more hours than usual, and I arrived home well after dark. A mechanic found that I had a defective ignition coil, and with a new one, I got many more miles from the old car.

Back to Manhattan for Senior Year

Back at Manhattan, I picked up where I left off, running the student chapters of IRE and Eta Kappa Nu (HKN). Choosing new members for the latter was a serious matter for students and faculty. It was important for high-achieving electrical engineering students to be inducted into BKN as it was the highest honor available for such students at Manhattan. Some years later, the college had a chapter of Tau Beta Pi, the national engineering honor society open to all engineering disciplines. My senior classes included two total dogs, electromagnetic wave theory—really Maxwell's equations—and microwave lab. The former was loaded down with high level, mostly unfathomable, mathematics, and the latter was loaded with faulty lab equipment and results had to be faked. Our instructor for both classes was Chester Nisteruk, who I thought knew little more than I did about both subjects. I think I scraped by the former with a D and with a C in the latter. They were my worst courses. Ironically, the college made a brochure entitled "Electronics at Manhattan College," and it contained a picture of me at work in the microwave lab. It was really a fraudulent production as there was really no electronics, as we now know it, taught at Manhattan in my time. To the best of my recollection, neither vacuum tubes nor transistors were ever mentioned. Fortunately my longtime buddy Bob Foran, who had graduated in 1954, got me interested in computers. His older brother who was into tabulating equipment—forerunner of the computer revolution to come—pushed Bob, unsuccessfully, toward IBM and computers. However, his notion paid off for me, and I thank God computer science and engineering involved neither Maxwell nor microwaves.

I had developed a close relationship with two engineering classmates, Bill McEvoy and Bill O'Hara. They had grown up together in the Queens Village neighborhood. Both were veterans—McEvoy in the Marine Corps and O'Hara in the navy. We attended parties together, and they became acquainted with my longtime friends such as Tom Gaffney and my other Hayes classmates. I would visit the Bills at their favorite bar in Queens, and they would reciprocate. Our local favorites were the Riverdale Steakhouse and a small bar in Yonkers that was very popular with John Collier and his Fordham crowd. If you were doing nothing else on a weekend evening, you could always go to the Yonkers establishment with hope of finding friends there. O'Hara day-hopped from Queens, but McEvoy lived in an attic room off campus. His father had been a civil engineer and pushed him into engineering. I think he, like me, was happiest in a liberal arts setting. Some evenings, I would meet with him in his room to jointly prepare for upcoming quizzes and exams. Invariably I found him unprepared and trying to figure ways to beat the system. Sadly, McEvoy dropped out of Manhattan in senior year and went nights to Hofstra University where he got an engineering degree. O'Hara, much less gifted intellectually, went on to get his degree with me.

I stayed in touch with both Bills and attended their weddings. Bill McEvoy married Edith and had a reception in a banquet room in a Netherlands legation facility in Manhattan. A job took Bill and Edith to Concord, Massachusetts, where they lived for many years. Bill vowed that he would someday get his engineering degree from Manhattan College, using his credits from Hofstra. The college fought him off for years but finally relented, and his death was reported in the alumni magazine as BEE, class of 1956. Bill O'Hara married Anne, a very cool and beautiful blonde, and their wedding reception was at Forest Hills Country Club where the U. S. Open tennis tournament was held. Bill and Anne moved to Connecticut where he owned and operated a very successful electronic parts distributorship. The O'Haras had a custom retirement house built in Florida, but Bill did not live to enjoy it. The O'Haras attended my wedding, but not the McEvoys, and I don't know why they missed it. Both Bills were a lot of fun and, as fellow veterans, helped me get through some bad days in engineering school.

After breaking off my relationship with Sheila, I was interested in meeting new girls. Tom Gaffney and some friends had found a dance hall in the East Bronx that was frequented by working girls, many of whom were employed by the telephone company. After attending a few sessions, I met a very charming girl who happened to be an immigrant from England. We began dating, usually with Tom Gaffney and Ann Sager. Tom had met Ann at a dance at the hospital where his mother was working. She was a nursing student, and she and Tom would soon marry.

When I went to call for my new date for the first time, I found that she had her own apartment, an ideal setup I thought. However, as it was my desire to leave New York after graduation, she realized that wedding bells were not likely for us. I had to agree that we should break up so that she could find a more suitable target for her wedding plans, so we did. For the rest of my college years, I dated a succession of great girls, but again with no long-term notions. I especially recall dating my sister's friend Peggy Lennon. She was one of the beauteous Lennon sisters of the Bronx, not the singing Lennon sisters of *The Lawrence Welk Show*. Speaking of Lawrence Welk, I had absolutely no appreciation for his musical variety show although it was very popular with my parents. Without a steady girl, I didn't even attend the 1955 Manhattan Engineers' Ball, and the ball was the social highlight of the college year for engineers. It was held every November in the ballroom of the Hotel Commodore, conveniently located adjacent to Grand Central Terminal on the east side of Manhattan

I Became a Hackie

Since my GI Bill funds had covered only three years of college and my scholarship funds were meager at best, I needed to find a part-time job. I spoke to Dad about driving one of his cabs, and he helped me get my "hack" license—as Driver 38787. Apparently, taxicabs had been called "hacks" at sometime in the past because licenses were issued by the Hack License Bureau of the Police Department. At first, I drove during the day shift but traffic was murder, limiting fares and tips. When Dad told me he often had cabs idle on weekend night shifts, I switched to Friday and Saturday nights. This worked out very well for me as I averaged $30 per night when that was a lot of money. I had no serious problems driving and stayed pretty much in midtown Manhattan. I had some fares to the New York airports, particularly to Idlewild, now JFK. These were not very lucrative as you had a long wait there for a trip back to Manhattan.

I had one call to Harlem, and as I dropped my fare off, a man came running up to me and begged me to take his friend to the hospital, saying that he had been stabbed. He pushed his friend into the cab, tossed some money at me, and told me where the hospital was. I wasn't happy with this situation, but what could I do; the man was already in my cab and possibly bleeding to death. I set off eastward toward the hospital on Harlem's main drag, 125th Street. It was a Saturday night, and I quickly became mired in heavy traffic. I could see the hospital about a block away when my passenger jumped out of the cab and, holding his abdomen, started running toward it. I pulled over and checked the backseat for blood and found none. I quickly headed back downtown.

Other than traffic, the only drawback to being a cabby in midtown Manhattan was parking. When you needed to make a pit stop for food or toilet, you had to go to a restaurant. You might have a favorite stop as I did, but if you couldn't find a place to park you'd have to go on and try somewhere else. One of my favorite spots was the Bickford's cafeteria on Eighth Avenue, between Thirty-fourth and Thirty-fifth Streets. It was just off the Theater District, so in addition to cabbies, it was popular with writers and bums. If I couldn't find parking there, I'd try another of the many Bickford's in Manhattan, or one of the Horn & Hardart Automats. I actually preferred the latter, but finding parking near an Automat was just about impossible. I usually had something in the cab with me to read while on break and enjoyed both reading and watching the world go by, particularly at a Bickford's where a good meal could be had for twenty-five cents.

I had only one trip to Brooklyn, and it was an adventure. I had no idea where I was going, so my fare directed me down one of the main streets. It became obvious to me that we were in a black area when he had me stop and pick up acquaintances along the way. Eventually, my original fare left me with a good tip, and I continued along picking up and dropping off and collecting money along the way. Now this was illegal as I was running what's called a "jitney" service. I was anxious to get rid of my fares and get back to Manhattan but had no idea how to do this. Finally, I was left with one man who wanted to go to Manhattan. I told him I would be happy to oblige him if he knew the way, and he did. The gods were with me as I never had another call to Brooklyn.

I wondered why the regular cabbies didn't work my lucrative night shifts and found that many took the weekends off if they'd made their nut during the week. Flushed with money, they'd often be found at the local New York racetracks on weekends—Jamaica and Aqueduct—and even distant Saratoga during their August meet. Having established a continuous source of funds, I was able to cover some of my tuitions and books, with plenty left over for partying, but not on weekends.

My final memory of my days as a "hackie" occurred on an eventful St. Patrick's Day, Saturday, March 17, 1956. It was a snowy day and heavy snow continued into the evening. Most cabbies had headed back to the barn, but I kept on working. It was heavy going, and the push-button transmissions in our Plymouth cabs made getting out of snowdrifts very dicey. Wouldn't you know it; I picked up a fare who wanted to go to a location in the far northeastern part of the Bronx. As I set out with him, fearing abandonment of the cab in the wilds, the sky suddenly cleared, the stars appeared, and it became a spectacularly beautiful evening. That is, until I heard the far-off sound of thunder, but nothing more. I made it safely to the Bronx and back, but things were very quiet and I called it a night earlier than usual. Unknown to me, that very evening Kitty Carlin, her sister Patricia, and friends, had left NYC for Europe on board the Italian liner, *Cristoforo Colombo*.

On to Graduation

A priority during senior year was interviewing and awaiting job offers. I was sure that I'd receive a job offer from GE, no interview required, and I'd be back to Syracuse. However, the offer was for a paltry $438 per month to start. I was put off by the low offer, so I decided to seriously look elsewhere. I was interviewed at a few electronics firms on Long Island but did not indicate interest, so no offers were forthcoming. Because I was interested in computers, I naturally gravitated toward IBM at Poughkeepsie, New York. I was very impressed with what I saw at IBM and was pleased to receive an offer to start at $455 per month. I had heard that RCA—then Radio Corporation of America—was into computers, so I visited their facilities at Camden and Moorestown, New Jersey. At the time, RCA was one of the country's leading industrial companies and would reach a billion dollars in sales before IBM. I could see that RCA's computer operation was a mess compared to what I'd seen at IBM, and that spelled opportunity to me. I received a job offer from RCA to start at $470 per month, plus $600 cash to show up. Since I was inclined to go south to work and for more money, I readily agreed to go to work for RCA. After I'd accepted RCA's offer, GE upped the ante to $468 per month, but for me it was too little too late. Apparently GE had poor success with their early recruiting that year and upped their initial offers to everyone. They did get acceptances from a number of my classmates. As an additional inducement, IBM put everyone who accepted their job offer on immediate half pay. They got many takers among my classmates, among them Bob Tomasulo, a close friend and colleague. It was a very good year for the engineering graduates of Manhattan College.

Earlier I mentioned favorite hangouts and neglected to mention the ever-popular Pinewood Bar and Grille, which was located just off Broadway and practically next door to the college. In addition to serving nickel beers, the Pinewood offered a selection of sandwiches. Several of us had been out on the town one evening and stopped at the Pinewood on the way home. There, for the first and only time, we met some of the Christian Brothers of the college faculty who had also stopped by for a few brews. I'm sure both groups were surprised to find each other at the Pinewood, but we chatted and parted amicably; and I thought no more about the encounter. After some time had passed, I was asked to come to the office of Br. Amandus Leo, dean of the College of Engineering. Br. Leo was a kindly man and eventually broached the subject of the meeting. It had been reported to him that I might have a drinking problem, that I might in fact be an alcoholic. He assured me that the college was ready to provide any professional help that I would need should I decide I needed it. Naturally I was shocked and emphatically denied the inference. Thus, I failed to take advantage of this well-intentioned early "intervention" and continued my binge drinking.

In senior year, I had no classes on Wednesday afternoon. I had always been interested in the theater and knew that matinee performances were given on that afternoon. One afternoon, I took the subway downtown, walked over to the theater district, and began checking out the offerings. I had read glowing reviews of Jean Anouilh's *The Lark* and decided to give it a try. For $3 I was able to get a seat in the very first row of the mezzanine. I was absolutely thrilled by the extraordinary performances of the cast—Julie Harris as Joan of Arc, with Boris Karloff, Joseph Wiseman, and Christopher Plummer as her tormentors. For a Broadway first, I could not have done better. Thereafter, I went every Wednesday and took a seat at whatever show was available. Broadway was never better than it was during 1955 and 1956, and I was fortunate to be a part of it, if only as an avid fan. I saved many of the original playbills from those years and used them to build the show listings in appendix K. My all-time favorite drama was, of course, *The Lark*. Julie Harris won that year's Tony award for her performance. I've seen some excellent musicals over the years, but I still think 1955's *The Pajama Game* is my favorite. The performance of Eddie Foy Jr. as the "time-and-motion" man was absolutely delicious as were the big production numbers "Steam Heat" and "Hernando's Hideaway."

One day after theater, I had a peculiar meeting. Mother had a close friend whose name was Bea Gormley. Bea was also from County Tyrone, but I think Mother first met her in NYC. Late in life, Bea married Ralph Florio, a New Yorker with a nice second-floor apartment in a converted Manhattan townhouse. Ralph was a big and hearty man who enjoyed cooking for Bea and her friends. He visited us in Rockaway and was usually available to meet family when they arrived on the boat or plane from Ireland. He often treated the new arrivals to lunch or dinner. We always wondered what Ralph did for a living, but the subject was never broached when we were with the Florios. Well, one Wednesday after leaving the theater, I walked toward Times Square and the subway. Someone came out of a shop I was passing and almost knocked me over. It was Ralph, and he was carrying a strange little black bag. When he realized who I was, he said, "Jimmy, what are you doing here?" He was very embarrassed, and so was I. Our meeting became a sort of secret within the family.

A big springtime event was the annual boat ride to Bear Mountain State Park near West Point, New York. I had been there a time or two by car with the family. The Hudson River Day Line boats left from a pier near West Forty-second Street and made stops at 125th Street and at Yonkers on the way up the river. At each stop, Manhattan students and friends could be seen lugging cases of beer and other goodies to the boat. It's a wonder the boat didn't sink under the load. Upon arrival at the park, places were staked out on the field, blankets laid, and games begun. For the hardier souls, there was hiking to the top of the mountain. The park had a fine old stone inn with restaurant that offered relatively fine dining and a cafeteria for quick snacks. These were hardly patronized by the crowd who had mostly brought their own food and drink. This became a problem because in later years, the state police would not allow any beer or liquor to be taken off the boat—you had to buy it at the inn. It was quite a sight to see perfectly good cases of beer and bottles of booze go over the side of the boat as the police enforced the prohibition. Some young women were able to hide booze in their bags as they were not inspected by the police. It became a rather dry and unpopular event thereafter.

One morning while I was asleep in my attic room, I was awakened by a crash in front of the house. I went to the window and saw that a car had crashed into my Chevy that was parked at the curb. To the west of our house, 261st Street went up a steep hill to Riverdale Avenue. On the hill were a number of shops, including Elsner's popular deli and butcher shop. Apparently the car's driver had parked on the hill to run into Elsner's and did not cut his front wheels into the curb. The brakes didn't hold, and the car rolled down, crossed the street, and into mine. I was very distraught when I learned that because I'd been involved in an accident I had to buy exorbitant insurance. I decided that since I was leaving NYC, I would forgo this expense. I then received notice that for this folly, I was to surrender my registration tags and my driver's license. But by then, I was on my way out of town.

The concluding "exercises" for the 1956 graduates consisted of the obligatory Baccalaureate Mass on Sunday, June 10, and commencement on Tuesday, June 12. I remember commencement day very well; it was bright and beautiful. In cap and gown, I marched onto the college quadrangle with my fellow EEs—proud, yes, but more relieved than anything else. I received my bachelor of electrical engineering degree from the hands of Francis Cardinal Spellman who had also done the honors at my high school graduation. The commencement address was given by Bishop Wright, of Worcester, Massachusetts, who had received an honorary degree. Fortunately there was some shade on the quadrangle as it was a warm day. So after almost six long years, I had finally finished college and gotten my "sheepskin." Some wise guy said that if it were a real sheepskin, maybe you'd be able to make a shirt out of it. After the ceremonies, the family headed home for a small party for the first college graduate of our O'Donnell clan. Directly after that, I packed up for my trip to Camden. Because I liked to travel "light," I gave away a lot of stuff that I later wished I'd kept. Foremost among the missing items were my autograph book with all the Yankee greats of the 1930s and 1940s and my collection of juvenile fiction.

While I was preparing to head south, my parents and sister Maureen were off to the "Old Country." They had gone over on the SS *United States*, the fastest of the transatlantic ocean liners, to attend cousin Alfie Quinn's wedding. Alfie was the eldest of Uncle Joe Quinn's sons still living in County Tyrone. He went on to become an IRA (Irish Republican Army) stalwart, and that brought a lot of grief on him and his family. Returning to New York on the same ship from the Irish port of Cobh (for Cork city), they ran into a late summer storm at sea. It was a very rough crossing, and many passengers were injured. The heyday of the ocean liners was drawing to a close, anyway, and my family swore they'd never take a ship again, and they didn't.

As for me, farewell, Riverdale, hello, Camden!

Dad with his 1948 maroon Plymouth, circa 1950.

Mother and the girls boarding their 1950 flight to Ireland on American Overseas Airlines. Mother and Maureen are on the right near the top of the stairs. Patricia is on the tarmac, third from right.

The main terminal at Idlewild airport, now JFK, 1950.

At Cornamaddy, 1950. Mother and Patricia ride the fenders, and Maureen is on the bumper.

Mother and the girls with my grandfather Patrick Quinn, 1950.
He died the following year.

At home in my attic room, Riverdale 1951.

I have a nice corner spot in the barracks, 1951.

We made a trip to Silver Springs while the family visited me in Florida.

My advanced radar class at Biloxi, 1952. My New York buddy
Norman Thierer is standing to my right.

On the beach at Biloxi, February 1952.

I was ready to defend the United States against all enemies,
foreign and domestic.

For drinking and dancing with the locals, a typical Mississippi
honky-tonk.

I'm standing in Jackson Square on one of my many weekend visits to New Orleans.

The Manhattan College quadrangle is beyond the arch.

Ultra high frequency techniques are employed by student as standing waves are observed on latest type transmission line in the course of a series of impedance measurements.

I was pictured in a promotional brochure for electronics engineering at Manhattan College. The microwave equipment shown in the picture was about all the electronics they had in the early 1950s.

Moorestown High School in the 1950s.

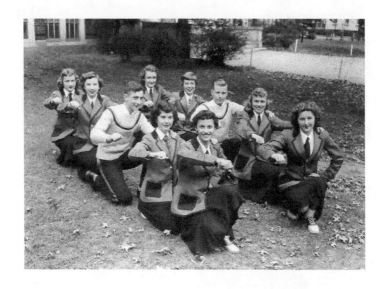

Moorestown cheerleaders. Kitty Carlin is second from last on the left side.

I'm working with a transit at surveying camp, summer of 1953. We wore our straw "boaters" to camp.

Classmate and golf buddy Eugene Donovan uses the splendid shaving facilities at surveying camp near Peekskill, NY.

Kitty Carlin and her sister Joanne, mid-1950s.

Cousin Mike Quinn (far right) proudly shows off his new convertible. In the backseat are Anna Ward, left, and my sister Patricia at the right. Sadly Anna, one of my sister's closest friends, died of cancer at an early age.

Mid-1950s family grouping in the Riverdale house. My sister Maureen is on the right; Patricia is on the left.

On a trip to Cincinnati with Aunt Kathleen Quinn, we stopped to visit Aunt Alice (Sr. Jane Frances Quinn) at her convent in Somerset, Ohio. It was my first long trip in my 1949 Chevrolet.

My "hack" license.

En route to Naples, Italy, March 1956. On board the *Cristoforo Colombo*. *Left to right:* Patsy Gross, Kitty, and her sister Pat.

The group goes "Dutch." *Left to right:* Wanda Curry, Patsy Gross, Herb Wirth, Noel "Curry" Wirth, Kitty, and Pat.

Kitty and Pat Gross catch some rays on Mount Pilatus, Switzerland.

"LIBERTÉ"

Art deco poster for the French Line's S.S. *Liberte*.

Patsy Gross and Kitty relax during the trip home on the *Liberte*, July 1956.

Commencement day at Manhattan College, June 12, 1956. I'm waiting, with my classmates, to go onto the quadrangle to pick up my diploma.

The County Tyrone Society Ladies' Auxiliary, c. 1956. My sister Patricia is second from left in the front row.

PART 4

RCA Years (1956-1971)

Welcome to Camden

Between my June 12 commencement and my start of work at RCA on Monday, June 25, 1956, I did some shopping. I had a very nice collection of 45-rpm jazz albums, and I still think the extended play 45s were the best format for jazz and popular music. However, for classical music, the new 33-rpm LP (long-playing) format was taking over, and gradually new jazz and pop releases were moving to it. For graduation, I had gotten a beautiful Magnavox hi-fi (high-fidelity) console record player, one that played 33/45/78-rpm records and had a fifteen-inch woofer and an AM/FM tuner. For some new music for my hi-fi, I went to the famous Sam Goody's record shop near Times Square and bought my first LPs. Goody's was about the only place to go shopping for LPs in those days, and you could test play them in one of their sound booths. My initial purchase included two albums that are still among my favorites: "I'm Wild Again" by Frances Faye and "Matt Dennis Plays and Sings." Years later, I transcribed both to CD format and often play my favorite tracks.

In reporting to RCA, I didn't pack much since I would be staying at a hotel at first and didn't know where I would be going thereafter. Consequently, I had to leave my new Magnavox and my record collection in Riverdale. On Sunday, June 24, I drove down to Camden in my aging Chevy via the almost-new New Jersey Turnpike. That evening I checked into the Walt Whitman Hotel—Camden's finest and Camden's only—and found that classmate Jim Maguire and I were rooming together. Jim and I hadn't socialized very much at Manhattan, so sharing a room together was quite fortuitous. I do remember that we got together with several other Manhattan grads and set out on the town. Maybe on a prank, we got steered to a gay bar at Airport Circle and quickly left. We did stop somewhere for a few brews and headed back to the hotel. Sometime during the night, Jim and I were awakened out of a sound sleep by the night watchman who admonished us for failing to lock our door. In 1956, Camden was very much the center of things in South Jersey. It had the first-run movie theaters and the best restaurants. The Walt Whitman Hotel, now the site of Camden County College, had a good restaurant as well as a very active office of stockbroker Merrill Lynch. The office had an arena seating for those who liked to watch the stock ticker, and many RCA people spent their lunch break there.

On Monday, we reported to the college relations office in the engineering building on the Delaware River at Delaware Avenue and

Cooper Street. That office handled college recruiting and, by design I
suppose, featured a number of attractive unattached girls. We were greeted
by section manager Robert Haklisch, filled out a lot of paperwork, and
got our assignments. At that time, each employee starting from college
did a rotation through four five-week assignments. At the end of the
rotation, he could request assignment to one of the four or somewhere
else he felt would be congenial; but you were not guaranteed your first
choice. I drew assignments as follows:

1. Moorestown, NJ (June 25-July 28)—radar systems
2. Cherry Hill, NJ (July 30-August 31)—color television
3. Harrison, NJ (September 2-October 5)—vacuum tube applications
4. Camden, NJ (October 8-November 9)—computer systems

Although I was determined to get into computer systems and was
impatient with the idea of rotating through places of little or no interest,
I had to go along with the program, and it's a good thing I did.

I believe that Jim Maguire's first assignment was in Camden, so
we decided to find a furnished apartment that we could share, and we
did. It was in a three-story building on the White Horse Pike in West
Collingswood. It was a good location, close to work and to all the
hot nightspots along the pike in Oaklyn. We had been there for a few
weeks when the landlady asked us to move out, saying that there had
been complaints about noise late at night. Actually we were not the
perpetrators. We'd seen the young couple on the floor above us bringing
home cases of beer on several occasions. They delighted in saving the
empties and dumping them all together down the trash chute that went
to the basement. Needless to say, it bothered everyone whose apartment
abutted the shaft. Fortunately for us, it was about the time for changing
assignments. Jim left to go elsewhere, and I was able to move in with
other RCA trainees who were living in a two-bedroom apartment at
111B East Coulter Avenue in Collingswood. My roommates included
Bob Lichtenberger and Dave Mackin, who were both ROTC (Reserve
Officer Training Corps) students, and Jim Maguire when he was on local
assignment. Bob was from Manhattan College, but I have no idea where
Dave went to school. Bob was an excellent cook, so living was good until
Bob moved on, probably into the military. Dave was called to navy duty
and some of us went to visit him at the officer's club near Washington.
The Coulter Avenue apartment was to be my home for several years.

It was very pleasant working in the almost-rural setting that was RCA's relatively new plant in Moorestown. At the time, they were working on large radar-based systems for the U.S. Navy. We often worked overtime and were paid for it; that made me happy. I remember looking out the window one evening and across the neighboring farm fields I saw two Negroes walking down a dirt road carrying watermelons. It brought back memories of my time in "Dixie," and I called my office mates to the window to share the timeless scene. I don't remember what kind of assignments I had but recall that I worked with Bob Buford, a year or two my predecessor at RCA. Bob was good to work with, and I was shocked to learn some years later that he had killed himself.

I Meet Kitty Carlin

What I did on the job at Moorestown was minor compared to a life-changing meeting that occurred there. One day in the cafeteria, I saw a girl carrying her tray and walking back toward the tables. I thought she looked just great and got up to ask her if I could sit with her. Unfortunately, she moved in with a group of coworkers before I could ask. When she left, I followed her back to her office where we chatted. Her name was Kitty Carlin, and I invited her to go out with me to a performance of *The King and I* at the Camden County Music Circus the upcoming Saturday, June 21; and she agreed. When I called for her at her home in Moorestown, I was met by her very suspicious little sister, Joanne. After the show, we stopped for drinks at the Honey Dew Lounge in Oaklyn. But as Kitty had no ID, they wouldn't serve her, so we left. We were to see many more shows at the music circus, which was located in Cherry Hill between Brace Road and Kings Highway. Adjacent to the theater was a very popular hangout called the Farm. We were there one summer night when a storm knocked out power, and the packed place was plunged into darkness. Candles were lighted and business continued apace, but without air-conditioning the old barn was unbearable.

I learned that Kitty had just started at RCA the week we met. She, her sister Pat, and some friends had recently returned from a four-month sojourn in Europe, crossing the Atlantic by ship both ways. In March of 1956, the group landed in Naples, Italy, on the Italian Line's *Cristoforo Colombo*. From Naples, they traveled through most of Western Europe, missing only the Iberian and Scandinavian countries. Kitty seemed most taken with Vienna and all of Austria. The country had only recently recovered its independence from occupation by the allied powers. *The Third Man* is an excellent movie that shows what postwar Vienna was like. The group returned to the United States in July on the French Line's S.S. *Liberte*. When Kitty started at RCA, I had only two weeks remaining on my assignment at the Moorestown plant. We had to meet within those two weeks or most likely would never meet. Fate put us together.

For our next date, I wanted to take Kitty to see *My Fair Lady* on Broadway. The show had opened just four months before, and tickets were hard to get. My friend and classmate Bob Tomasulo had an uncle who was a ticket broker, and voila, Bob and I had tickets for the Saturday matinee. Kitty was thrilled to be going to the show but less so when I told her I would pick her up at 8:00 a.m. I explained that I had promised to take my sister Patricia to Idlewild Airport for another trip to Ireland. She too was going over for cousin Alfie Quinn's wedding. When we got to Riverdale, we found that some of Pat's Irish "lads" had come to take her to the airport, so our early departure turned out to be totally unnecessary. Pat stayed in Ireland for six weeks, mostly in and around Dungannon and Pomeroy, NIR. She wisely flew back to New York.

We met Bob Tomasulo and his date at the Mark Hellinger Theater and found that we had truly exceptional seats in about the fifth row of the orchestra. Actually, they were "divan" seats, like sitting on a sofa. The show was wonderful, and afterward, Kitty and I had dinner in a restaurant on Times Square. Bob and his date had left for Brooklyn. After dinner, we went to Greenwich Village where we met the Bills—McEvoy and O'Hara—and their dates. I'm not exactly sure where we spent the evening, but in any event, it was very late when I dropped Kitty off in Moorestown. It had been an exciting but exhausting day for us. Kitty celebrated her birthday the following day with her family. She was twenty-three years of age.

On Monday, July 30, I started my assignment with color television engineering, then located in RCA's new facility on NJ Route 38, next to the Cherry Hill Inn. That location—now the site of a movie theater and shopping complex—was then in Delaware Township, later renamed Cherry Hill. The assignment was interesting from a historical perspective. Just a few years before, this group that developed the color television system that the FCC (Federal Communications Commission) had chosen over that of CBS (Columbia Broadcasting System) for the United States standard. It was a close call by the FCC, but we were thankfully spared the clunky rotating disk design that CBS proposed. My duties were limited to designing and setting up test signal generators for use in the laboratory. I did attend some meetings where cost savings of pennies were hotly debated. By 1956, television sets were no longer manufactured by union workers in Camden. Color set production had been moved to the Midwest, and engineering would follow a few years later. The company was only interested in colored TV, so black-and-white sets were imported and sold with the RCA "Nipper" logo.

While I labored at Cherry Hill, Kitty and I dated regularly. That summer, we saw many of the shows at the Camden County Music Circus. The shows were performed "in the round," and though the performers were not Broadway caliber, the shows were a lot of fun. I fondly remember *Wish You Were Here*, which was set in the Catskills, New York's "Jewish Alps." Eddie Fisher, Elizabeth Taylor's spouse, had starred in the New York production. One scene was set around a swimming pool, and the patrons in the first rows got soaked from the splashing. After the show, we usually stopped at the aforementioned Farm for refreshment. We also went to first-run movies then shown in Camden's Stanley and Samar theaters and to the new foreign films shown at the Ritz in Oaklyn.

In the fall, Kitty bowled with the RCA league and her team won a trophy—where is it now? We occasionally bowled together and Kitty was definitely the better bowler. We found time to attend the Pennsylvania-Navy game at the historic Franklin Field, also home field for the Philadelphia Eagles. A favorite outing was to the Pub at Airport Circle in Pennsauken. That establishment widely promoted its drinks as being a full 1½ ounces, which didn't seem like a lot to me. We did enjoy their famous hearts of lettuce salads and grilled steaks. Many dates ended up at the Honey Dew Lounge in Oaklyn. They had a great live swing band for listening and dancing, and their trumpeter could blow with the best. The highlight of the evening was their full-blown rendition of "Cherry Pink and Apple Blossom White." *Perez Prado*'s 1955 orchestral recording of the song had reached number 1 on the *Billboard* charts.

Off to Harrison and Back to Camden

On Monday, September 2, I started my third assignment, this time at RCA's vacuum tube plant at Harrison, New Jersey. This was an old facility that made vacuum tubes for radio and television receivers and for high-fidelity sound equipment. The main building was about eight stories in height and occupied most of a city block. On each floor, hundreds of

woman sat at workstations assembling by hand the parts of the various
tube types being run. The assembled tubes were then placed in machines
where the base was attached and the bulb evacuated and sealed. I had
several made with my name inscribed inside. The tubes were branded for
use and sale by RCA or for other electronic firms. No one company made
all of the many types then in use, so vacuum tube manufacturing was a
cooperative enterprise. My duties in the lab were to design, build, and test
various amplifier designs for something new on the horizon, stereophonic
sound. This was a sort of circling the wagons in the face of the looming
onslaught of solid-state semiconductor devices—principally transistors.

During this assignment, I stayed with my aunt Kathleen and uncle
Mike Quinn at their home at 636 Valley Road, Clifton. Their house sat
at the foot of Garrett Mountain and overlooked the flatlands to the east,
clear to the skyscrapers of NYC. The setting sun reflecting off the distant
buildings made a dazzling sight. I had the first-floor room at the back of
the house. I don't remember which of my cousins were living at home at
the time. I do remember meeting Eileen's boyfriend Joe Dodd, who was
selling shoes in Paterson on a part-time basis. I bought an expensive pair
of Nunn-Bush black dress shoes from him, but they never fit me properly,
so I gave them to my classmate Jim Maguire. During the summer, the
Hungarian Revolution against the Communist government was under
way. One of my coworkers had recently emigrated from Hungary and
was very concerned with events back home. During coffee breaks, we'd
go to a nearby restaurant, read the newspaper reports, and commiserate
with him. Sadly for him and his people, the Russians crushed the revolt.
Lunchtime entertainment was offered in RCA's large plant cafeteria.
The star of the show was an Elvis impersonator, and the lunch crowd
of women cheered loudly during his act. The real Elvis Pressley was an
RCA recording artist who made a major contribution to the company's
bottom line in the 1950s. Weekends, I continued to see Kitty Carlin
during my Harrison sojourn. I particularly remember driving back to
Clifton on Sunday nights while listening to Jean Shepherd, the greatest
storyteller ever. His show was broadcast over station WOR, New York's
most powerful. One night, he talked hour after enthralling hour about
one thing, the great snowy owl.

On October 8, 1956, I started my final assignment, which was in
Camden's Bizmac Engineering, Bizmac being RCA's branded large-scale
computer system. I resumed residence in the group apartment in
Collingswood, and it would be my home until July of 1959. Finally settled,

I was able to make trips to Riverdale and take my prized Magnavox, along with my books, records, and desk south to Collingswood. Now that I was back in South Jersey, I was able to spend more time with Kitty. I learned that she was a 1951 graduate of Moorestown High School, where she had been a varsity cheerleader. She and her friend Patsy Gross had spent two years at Dunbarton College of the Holy Cross in Washington. I asked how she chose Dunbarton, and Kitty said that she and Patsy had come across the college's brochure in the town library. The brochure and the idea of being in Washington were appealing—not too close and not too far. I'm sure that they had a wonderful time partying with the Georgetown, and other, boys. Unsure about the benefits of continuing at Dunbarton, they left after the spring term of 1953. To improve their prospects of gainful employment, the girls enrolled in Philadelphia's Taylor School of Business. From Taylor, Kitty decided to test the job market by applying at the Campbell Soup plant in downtown Camden. It was her first and last interview, as she was hired on the spot to start in February of 1954. After working at the soup company for two years, Kitty and her friends went to Europe, and upon return, she started work at RCA where we met.

Now that Kitty and I were a couple, I finally had someone to take to the Manhattan Engineers' Ball in November of 1956. As always, it was held in the Grand Ballroom of the Hotel Commodore and featured the Jimmy Caruso Orchestra. I don't remember who sat at our table, probably one or both of the Bills, Bob Tomasulo, and perhaps, Tom Gaffney who was back at the School of Business from his army service in Korea. Four years later, we attended the 1960 ball, again with Jimmy Caruso and company. In October, Kitty finally met my family—sisters, parents, aunts, uncles, and cousins—at the wedding of Cousin Barry Quinn to Dolores McNenney. After the reception, we went to the Quinn home on Valley Road and sat outside to take in the great view eastward to NYC. Kitty and the family went well together, and I knew that we were into a serious relationship.

My First Project

This part of my story may be too technical for some readers, that's tough. I'm a systems and software engineer, and this is my story, so here we go.

At work, I'd finally gotten where I hoped to spend my career, in computers. J. Wesley Leas was chief engineer and welcomed me warmly. Over the years that I was to work for Wes, I found that my initial approval of him gave way to something less. I believe that it was my own inexperience that accounted for this change, because I sorely missed him when he was gone. From the outset, I was assigned to the Transcribing Card Punch (TCP) component of the overall system. Its function was to read a magnetic tape and punch the data into Hollerith cards at the rate of 150 cards per minute. At the time, IBM's best was 100 cards per minute. The electronics for the TCP were housed in two double six-foot racks, one for the power supply and the other for the control electronics and magnetic drum storage device. While the electronics were implemented using transistors, the huge power supply had been designed so that vacuum tubes could be used if transistors proved to be inadequate. Joe Brustman, a German émigré, was the original designer and had little faith in transistors.

At the time I joined the project, Jim Palmer was the lead engineer, and his boss was Gary Chien who was responsible for all punched card equipment—readers and punches. Jim and I worked well together, and he quickly left the TCP to me while he helped out on the more problem-plagued card reader, a vacuum tube design. Even though I had never even seen a transistor before, I plunged right in. In the heady days before all of the TCP problems had been identified, Jim Palmer, Chuck Propster, and I had an article published in the December 1957 issue of *Electronics* magazine. It was optimistically titled "Puncher Transcribes Computer Output" and featured a picture of me (unnamed) alongside the equipment racks. By the time the article appeared, Chuck had left for GE Computer Systems in Arizona, and I was all alone with the project.

I soon found that the TCP had some major problems, electronic and mechanical. Ray Bove, our lead mechanical engineer struggled with the mechanism that he had inherited from Bob Sinn, its inventor. I found the electronics to be a patchwork of transistor logic that I was able to muddle through and fix. But, the whole design was based upon three transistor types that had been designed for use in RCA's new portable radios—RF (radio frequency), IF (intermediate frequency), and AF (audio frequency). The mechanical design had the cards travel sideways through the punch mechanism. That meant that up to eighty punches had to be activated at one time, because there were eighty columns in a punched card. To generate the power needed to energize that many punches, the designers had put eighty AF transistors in parallel in a faulty design such that when one failed, it eventually caused the remaining transistors to fail. The failure mode was such that the caps were blown off the transistors and hit the inside of the rack door, sounding like machine-gun fire. As I stood appalled at this disaster, I could hear the maintenance guys talking among themselves and saying that they saw the TCP as a moneymaker for them—lots of overtime to keep it running.

My first order of business was to solve the problem of the parallel transistors. At the time, there were very few transistor types available. I solicited all transistor companies, asking for a power transistor. Many companies were working on such a device, companies now long gone from the industry. Finally I got samples from a company whose name escapes me. I was able to make a design with a single power transistor card that replaced the AF transistor cards. This improvement reduced the space required for circuit cards and reduced demands on the unit's power supply.

However, I wasn't finished with the punch power problem. We'd run a tape and compare the output cards with the input data. Ray and I noticed that holes were missing when it was necessary to punch many columns at once. Naturally, I blamed him and he blamed me as we tried to figure out why this was happening. One day in the lab, while the system was running, I happened to touch the electronic rack and the separate punch unit at the same time. I got a mild electrical shock, meaning that there was a voltage difference between units supposedly at the same potential. I got a big piece of grounding strap and connected it to both units. You could audibly hear the punch tempo change as it started to punch all of the holes. Problem solved. I went back to the prints—engineering drawings—to determine why this problem had occurred. I was amazed to find that while there were eighty wires to the punches, there were only six or eight wires for return of the current from the punches. This was a major flaw in the original design and easily corrected once it was recognized. A ground strap was incorporated into the cabling between the two pieces of equipment.

Once I realized that fundamental problems existed within the design I inherited, I began to question other elements of that design. I had been curious as to why I had a whole rack full of capacitors—some 2.2 farads—to supply current to enable the punches. This was an unbelievable amount of capacitance. Furthermore, the capacitors were in a cabinet, not in the punch unit where the current was needed. I did a few calculations and found that while a lot of current was needed to pull the punch solenoids, much less power was needed to hold the solenoid during the punch cycle. Thus I was able to reduce the capacitance to 0.47 farads, an almost fivefold reduction. Ray and I came up with a small capacitor bank, which fit within the punch unit, where it should have been from the outset. Better yet, I was able to eliminate a full rack of electronic "junk." This made the TCP look far more state-of-the-art, comprising a single small rack and separate punch unit.

While I was busy solving problems as they became identified, TCPs were being built for delivery to our customers with RCA Bizmac systems. I had two engineers working under my direction handling production problems and debugging electronic malfunctions caused by wiring errors or component failures. Oddly, at least to me, one of them was an engineering graduate of the University of Kansas, which I though of as a "cow" college. One day he asked me when we would be wearing our "engineer boots." I was at a loss for a moment and then recalled old time pictures of engineers in high leather boots. I guess everyone at Kansas had them, so he was quite disappointed to learn that we had no use for them back East. A major problem arose with the magnetic drum used to store records read from tape so that the data could be fed row by row to the punches. The drums would come in slightly out of round or the bearings would fail, causing our magnetic pickup heads to strike the drum surface and rendering the tracks unusable. They were made for us by a firm in Connecticut, and Ray and I made several trips up there to try to find ways to improve drum reliability. Over time, we did get the drum problems under control so that they were no longer showstoppers.

Sometime in late 1957, I believe, we shipped the first TCP to OTAC—Ordinance Tank Automotive Command—headquarters in Detroit. Early computer customers ran acceptance tests before they signed off on the individual pieces of equipment that comprised the system. The TCP acceptance test was rigorous. It required that the system work without error twenty-four hours a day for two weeks. I started on weekly trips to Detroit with the many other RCA people involved in getting the Bizmac system up and running. We'd leave Monday morning and, usually, come back Friday afternoon. We always flew Capitol Airlines on the Comet, and the airline prospered so long as RCA had major problems in Detroit. Afterward, Capitol was folded into Eastern Airlines, which somehow became U.S. Airways.

While in Detroit with our maintenance guys looking over my shoulder, I continued to find and fix small problems with the electronics. But the real problems showed up when the endurance tests began. The heart of the mechanical system was a die, which constrained the punch motion to correspond with the card columns. After running for a while, the punches would "seize" in the die, and the entire mechanism failed.

We were allowed to make an on-site repair during the test, and a new die was sent out from Camden and installed by the maintenance guys that we'd been training. Once I'd solved the remaining electronic problems, I hoped to get off the Detroit run. However, for political reasons, I was ordered to go to Detroit so that RCA could say that their top engineers were on-site to deal with any problems. It turned out that the mechanical punch unit kept failing, and the TCP was not accepted by OTAC. Back in Camden, they scrambled to find another solution.

From Detroit, I had to go to New York City to sort out an electrical problem within the TCP at New York Life Insurance Company. It turned out that our power input wiring did not meet the New York electrical code. After on-site consultation with the company's installers, I realized that we had really bungled this aspect of the equipment design. It was a reflection of our total inexperience with such things as electrical codes. I was able to make a few relatively simple changes to fix the problem in various components of the New York Bizmac system and bring back the fixes to go into the equipment being prepared for shipment elsewhere.

Wedding Bells

While I worked on the TCP, Kitty and I continued to date, and I began to talk about getting married. By then, I had met her family and felt comfortable with them. In addition to her parents, Marie Gegan and Joseph Carlin, I learned that Kitty had four sisters—Marietta, Agnes, Pat, and Joanne—and a number of Carlin aunts—May, Etta, Peg, and Agnes—living nearby. It was later reported to me that her aunts had warned Kitty about getting involved with an RCA engineer because, in their experience, they were here today and gone tomorrow. One fall evening, on the way to a play in Philadelphia, I decided to make a fast stop at the Florsheim shoe store near the Wanamaker's flagship emporium on Market Street. I told Kitty to drive the car around the block, and I would probably be finished by the time she got back. It turned out that all the streets ran in the wrong direction, so that her short trip became a much longer trip. I impatiently waited on the corner, wondering what was keeping her. Kitty finally arrived on foot, sans car. She explained what had happened, that the car had died and that the police had pushed it out of traffic for her. She was frazzled, so I said, "Come on, I think you need a drink." After a drink or two, we went on to the theater and arrived as the show was well under way. The only thing we can remember about the show is that it wasn't very good.

Kitty felt we shouldn't rush into marriage because she'd had no time to save any money and owed money to her father for her fur coat. I didn't think her financial position was a serious deterrent to marriage. Being new to the area, I had no idea where to go to buy an engagement ring. At work, I learned that you always went to Jewelers' Row on Sansom Street in Philadelphia for the best deals. In December of 1956, while walking down Sansom Street, I met Kitty's sister Pat, probably the only person I knew in the whole city. That day I bought a platinum ring with a "modern cut" solitaire 1.1-carat diamond. Pat never mentioned that we had met until long after the fact.

Kitty and I announced our engagement that Christmas and went to Midnight Mass at Moorestown's creaky old Catholic church on Main Street. I think we arrived late and had to take seats in the front row. The arrival of the collection basket took me by surprise, so I fumbled about in my wallet and threw a bill into the basket as it went by. I realized that I had contributed far more than I thought proper, so when the basket came back I pulled out my bill and replaced it with a smaller denomination. Kitty was mortified and probably had second or third thoughts about her spouse to be. Anyway, we had a very nice Christmas with both of our families; everyone seemed pleased with the proposed match. We decided to be married the following April; little did I know what I was in for.

Once you agree to be married, the bride and her family go into overdrive, and the groom becomes a fifth wheel in the process. While the women worked on the wedding, I did what any red-blooded American man would do. In March of 1957, I bought a new car, my first. It was a 1957 Chevrolet two-door sedan in a two-tone paint scheme—blue and cream—and it had fins. I bought it from Randolph Chevrolet on Admiral Wilson Boulevard in Camden, a place I passed every day on my way to and from work. The car turned out to be a bit of a lemon, and I did not keep it for very long. I traded it in for a 1958 Peugeot, a beautifully designed car but with an underpowered engine and rotten tires. There will be more about it later.

Our April 27, 1957, wedding day dawned bright and beautiful. The wedding ceremony at Good Council Catholic church in Moorestown, New Jersey, was perfect and drew a good crowd. The reception that followed was held at Kenney's restaurant in downtown Camden. On my side of the wedding party were: best man Tom Gaffney, a high school colleague; brother-in-law Jim Fletcher; Bob Foran, a friend from grade school; and Bob Tomasulo, a Manhattan College colleague. Kitty had her older sister Pat as maid of honor, and as bridesmaids, my sister Patricia, her cousin Peggy Byrne, and her close friend Pat Gross. Kitty's "baby" sister, Joanne, was flower girl. My college classmate and roommate, Jim Maguire, attended the wedding as the guest of my sister Patricia. After the reception, we returned to the Carlin home, put on our traveling clothes, and said farewell as we set out on our honeymoon. When Bob Tomasulo's car wouldn't start, Bob Foran drove Kitty and I to New York City and the Hotel Biltmore for an overnight stay. It had been an ungodly hot day, reaching well up into the ninety-degree range, so the air-conditioned hotel offered blessed relief.

The next day, we took the shuttle bus to Idlewild, now New York International Airport, for our trip to Nassau in the Bahamas. Our plane ride in the state-of-the-art Douglas DC-6 was uneventful. However, upon arrival in Nassau, we were sprayed to kill any unwanted visitors to New Providence Island—our welcome to the Bahamas. We stayed at the beachfront Fort Montagu Beach resort on the plan, which included meals and beach privileges. Unfortunately, the weather was poor, and Kitty had little opportunity to work on her tan. We rented an MG sports car and explored the island, whereupon we met kids attending the Catholic high school. They certainly had a far more rigorous mix of studies and sports than was our experience. We also attended a native Calypso entertainment in the center of town; Calypso music and Harry Belafonte were very much in vogue at the time. For Christmas, Kitty was to buy me the marvelous LP record set whose title was "Belafonte at Carnegie Hall." I was unsuccessful in finding any up-to-date information on the hotel. It opened in 1939 and now seems to have fallen off the Bahamas hotel list.

When it came time to leave Nassau, we decided to take advantage of the low-cost liquor and bought a few bottles to take home. We were stunned to learn that there was a boarding fee for the flight and an overweight fee on our bags stuffed with booze. These unforeseen fees seriously depleted our funds. Upon arrival in New York, we took the shuttle to Manhattan's east-side airline terminal. From there, we had to carry our bags through the subway system to the bus terminal on the west side of Manhattan. We then took the Greyhound bus to Mount Laurel, New Jersey. When we alighted, we had only a dime between us. We used the dime to call Kitty's father to come pick us up. We had not one cent thereafter.

This is a good spot to say a few words about Kitty's father, Joseph Carlin, who was known to his many grandchildren as Pop-Pop. His family had moved from the farm into Moorestown after his father's death in 1906. Joe went through the Moorestown school system and got a job with the Wetherill Paint Company, a Philadelphia firm whose principals were the Matlacks, a Moorestown Quaker family. Joe saved his money and bought a car, which he wrecked about the time he married Marie Gegan, of Philadelphia in 1928. Many years went by before he had the money to buy a replacement. Joe eventually became Wetherill's sales manager and a supplier of paint "samples" for the family. Wetherill's fell on bad times when latex paint came into widespread use. The company is no more, but Kitty and I managed to save one of their wooden store signs.

A lifelong Democrat, Joe dabbled in local politics without much success in rock-solid Republican Moorestown. Joe was very active in the Knights of Columbus (K of C), a Catholic men's fraternity. When I first met him, he was serving as Grand Knight, leader of the Moorestown Council. Later—1958 to 1960—he served as state deputy, the highest KofC post in New Jersey. He got all of his Catholic sons-in-law to become Knights through the fourth degree; however, as an economic measure, I later resigned. Over the years, I got to meet many of Joe's fellow Knights and their ladies, a most congenial group. After retirement from Wetherill's, Joe served as New Jersey's executive secretary until his death. He worked from an office within walking distance of home, and I occasionally stopped in to see him there. In the family setting, Joe was very opinionated and frowned upon those who disagreed with him. But, he was always charming and a wonderful father-in-law. He made excellent drinks too.

Home Sweet Home

Back in New Jersey, we took up residence in my two-bedroom apartment on Coulter Avenue in Collingswood. My roommates had been transferred out of the Camden area. Before leaving, they had offered to paint the living room and hall a sort of butterscotch color while I was away. Kitty and I arrived to find that they had mistakenly used a glossy paint that was hard to live with, but we did. That June, Kitty's sister Pat was married to Richard Selkirk Davis, of Albany, New York. Dick was an RPI (Rensselaer Polytechnic Institute) engineer who had recently returned to RCA, Moorestown, after earning his master's degree at Michigan. Pat's was another wonderful Carlin wedding, and Kitty as maid of honor and in her yellow dress and hat never looked lovelier. Once again, Patsy Gross served as bridesmaid and younger sister Joanne served as flower girl, both were gorgeous. After their honeymoon in Hawaii, Pat and Dick took up residence in a second-floor garden apartment complex in Moorestown. In October of 1958, we had another wedding. My sister Pat and Jim Maguire were married at St. Margaret's in Riverdale. They had their reception at Patricia Murphy's in Yonkers. Since Jim was working at RCA, Harrison, they took an apartment in Bloomfield, New Jersey.

It doesn't seem that we had been in residence in Collingswood for very long when Kitty was laid off from her job at RCA. We had serious layoffs throughout RCA in the years 1957 and 1958. Sadly, many of the class of 1957 engineers taken in at Camden were laid off. One of them was working for me, and I hated to see him go. Kitty quickly found a job at Campbell Soup Company and for more money than she'd been getting at RCA. Kitty had worked at Campbell's before she'd gone to Europe but that was in the plant-based engineering department in downtown Camden. Her new job was not far from RCA and downtown Camden. While working at RCA, Moorestown, Kitty had taken the car, and I rode a bus to work. The bus picked me up a few blocks from home on Haddon Avenue in Collingswood and went down that street all the way to the center of Camden. From that point, I walked several blocks to my office in RCA's Building 10 on Delaware Avenue at the foot of Cooper Street. When Kitty moved to Campbell Soup, we could go together in the car—she could drop me off downtown, or I could drop her off. Since I was out of town quite a bit, Kitty often had the car to herself for days at a time.

With our focus on our new life together, we didn't pay a lot of attention to one of 1957's historic events, the launch of an earth satellite by the Russians. Sputnik's October launch stunned the world and exposed serious shortcomings in our scientific and military planning. Previously, it seemed that nothing could shake our national complacency and confidence in fatherly Ike, President Eisenhower. Ike's presidency was notable for its fiscal conservatism and scaling back of military spending. But Sputnik exposed the flaws in our outdated military planning and blew poor Ike out of the water. Vice President Richard Nixon and senate leader Lyndon Johnson were seen to be forward-looking leaders as they had been pushing for more funding for ICBMs (Inter-Continental Ballistic Missiles) and reconnaissance satellites. Sputnik set the stage for the Kennedy-Johnson slate's presidential election victory over Nixon in 1960.

The Carlin sisters and their husbands and half-dozen maiden/widowed aunts made a wonderful extended family. Kitty's mother, Marie, was absolutely charming and hosted a huge family dinner just about every Sunday of the year. Her father, Joe, presided at the head of the table and tried to keep everyone on board with the Democratic party of Roosevelt, Truman, and Kennedy. The rare defector was soundly admonished to sin no more. Marietta, the eldest of Kitty's sisters, lived in a garden apartment complex nearby while her husband, Dr Ed, was doing something in one of the Camden-area hospitals. They had a young son, Edward, so we usually went to their place for an evening of bridge. Ed decided to set up his obstetrics practice in Pennsylvania, so they moved to a small split-level house in Springfield, Delaware County. We often visited there, particularly on Christmas Eve, before we had children of our own. We also became close to Kitty's sister Agnes and her husband, Jim Fletcher. They had been married in 1951 and had two boys, Mike and Mark. At the time, they were living in a house in Moorestown, which Jim had built pretty much by himself. It was a nice, but small, two-story house on Irving Place. One of my early memories of married life was babysitting their boys one night when Mike fell out of bed and broke his collar bone.

After their marriage, Kitty's sister Pat and her husband Dick lived in a Moorestown garden apartment as he worked at the RCA facility in Moorestown. We often played bridge and went out to dinner with them. They later moved to a house in New Jersey's Levittown—now Willingboro. Kitty and Pat remained close to many of their married school friends and their husbands—Wanda Curry and Art Fitzpatrick, Noel Curry and Herb Wirth, Joan Colsey and Richard Heigh, and Evelyn Clancy and Ken Curtis. With them, we had the basis of a couple's bridge group that played for many years, even after some of the couples moved out of the area. On one memorable occasion, a very pregnant Wanda and Art played cards with Kitty and I at our Collingswood apartment. Wanda said she had to leave and went directly to the hospital and had her baby.

In the early summer of 1957 and again in 1958, we traveled to Cape Cod and stayed in small inns or bed and breakfasts, one year at Dennis and the next at Chatham. On both occasions, the weather was disappointing, so we spent lots of time drinking, dining, and shopping—really just enjoying the ambience. At Woods Hole, we found the Landfall, a wonderful restaurant overlooking the harbor. They served terrific steaks and excellent martinis, and accordingly, moved to the top of our list of favorite places. Alice and Dave Walton, friends of Kitty, lived near

Falmouth; and we visited them for a cookout and martinis. Dave used some charcoal seasoning on his burgers, and they tasted wonderful. Of course, I think food tastes better out of doors and near the ocean. We started using the seasoning and did for many years until it disappeared from store shelves—Kitty thinks it was suspected to be a carcinogen. In June of 1958, we went to the Falmouth Playhouse and saw *He Who Gets Slapped*, with Broadway's Alfred Drake playing "He." It was great to see Drake, star of *Oklahoma!*; *Kiss Me, Kate*; and *Kismet*, but the play was not very interesting. As the Kennedys became prominent and identified with Cape Cod, crowding made the cape a less inviting place to go, so we started going elsewhere.

Meanwhile, we ran into a problem on the home front. Our landlady became unhappy with our alleged heavy usage of her washing machine in the basement. She didn't mind it when the guys lived there as we didn't do as many washes, and she said we really didn't have permission to use the washer anyway. We had kept using it because we'd always used it; it had never been an issue before. That gave us the motivation to hunt for an affordable house. Since we were living in Camden County, our search was focused on what is now Cherry Hill. At that time, that's where all the new construction of large housing communities was happening. Kitty wasn't happy with the notion of living in Camden County. She kept nudging me toward her home county of Burlington, unknown territory for me. We found a new community we liked called Riverton Estates on Highland Avenue in Cinnaminson. We put a deposit on a three-bedroom, two-bath rancher with a one-car garage and no basement. It was to be ready for occupancy in the summer of 1959. The price was $20,500. I'm not sure what my salary was at the time, but it was well under $10,000. But we had saved Kitty's salary and, fortunately, had the funds needed to close on our new home and to furnish it.

Union Activities and Promotion to Management

Once back in Camden from RCA's various job sites, I became the lead systems engineer for the card equipment group, Jim Palmer having moved to another assignment. While I had been working up the engineering ranks, I was asked to run for election to the position of councilman in the engineer's union known as ASPEP (American Society of Professional Engineering Personnel). This union was loosely affiliated with engineers at a number of companies, including Boeing. Effectively, I became a shop steward, handling problems within my group of forty or so engineers, including formal grievances. I had a unique experience one summer, when the engineers went on strike. I had to lead my group in picketing on Camden's Delaware Avenue at the foot of Market Street. This was a hot and busy spot, with heavy truck and train traffic on the avenue. My vice councilman was a former Argentine naval officer named Carlos Monteret de Villars. Carlos was unhappy that RCA put him on the payroll as Carlos de Villars rather than his true name of Carlos Monteret. Carlos was not happy with the idea of losing pay while picketing in the hot sun. He kept asking, "Jeemee, why we do this?"

The strike was soon settled, and I suppose my diligence came to the notice of union bigwigs. I was invited to join the Camden Grievance Committee and did so. From this vantage point, I was able to see the whole nitty-gritty workings of union-management relations. Thereafter, outside of the formal structure, the engineering management came to me many times for advice on how best to handle problems at lower levels. Basically, engineers in management positions had no training or experience with union issues. I tried to serve both sides until, in no time at all as you will see; I was promoted out of the bargaining unit.

By now, Bizmac Engineering had become RCA Electronic Data Processing (EDP) and work had started on a new system to be known as the RCA 501. As we prepared for the next generation of punched card equipment, a graduate student worked all summer on a mathematical study of our punch motion. Before returning to Lowell Institute, now part of Northeastern University, he submitted his final report, which concluded that our punch design could not be made to work. Thus, theory confirmed our field experience, and the mechanical design was abandoned. Over time we had acquired and studied the IBM card punch. It was a simple elegant design, but slower that ours; however, it worked. In consultation with the overall systems software team, I developed the specifications for the new 501 card reader and punch. My punch specifications assumed a mechanical design along the lines of IBM's.

After a change in our group's upper management, it was decided that we would go outside for a new punch mechanical design. Fred Pfleger, my immediate supervisor, and I began to attend meetings with the mechanical engineers at Bridge Tool and Die. One of their claims to fame was that they had on staff the "famous" Dr. Schussler. He was ex-Luftwaffe and allegedly the inventor of the American air-cooled .30-caliber machine gun. He kept several such guns in a locker in his office. Schussler, to my mind, was the prototype for the mad scientists in such future films at *Dr. Strangelove*. He and I could not agree on the basis for a new design. He wanted a big, strong mechanism for high-speed operation. I wanted a small light mechanism that could easily be controlled by system software. Fortunately for me, a promotion opportunity opened in another component, and I left the card group. The group went on to make a punch unit with a Bridge mechanical component; it too failed. Finally, RCA made the wise decision to buy punched card equipment from IBM and incorporate their proven designs into our electronic systems. I had spent four intense and interesting years with the card group, only to see that our efforts had been in vain. RCA just could not build mechanical subsystems, and unfortunately, I was to be involved with them more than once again. Despite differences at work, Kitty and I remained friendly with the Pflegers (Fred and Mary) and often exchanged visits.

Late in 1960, RCA's computer division underwent a major reorganization into four business units: commercial computer systems, government computer systems, data communications, and advanced development. Effective December 1, I was moved to the data communications department and promoted to the position of leader, design, and development engineer. Leader was the first rung of the engineering management ladder. You had a dual responsibility, that of lead engineer and project manager, along with management of your group of engineers and technicians. At age twenty-eight, I was in management, possibly the youngest to reach that position. My group was responsible for design and development of data communications equipment for independent use and for use in conjunction with RCA computer systems. My immediate supervisor was Jim Palmer, and his boss was Dave Nettleton who was engineering manager for the entire department. Within two years, both Jim and Dave would leave RCA to start their own companies, leaving me high and dry. One really good thing about the leader position was that, although you were no longer in the union, you received any benefits the engineers earned through collective bargaining. The principal benefit of interest to me was paid overtime. Another benefit was that you could return to the bargaining unit at the rank you had when you left it. This kind of gave you a safety net should you fail as a leader. After I'd been in management for a while, RCA decided to implement a management training program for all disciplines—engineering, finance, personnel, and so forth. Though a little late for me, it proved to be an excellent program. Naturally, the fine training crew was all laid off on the next business downturn.

My New Assignment

When I took over the data communications group, it was responsible for a piece of equipment than was in production in the factory and for two projects in the design phase. The former was the TAT/MUX (a transatlantic cable multiplexer), and the latter were DaSpan (a paper tape transmitter-receiver) and SCHARQ (single-channel automatic error correcting unit). Our office and laboratory spaces were in a facility RCA leased on U.S. Route 130, in Pennsauken. TAT/MUX was already in operation at several transoceanic telegraph companies. DaSpan was intended to function with a modem and use the national telephone network for data transmission. SCHARQ was intended to do error detection and automatic correction on the long-haul radio telegraphy facilities of NSA (the National Security Agency). It turned out that those who had conceived of these systems had moved on, leaving behind problems for me to solve. Fortunately, I inherited a fine group of young engineers; among them were Carroll Hughes, Jim Fayer, and Ed Marshall (electronics), and Norm Lincoln (mechanical). Oddly enough, Carroll had been a minor league pitcher for the New York Yankees.

TAT/MUX was an interesting subsystem. At the time it was deployed, long-haul data (or record) transmission was via telegraph signaling over radio or transoceanic cable circuits. In this business, Western Union was a key player, but not one of our customers. Radio was used where cable was not available but was subject to the adverse effects of sun spots or flares, so cable was the preferred medium. However, cable circuits were overbooked in the booming postwar years, and new and improved transoceanic cables were a few years off. Many users shared a cable via channels separated by what's known as frequency (analog) subdivision. Each channel used a bandwidth of about 300 hertz and achieved a data transmission rate of 60 wpm (words per minute). It was found that many of the cable users did not need the full 60 wpm capability so that if the channels could be subdivided further, say to 15 wpm, more users could be accommodated. No further analog subdivision was technically feasible, so digital time-division multiplexing was needed to derive the 15 wpm subchannels. Each TAT/MUX cabinet contained the circuitry needed to subdivide four or eight channels, I don't recall which. In the days before microprocessors and integrated circuits, this subdivision task was accomplished by the brute-force use of discrete transistor logic cards stacked in drawers, with lots of front panel lamps so that the user could monitor circuit activity.

TAT/MUX was a marvel of the early transistor age and as I became involved with it, some problems immediately surfaced. I was somewhat surprised, as the equipment was supposedly out of the design phase and into the production phase. The most insistent complaints were coming from the factory. As all young managers were wont to do, I attributed the factory problems to their own incompetence. To keep them off my back, I sent over to the factory one of my most available engineers. I don't recall what he did, but he did buy me some time. Eventually, I reluctantly went down to the Camden plant myself. What I found was appalling. They were having trouble getting any units out the door. First of all, they were running a forty-eight-hour equipment burn-in test with continuous output to a bank of Teletype printers. The test, as it stood, was wearing out the printers; and the factory had been scrounging for printers and printer parts all over the country. I was able to quickly solve that problem by changing the test specification so that the printers were run for just five minutes every hour. Many of the factory-built circuit cards were failing when installed into the equipment racks. This problem would require more extensive investigation, so I had the factory personnel put the bad cards aside for us to look at later. That way, they would be able to ship some units every month and make their production target.

I realized by now that I'd have to address TAT/MUX problems myself, rather than push them off onto my staff. Accordingly, I decided to visit the installation at RCA Globecom in NYC. There I found two concerns. Firstly, the small incandescent lamps on each unit were failing faster than they could replace them, so they could never be sure that a component was working. There were some two to three hundred lamps per unit. When I checked the lamps' specification, I found that they were rated for one thousand hours. That meant on each rack, two or three lamps failed every eight-hour shift. I was able to find a twenty-five-thousand-hour lamp that fit into our lamp socket, and that took care of the problem. More disturbing was the news that one of our units had burst into to flames at Tropical Radio in Panama. There had been no fire in New York, so we assumed that the Panama fire was a freak occurrence. This turned out to be a bad assumption.

In those years, RCA was in a mode of hiring retired admirals for innocuous positions in order to curry favor with the navy. Admiral Waller, reputedly a WWII carrier commander, was given the job of Pennsauken plant manager. I often stopped by to chat with him because he was lonely and had no idea what was going on in the building. When union representatives started harassing him about safety violations, he turned to me for help. Having never dealt with unions, he was very concerned that he would cause a strike and disgrace himself. I explained the union facts of life to him, and we were able to find a way to solve the legitimate safety problems to everyone's satisfaction.

Life in Cinnaminson

As mentioned earlier, we had purchased a ranch house in Cinnaminson with a move-in date of July 1959. Just prior to closing, a strike hit PSE&G (Public Service Electric & Gas). Our builder's crew said no problem; they would rig up a temporary electric service from the neighbor's house. Moving day started hot and grew hotter. I had thoughtfully laid in a supply of beer and stashed it in our new refrigerator. Although we had engaged movers, we expected visitors throughout the day and wanted to have plenty of beer and ice available. About noontime, I went to the refrigerator and found only warm beer and no ice. Furthermore, the refrigerator seemed to be laboring. I began to investigate our electric service and found that the on-site crew had given us a single hot wire connection to the house next door. This meant that the return current was by way of the earth, a non-metallic path. When I finally got around to measuring our voltage at the refrigerator, I found it to be 80 rather than 120. At that low voltage, we were in danger of burning out the motor in the refrigerator, so I had to turn it off. Consequently, we had no cold refreshments during that hot day and more to come.

We had another surprise prior to closing on the house. When we'd signed the sales agreement in the winter, it was understood that we would have a septic system for our waste water. During the intervening months, the township decided that a sanitary sewer would be run down Kathleen Avenue and that we would have to pay an additional $2,000 for it at closing. We were shocked at the time, but it was really a blessing as people were forced to pay $10,000 or more some years later. Our property was nominally one-half acre and had a steep driveway that fortunately lay to the south. Our house was the last to be built for some time, so the next lot held a large mound of soil that had accumulated as a result of grading the completed properties. The mound was to serve as a ready source of top soil until building resumed a few years later. I had no idea about the boundaries of our property, so I just mowed everything from the back of our house to the tree line. It was a lot of work, even with my new rotary power mower. Because our property was so open, we decided to install a post-and-rail fence to create some sense of intimacy. Our ground was hard-packed clay studded with rocks. The fence installers took forever as their power auger bits kept breaking on our rocky soil. Upon moving, Kitty had resigned from the Campbell Soup Company to become the lady of the house, so she was present to monitor the fence project. When the fence was completed, I learned that I'd been mistakenly mowing a full acre. I loved that fence. Kitty quickly adapted to her new surroundings and in no time, started having our groceries delivered from the Beitz market in nearby Palmyra.

Our house had three bedrooms, two baths, galley kitchen, dining room, living room, and study. The last three rooms listed had large windows, and we overlooked the open area to the north and east, toward Highland Avenue and U.S. Route 130. I didn't know much about the town and had really only been there once before buying the house. Kitty attended the RCA Moorestown's 1956 Christmas party at the Riverton Country Club. It was held in the old clubhouse on Park and Thomas Avenues, Riverton. I was to pick her up after the party, and she had given me directions from Collingswood. A fog closed in on the night of the party, and I struggled up Route 130 and after much trial and error, stumbled onto the clubhouse parking lot. Weeks later, in daylight, I could see that Cinnaminson was almost rural, with lots of orchards and farmland.

In 1960, Cinnaminson celebrated its centennial, 1860-1960. Herman Denneler was mayor, and we had police department of six. Cinnaminson in 1960 had a small-town feel with local shopping at small stores in Palmyra and Riverton. But even then, the town had lots of motels, probably going back to a time when U.S. Route 130 was a major gateway to Philadelphia. Local amenities included two movie theaters in Palmyra, one of which was a drive-in with electric in-car heaters. The Palmyra Savings and Loan offered 3.5 percent interest on savings accounts. Fortunately for us, we had Ed McMahon's Esso service station right nearby on Highland Avenue. Ed and his mechanic took care of our cars for years after we moved away. Winters TV on Cinnaminson Avenue kept our television set working and had a tube checker for the do-it-yourselfers. By 1960, Owen Chang, MD, had established his medical practice; and the Hathaway Lodge, a bar and restaurant, was in its thirty-first year. Best of all was ice cream at the nearby Millside Farms in Delran. This dairy occupied both sides of U.S. Route 130 and caused a shutdown of the road when the cows were moved from the pasture to the barn located on the southwest side of the highway. Sadly, the many peach and apple orchards, and farms, would soon give way to houses.

The summer of 1960 found Mother and my sister Maureen off to Ireland again; this time with former neighbor Charlotte O'Brien in tow. Apparently Maureen was invited along to be escort and driver. It was surprising to me that, after their horrible trip home on the SS *United States*, they again took a ship to and from Cobh, Ireland; this time on the Cunard liner *Mauretania*. Mother and Charlotte had a good time together on the ship but went their separate ways ashore, Mother and Maureen by train to County Tyrone and Charlotte to her home county Wicklow. Tommy Donnelly, a family friend, took them to Belfast to rent a car. Thereafter, Maureen had the usual driving problems and misadventures that beset Americans driving on the opposite-handed, narrow, and unmarked Irish roads, with no automatic transmission. For Maureen, at least, the visit was too long, and she was happy to be on her way home to New York.

Cinnaminson in the 1960s seems, in retrospect, to be one long party. Our friends Mickie and Hank Smith hosted many parties at their home on Woodside Lane. There, we did a lot of dancing to the funky music of Trini Lopez. While Hank had been a Democratic committeeman, he was inexplicably an avid member of the far-right John Birch Society, and he had targeted me as a potential convert. Although I never became a Bircher, I did serve as a Confirmation sponsor for his son Hank Jr. That was a fabulous event held at the new Church of the Sacred Heart in Riverton. The bishop allowed the new church only after our cranky old pastor—Fr. Quinn—passed away. Our near neighbors the Grays (Tom and Nancy), the Pollocks (Ron and Phaedra), the Singleys (Ben and Marilyn), and the Kellers (Charlie and Adrienne) were among the hearty partiers. Sadly, alcohol became the downfall of several of my friends and led to their early deaths. Only Mickie, Ron, and Phaedra were still living in the area at last count. Ron Pollock was a chiropractor who served Kitty and I until his retirement.

When we moved into our house, Francis "Luke" Brennan, a farmer, was past mayor and past president of the Cinnaminson Democratic Club. Luke was later elected sheriff of Burlington County and served in that capacity for decades. Nick Batlis succeeded Luke as mayor and club president. Kitty became active in local politics and eventually served as vice president of the Democratic Club. In the 1961 primary election, she was the unopposed choice for the Democratic County Committee. Her election was certified by the county clerk shortly thereafter. With Luke out of the way and Nick often busy elsewhere, Kitty was effectively the local Democratic chair until we left for California in 1962. I continued my affiliation with the Republican Party and, for that reason, was appointed to several commissions—health, shade tree, and port—by the democratic leaders to show bipartisanship. I was a Teddy Roosevelt Republican and hoped for another Teddy to come along and enervate the party. It never happened, so I usually voted with the Democrats. Most Democratic Party business seemed to be conducted in a new gin mill on Route 130, presumably owned by one of the party stalwarts. It seems to me that it was called the Country House and that the building has since been torn down and replaced, most likely, by a bank.

We were all very excited by the 1960 presidential campaign of Senator John F. Kennedy. Even the John Birchers decided to play down their anti-Democrat campaign in case he turned out to be a closet Bircher. A group from Cinnaminson went to the Willingboro Plaza shopping center where he was due to make a campaign appearance. We waited for hours, but no Kennedy, so we left very much annoyed. During the campaign, there was little overt anti-Catholic bashing of Kennedy. However there was much well-founded suspicion that his Pulitzer Prize book *Profiles in Courage* was written by staffers hired by his father, the "evil" Joseph P. Kennedy. I read the book and enjoyed it very much. We were happy to see Kennedy get elected by a very narrow margin over Richard Nixon. As we moved into the early 1960s, we looked forward to many years of peace and pleasure with our man and his family in the White House. Sadly, our hopes were to be dashed in many ways. It was a time when it became very apparent that our vaunted CIA (Central Intelligence Agency) was an expensive paper tiger.

A big local event every year was Riverton's Fourth of July parade and evening fireworks. The fireworks drew a crowd from all over the area, so it was important to get down to the riverbank early and stake a claim to a good spot. The fireworks were fired off from the pier of the Riverton Yacht Club, and the west wind blowing off the Delaware River sent much of the debris back onto the spectators. On a hot July night, you might well come away looking like a chimney sweep. Naturally, we had liquid refreshment before, during, and after the show. One year, we were all invited back to the home of one of Hank's neighbors for a cool dip in their swimming pool. I remember the pool having a kind of oil slick afterward. Partying was not limited to the home front.

In the summer of 1961, Kitty and I decided to motor down to Florida in our Peugeot and rendezvous there with Kitty's sister Marietta, her husband, Ed Sullivan, and their children—Edward, Maureen, Anne, and Joseph. Kitty's younger sister Joanne went along as a babysitter. Along the way, Kitty and I spent a night and a day at Colonial Williamsburg, our first visit to that tourist venue. I had dumped my dog of a 1957 Chevy for a used 1958 Peugeot. The car was a four-door sedan with a four-cylinder engineer and beautiful coachwork. The clever trunk design allowed easy access to the spare tire without having to move anything else. That was a good thing because we had tire problems coming and going—a southbound flat tire at night alongside Virginia's Great Dismal Swamp and a northbound blowout somewhere in North Carolina. I hadn't realized that the French-made tires were pretty well shot when I bought the car. In those days, there were no broad interstate highways, and our narrow car didn't quite fit into the existing grooves in the roadways we had to take. So it was a rough ride, and we were glad to finally arrive at Miami Beach.

It was so hot in Miami that I stayed in the hotel room until it was time to go out to dinner. Kitty, her sister Joanne, and the Sullivans all spent much of the day at the poolside. Consequently, we didn't feel like doing any sightseeing, and we didn't. One evening, we all went to visit Ed's cousins, the Duffys, at their mansion in Coral Gables where we had a very bizarre cookout. Their very well-to-do parents had died and, hardly more than teenagers, they were left on their own in a big old Florida classic home. One evening, we all went to dinner at Pier Sixty-six in Fort Lauderdale. The meal and the ambience were outstanding. Except for the aforementioned blowout, the northbound trip was equally rough but otherwise uneventful. Back to work.

A big local event in 1961 was the opening of Cherry Hill Mall in October. It was the first large indoor shopping center east of the Mississippi River and soon attracted busloads of visitors. The mall is widely held to be the cause of the 1961 renaming of what had been called Delaware Township to its current name of Cherry Hill. The town fathers wanted a unique name and ZIP Code, and the United States Postal Service suggested this name because of the mall's proximity to developer Eugene Mori's old Cherry Hill farm, which sat on a hill across NJ Route 38. However, by 1961, Mori's farm had already given way to the Cherry Hill Inn and the large and very modern RCA Cherry Hill campus. Mom-Mom Carlin went to the mall on opening day and came away with a free potted cherry tree. We planted the cherry tree in our backyard where it became food for the local rabbits. Kitty didn't get to go to the mall with her mother because she was working. Through Kelly Girls, she had taken a short-term assignment at McAndrews & Forbes, an importer and processor of licorice root. It was one of the three Camden firms then listed on the New York Stock Exchange, the others being RCA and Campbell Soup.

During the Wesley Leas years, RCA Bizmac Engineering had wonderful Christmas parties at the grand ballroom of the Cherry Hill Inn. The ballroom was on the inn's lower level and reached via an elegant wide stairway. Kitty and I often went to the inn on our own for fine dining and dancing. In its heyday, the Cherry Hill Inn was one of the top-ranked inns in the country. The hill upon which it sat was the actual Cherry Hill that gave its name to the township. Eugene Mori had owned the cherry orchard and adjacent land that became the inn and the large RCA facility, both built during the early 1950s. The Moris were probably the wealthiest family in South Jersey. Sadly, both facilities are no more.

DaSpan and SCHARQ

Back at RCA, with TAT/MUX problems somewhat under control, I was able to spend time on the other projects I inherited. I worked with Ed Marshall, the lead engineer, on DaSpan to resolve a number of mechanical and packaging problems. DaSpan used the same basic circuit cards and drawers as TAT/MUX but incorporated special electronics for the adjunct equipment, including a paper tape reader/punch. We basically just attached the purchased mechanical device to the top of a short equipment cabinet. We were just building two units on a tight budget, so an elegant design was not in the cards. With the help of mechanical engineer Norm Lincoln, we were able to ship on time and within budget. My team and I received lots of "attaboys" on that job. DaSpan was a very primitive step on the long road to high-speed data transmission. We were forced to connect to the telephone system via clunky Bell's "data subsets" as modems were then called. In 1961, *The RCA Engineer* published my article entitled "The DaSpan Hi-Speed T-R Unit." My picture and biography were better this time. It turned out that our work would never have been recognized if I hadn't written articles for publication, and more were to come.

The SCHARQ project was far more serious as we had booked an order for some thirty or so units. When I inherited the project, we had two prototypes under way for delivery and verification of the design. The customer was NSA, the National Security Agency at Fort Meade, Maryland. I helped lead engineer Carroll Hughes complete the design and get the prototypes ready for delivery. SCHARQ was transistorized terminal equipment that used a redundant five of eight Morse codes to permit forward error detection and correction. When the prototypes were on-site at NSA, we were finally able to see the power of this code. The SCHARQ was connected to a very error-prone incoming radio circuit from Johannesburg, South Africa. With SCHARQ, the garbled, unreadable text transmission became virtually error-free and totally readable. Naturally, actual NSA transmissions would have been encrypted, making the need for error correction even more pressing. So after that successful demonstration, Carroll and I left Fort Meade with a happy customer and with pride in our accomplishment. Then disaster struck.

As I lay in bed one Saturday morning, I got a shocking phone call from Chuck Breder, one of our administrators at RCA. A fire had broken out at Fort Meade in the locked room containing our SCHARQ terminals. NSA runs a highly secure facility with marine guards who had broken into the room to put out the fire. I couldn't imagine what had gone wrong, but I remembered the reports of fires in the TAT/MUX. Both equipments used similar technology and therefore, must have a common flammable element but what? One component that was unique and common to the equipment was a "potted" resistor-capacitor module, which was used to create "flip-flops," transistor storage elements. These R-C networks were laid on a ceramic substrate and encapsulated in plastic, then filled with a "potting" material to create a solid capsule, which could be soldered to a printed circuit board. Something made me suspect that the module was the "firebug." When I put a match to one of the modules in the lab, it burst into a hot flame, much like a blowtorch. It was easy to see that it could set fire to its neighbors and burn up anything flammable in an equipment cabinet. I had found the culprit. I quickly checked our RCA specification for the module and found no requirement for a nonflammable potting compound. Our vendor quickly made up some modules with a nonflammable compound. I tried to ignite them in the lab, and they wouldn't burn; problem solved—or not?

The government called for a formal inquiry into the fire at Fort Meade, and I was to be on the "hot seat," pun intended. In thinking about the problem, I convinced myself that the source of ignition was a fine crack in the substrate within the module. Even at the low voltages we used, a hairline crack could cause an electric arc with sufficient heat to ignite the highly flammable potting compound. Unknown to me at the time, the board of inquiry was chaired by NSA director Admiral Frost himself. Had I known, I would probably have wanted an attorney or a rabbi with me. I explained to the NSA board how the fire occurred and what we had done to fix the problem in the SCHARQs to be delivered under the production contract. The board seemed satisfied and proceeded to sweep the whole thing under the rug. It seemed that the sole-source procurement, itself, might not have been by the book.

Back at RCA after slaying the NSA dragon so to speak, I was confronted with a major management concern. It seemed that some of our TAT/MUX equipment was on a ship bound for Taiwan. Management—particularly Wes Leas, our department head—was concerned that if the equipment were installed in Taipei and burst into flames, it could cause a serious fire in a city of ramshackle wooden buildings. I had to decide if the equipment should go to Taiwan, or be put off the ship when it stopped at Honolulu. I had concluded that the substrate fractures were an early life problem and that the TAT/MUX equipment had sufficient factory burn-in testing to weed out any defective modules. Thus, I recommended that the equipment proceed to Taiwan, and management was happy to go along. Fortunately for all of us, no fires were reported to us from Taiwan.

Unknown to me, my time with the exciting SCHARQ and TAT/MUX projects was coming to a close. As a final tribute to SCHARQ, Carroll Hughes and I prepared a technical paper for the AIEE (American Institute of Electrical Engineers), and it was published as a conference paper of the January 1962 Winter General Meeting in New York, with the title "Design of a Single Channel ARQ System." Actually, I presented the paper in person at an AIEE sectional meeting in Utica, New York, the previous fall. RCA had staffers to help authors prepare for such formal presentations, and they were a godsend to me, a total beginner. Kitty joined me for the several days of the Utica conference. While I hung out with the techies, she enjoyed the amenities of downtown Utica. Our hotel hosted a morning radio talk show, and Kitty was interviewed on the air. She was asked the usual questions: what brought her to Utica, where she was from, how long she'd been married, and how she was enjoying her visit to Utica—queen city of the Mohawk River. I don't remember much about our visit, but Kitty remembers staying in an old hotel whose ventilating transom was painted shut, never to be opened again.

Off to Europe and Ireland

By the booming 1960s, worldwide record traffic—Telex and telegraph—had grown immensely. As was the case for the telephone, central offices arose to allow record traffic to be routed from its origin in one country to its destination in another. Consequently, huge telegraph switching centers arose at major world gateways such as London. These centers employed electromechanical equipment to receive, switch, and retransmit record traffic. The growing airline reservations systems also used telegraphy for reservations. These applications desperately needed upgrades to computer-driven electronic switching. Our department of the computer division had built such a system—AUTODIN—for the defense department's logistics service. We hoped to exploit our solid position in computers—BIZMAC—and record transmission—TAT/MUX and SCHARQ—by tackling the switching systems application. With marketing's Tom McKee, I developed a proposal for such a system for United Airlines. Tom and I went to Denver to present the RCA approach to the airline's technical staff. Actually, I presented my proposal while Tom sat and watched. The customer seemed wary, but Tom and marketing loved my pitch and asked for me to go to London to study the problem firsthand.

I quickly got a passport and set off in March of 1962, with marketing's Don Cianto, on my first trip overseas—as an adult. After an overnight flight to frigid London, we checked into the old Park Lane Hotel on St. James' Park. I noticed going into the hotel that all the plumbing seemed to be on the outside of the building. We were assigned to rooms on one of the highest floors. As we settled in and tried to wash up, it seemed that we had no hot water. It was immediately apparent to me that the water wasn't going to stay hot while climbing up the outside of the hotel in frigid weather. We asked for rooms on the lowest possible floor and were moved down to the third floor. When I got into a bath filled with slightly warm water, I was beset by cold winds howling through the large bathroom, probably a converted bedroom. Bath towels had been hung on what appeared to be a steam rack to keep them warm and promote drying. Here, again, appearances deceived; there was little warmth from the rack, and the towels were never to dry. Welcome to England.

When we finally ventured out, we were steered to a restaurant that was fairly good and then to a nightclub with an excellent floor show. To ward off the chill as we taxied around London in drafty and unheated cabs, we acquired a bottle of Scotch whiskey, which Don carried in his attaché case.

Liberal doses of Scotch kept us alive and fairly mobile. Next day, we went to meet with the technical staff of the UK GPO (General Post Office) to see the world's largest record switching center. It occupied a vast room and though it was semiautomated, there were lots of people there to keep it all going. I was particularly amused by the hookup to the Vatican. All traffic to and from the Vatican used old-style Morse telegrapher's keys such as you'd see in American Western films. The operators even wore green eyeshades. While the Brits were polite, I didn't get the feeling that they were ready to go for RCA's electronic solution.

We quickly wearied of frigid London and of marginal food, and we set off for Rome. As soon as we got on our Alitalia flight, I knew our fortunes had turned for the better. The food and wine on board were great. Our hotel, the Albergo Ambasciatori on the Via Veneto, had nice warm bathrooms with hooded towels the size of togas. That hotel today is now the five-star Hotel Ambasciatori Palace and has double-occupancy rooms for €350 per night. On our arrival, Rome was cold and windy but sunny, so all would be well.

Our meeting with the technical staff of ItalCable was friendly, but they didn't invite us to see their shop. They did take us to a splendid and very popular lunchtime place called GiggiFazzi's. It was a sort of indoor-outdoor restaurant so that while eating, you could watch all of Rome pass by. It was wonderful. Our Italian host and his wife took Don and I out to dinner at a popular restaurant-nightclub called Da Meo Patacca. The place is a going concern, even now, as I write. It was housed in a converted stable and was quite an enjoyable experience, except for the language problem—our host's wife spoke no English and his was rather limited. Next day, ItalCable sent us a car with an English-speaking guide, and we did an incredible one-day tour of the city. I don't remember much about what we saw except for the catacombs and St. Peter's Basilica. We asked our guide, a university student, to join us afterward for a drink and he did. That night as I lay in bed, I was awakened about 1:00 a.m. by a phone call. It was my boss Jim Palmer advising me that I was to proceed immediately to Los Angeles to take over the engineering operation there. I told Jim that I would go, but that I had some stops to make, one of which was my first visit to Ireland in almost thirty years.

My flight from Rome to London was absolutely thrilling. It was a beautiful sunny day as we flew out over the Tyrrhenian Sea and the island of Corsica, then north right over Genoa. It was such a clear day, and we were flying low enough that I could plainly see the famous pillar at the

entrance to the Port of Genoa. We then gained altitude as we flew over
Italy and the Alps into France, with a view of Switzerland where the
Rhone River flows from Lake Geneva. As we crossed the French coast, we
got a fleeting view of Mont Saint-Michel, the medieval church and town
perched on a sometime island in the English Channel. As England came
into view, we saw the famous white cliffs of Dover. What a wonderful
overview of Western Europe, an airplane ride of a lifetime. In London,
Don Cianto and I parted. He planned to stay in London for a few days
but came down with the flu and was bedridden for most of the week. He
unwisely chose to stay in the frigid Park Lane Hotel and suffered the cold
until they finally came up with an electric heater for alongside his bed.

Now on my own, I flew from London to Dublin where I rented a
car from Hertz. I had a bit of trouble with the car rental people at the
airport, but pulled rank on them—RCA owned Hertz. I stayed that
evening in Dublin and had a fine dinner at one of the big hotels. Next
day, I set out for County Tyrone and got caught in Dublin's morning rush
hour. Dublin's traffic and hills worsened my struggle to shift gears on my
right-handed Morris Minor automobile. After crossing the border into
NIR, I passed a farmer who waved as I flew by. To me, the wave seemed
to be a welcome home, and I immediately felt a surge of affection for the
ancestral homeland. Mother had told me to go to the town of Pomeroy
and ask for directions to the home of "Yankee Pat" as my grandfather
had been called. I found my way to town, but there were very few people
about. I finally came upon an elderly man, and after some difficulty, Pat
had been dead for some years, he sent me off in the right direction. I
arrived at the "new" Quinn homestead in Cornamaddy, now the home
of Uncle Joe and Aunt Minnie, to a warm welcome.

Ireland in 1962 seemed to be about seventy-five years behind the
United States. Uncle Joe's new house did have running water and an
indoor toilet, but no electricity. It was very cold and it was St. Patrick's
Day—the only heat in the house came from the kitchen stove, which
burned turf (peat). Joe took me into his parlor where he had stashed a
bottle of whiskey, and we had a belt to ward off the chill and honor the
day. It was probably my first taste of Irish whiskey, and it seemed to be
very rough. Thereafter, I stayed with Scotch as Joe took me around to
all the pubs and introduced me to his friends and relatives. One of the
pubs was Boyle's in Cappagh village. Legend has it that Mother and
Alfie Boyle had been an "item." When I asked to use the men's room in
Boyle's, I was sent out the back door and into the fields behind. Boyle's

and all of the other pubs were cold and damp, with turf fires the only source of heat. Turf smoke wasn't unpleasant, but seemed to stick to your lungs and your clothing. The exhausting day ended with a very late visit to the home of Aunt Celie Quinn and her husband, Johnny Donnelly. They were sleeping when Uncle Joe pounded on their door. Roused from sleep, they quickly set about making a fire and putting on the kettle for tea. The turf must have been damp because the smoke gave me a severe coughing fit. I toughed it out and had a pleasant visit. I wouldn't see Aunt Celie again for some thirty-seven years. Back at Joe Quinn's, I had the unpleasant prospect of sleeping in a frigid bed, which they tried to warm with bricks heated on the kitchen stove. Unfortunately, the warmth soon dissipated, and I shivered the night away.

Next day, I went to mass with Joe and Minnie at Altmore Chapel, my mother's church, located not far from her Altmore School. The churchgoing custom was very strange to me. The women wore black shawls and went directly into church and sat on the left side, while the men lingered outside talking and smoking. When the mass started, the men rushed in filling pews from the back toward the front on the right side of the center aisle. Even when it appeared to me that a back pew was filled to capacity, a latecomer would push in rather than go forward a few rows. I don't remember much else about the mass other than the rich odor of cow manure tracked in on muddy boots.

Later that day, I had Uncle Joe lead me over to the O'Donnell homestead at Gort in the Brantry District of southeast Tyrone. As instructed by Mother, I left Joe in the car and went up and pounded on the door. Dressed as I was in a long coat with a fedora on my head, I looked and acted like a British copper. After a long while, a woman came to the door, and I said I had come about the "poitin," an illegal home brew. She looked me over and said something like "You're an O'Donnell, aren't you?" I asked how she knew, and she said it was the eyes, apparently I had O'Donnell eyes. I don't recall which of my two resident aunts came to the door, but I was swept in on a tide of amazement at my unheralded arrival. I called to Uncle Joe to join us, and they laid a "tea" for us. We had a pleasant visit with Aunts Maggie and Cassie and Uncle John. Uncles Eddie and Pat were at work in the fields, so I didn't see them on that visit. Next day, when I turned my Hertz rental car in at Dublin, the attendant was appalled at the dented fenders and broken taillight incurred in the dark on narrow roads. But he didn't dare to charge an RCA "executive." I don't remember anything about my return flight from Dublin to London and on the USA.

On to California

RCA's computer division had set up a West Coast engineering operation in an available office and a laboratory space in RCA's microwave engineering department. The building was just off Pico Boulevard on the border of Santa Monica, a very good location. The original idea was to use West Coast talent in support of our big ComLogNet logistics system for the U.S. Air Force's Sacramento depot. The West Coast team had designed and built some of the components of the system. As work on ComLogNet wound down, many of the engineers were expected to transfer to the Camden area so as to stay with their products, while others were to be laid off. The remaining staff was working in support of EDGE—Electronic Data Gathering Equipment—which had been installed as part of RCA's computer systems at two California Lockheed plants—Burbank and Sunnyvale. Ralph Montijo, a very talented and persuasive person, managed the operation.

During late 1961 and early in 1962, I had made a number of trips to the operation in West Los Angeles. It was quite a thrill in those days to go by plane to California as the Boeing 707 jetliner had just come into service. On clear days, we would drop down to get wonderful close-ups of the Grand Canyon, something no longer permitted. In each case, I was accompanied by an administrative troubleshooter, and we became a team. Our task was to investigate what was being done technically and financially in that remote operation, which had little executive oversight. I was also asked to evaluate the work of the individual engineers and to induce the good ones to transfer to ugly old Camden.

One of the senior engineers that I had been told to recruit asked my colleague and I to dinner at his home. He and his wife—I'll call them Mike and Madge—were very much into martinis and by the bucketful. We guests sat around his living room and did our dead-level best to keep up with our hosts. Madge finally decided to cook dinner on a hibachi pot that she set in the middle of the coffee table. She was able to assemble dinner without straying far from the martini bucket. Mike and Madge did accept the transfer to Camden, but it did not work out for Mike. He was overcome by his alcohol addiction and left RCA to accept work wherever he could find it, usually in hardware stores. Madge became an editor of the *Burlington County Times* and seemed to be doing well. Eventually, I lost touch with them.

Management back in Camden was under pressure to shut down the West Coast engineering operation as its spending rate was way over budget. Operations—engineering and manufacturing—needed engineering support to get the apparently marginal EDGE product out the door from the RCA plant in Van Nuys. The financial types felt that Ralph Montijo had a vested personal interest in keeping the operation going—his family was from Mexico. The compromise was that Montijo would be transferred back to Camden, and I would replace him in Los Angeles. I was the logical choice as I had been to the plant on many occasions and knew many of the staff. My assignment was open-ended and dual edged; I was to close down the engineering operation while keeping good product flowing from the factory. Wow!

Accordingly, in late March of 1962, Kitty and I closed up our house and flew out to LAX (Los Angeles International Airport). It was her first jet flight. We picked up our rental car and drove up to Wilshire Boulevard and then west to where it ends at Ocean Avenue and the white statue of Saint Monica. Ocean runs along a high bluff, which overlooks the ocean and the Pacific Coast Highway. We were heading to the Miramar Hotel a few blocks north on Ocean. It would be our home for several weeks. I had stayed at the Miramar on all of my prior trips and found it very congenial. It offered bungalows as well as rooms and suites. While I went to work, Kitty hung out at the pool and played bridge with the airline pilots and crew who stayed at the hotel. We made a few trips down onto the beach where we read the famous Santa Monica signs that outlawed just about everything you'd want to do on the beach. After a few weeks, we got tired of hotel living and secured an apartment a few blocks north on San Vicente Boulevard.

Our second-floor unit in the Coral Gables apartments was furnished, but had no bedding, towels, and the like. These we picked up at the local Sears, and we essentially camped out until we flew back and had our stuff shipped out, along with our cars. The apartment had three bedrooms and two baths. The view from the front windows was of Pacific Palisades and the Santa Monica Mountains. The view from the rear windows was of the pool and its surrounding lush shrubbery. Quite a few airline hostesses lived in the adjacent apartments and could often be found at the pool, but unfortunately during the day while I was at work. Our fellow residents were a hearty-party group. It wasn't unusual for me to receive a call when at work, telling me that the party had started, chilled martinis were at the pool, and I was missed. Fellow dwellers included a surgical chief at Santa Monica Hospital and the superintendent of schools. One pool party was raided by the police, and those two escaped over the back fence. However,

I was caught red-handed, playing the bongo drums outdoors, a no-no within the city limits.

It was a pleasant ride of about fifteen minutes to work, most of it along San Vicente Boulevard, through Brentwood, and into West Los Angeles. There was a wonderful liquor store called Pearson's that I'd pass every day and where I'd sometimes stop for supplies. One day as I was pulling in, I saw Doris Day getting into her convertible to leave. She was the biggest star of the day, and I'd just missed meeting her. As I got immersed in my new job, I often found it necessary to go to the factory in Van Nuys. This required a perilous trip over the Santa Monica Mountains and into the San Fernando Valley on twisting Sepulveda Boulevard. This was only a two-lane road with no guard rails. Because of the fire hazard, no smoking was permitted while in the mountains. One of our neighbors had gone off the road and had been severely injured. She had quite a few pins in various joints. I didn't enjoy going to the valley as it was always much hotter than the Los Angeles Basin, ringed as it was by sheltering mountains. Returning to West Los Angeles, as you approached the mountains, you could see smog seeping over the hills from the basin. It was always good to get back to cool and foggy Santa Monica. Many years ago, the tricky Sepulveda crossing was replaced by I-405, the eight-lane San Diego freeway.

As we explored our surroundings, we finally made our way to Sunset Boulevard and the strip through Hollywood. It was quite exciting to take the famous road to the coast and actually see the sunset over the Pacific. In the early 1960s, there was a popular television show called *77 Sunset Strip*. It had a great cast and featured Edd Byrnes as "Kookie," a car hop—parking attendant. Next to number 77 was a nightclub called Dino's, a hangout for the leading Hollywood gossip columnists and their prey. Kitty and I often went there because it had a beautiful nighttime view of the entire Los Angeles basin. I don't recall seeing any stars there, but I might not have recognized them off the screen. Another favorite place to go was the movie theater in Westwood Village, near the UCLA (University of California-Los Angeles) campus. The owner-manager had a heavy German accent and came on stage to introduce the feature film. His parting words were that we were to do two things during the film—sit down and be quiet—or else. He would always draw a cheer, and then we would settle down to enjoy the film. We saw some sexy, off-beat films at the Westwood.

There was no shortage of great restaurants in the Los Angeles area. An early discovery was the famous "Restaurant Row," actually La Cienega

Boulevard in Beverly Hills. We made our first trip there to celebrate our fifth wedding anniversary. With so many restaurants in a row, we decided to start at one end and work our way down the strip, sampling each restaurant on successive visits. Consequently, on that first night we chose Lawry's "The Prime Rib" since it was at the near end. While waiting for a table, we sampled a few "Black Russians," a newly popular and deliciously potent vodka-Kahlua concoction. Once seated, we were shown actual samples of the various beef cuts available on the menu. We were to find this to be pretty much standard in the California restaurants and a great idea. I suspect that I had a steak and that Kitty had prime rib. We loved Lawry's and thought we'd be back many times. However, there were so many others to sample that we never made it back to Lawry's. One of the restaurants on La Cienega featured Polynesian food and drink. There we had our first mai-tais, rum-based drinks that came complete with fruit and little umbrellas. We also sampled some of the restaurants such as Chasen's and the Brown Derby that were frequented by Hollywood stars. Actor Ronald Reagan allegedly proposed to Nancy Davis in a booth at Chasen's. We went there with our starstruck neighbors Jan and Jim Cairns, so we hoped to see some stars, but alas, we didn't. I doubt that we could afford to go to some of those restaurants today; being on expenses was just great. The original Chasen's closed in 1995, and the original Brown Derby now decorates a strip mall. We recently saw an interesting 1997 documentary film, available on DVD, entitled *Off the Menu: The Last Days of Chasen's*.

Work and Play

Shortly after I arrived in LA, Ralph Montijo, the departing engineering manager, threw a farewell party for the remaining staff. In typical free-spending Montijo style, he had food flown in from Mexico City. It was a great party, and thereafter, Ralph did what he could to bring me up to speed. When he and his lovely wife left for Camden, I was on my own. I quickly found myself beset by many technical problems with the data collection units that we were trying to ship to waiting customers. Meantime, we were getting complaints from the existing Lockheed installations. Although primary engineering responsibility was in Camden, I was called upon to do whatever could be done to placate Lockheed and advance the cause. After intensively studying the mechanical design of the units and visiting the Lockheed plants, it became clear to me that we had a marginal design—one that would be hard to get through factory testing, and one that would not hold up in the field.

My staff of thirty-plus was equally divided between technical-engineers and technicians—and administrative personnel. The former were lead by Bob Bricker and the latter by Wayne Beeman. They and their teams were aware that I would be laying people off and consequently, were quite concerned about their jobs. I encouraged everyone to take time off as needed to find new jobs. Dress was very casual in LA, but I adhered to my East Coast wardrobe because it was basically all I had. I looked overdressed, and they, to me, looked underdressed. I had a wonderful secretary, Marie May, who stayed with me to the end. She did a pencil portrait of me that I still have. Every few weeks, I'd have a layoff and take the entire team out to lunch and drinks on me. We always went to the Blue Fox, one of LA's best, where I had a credit card that sent all the charges directly back to Camden. On one occasion, I got mad and refused lunch to an engineer on loan from Camden because he was not dressed properly. I didn't much like him anyway, and he soon left for home.

Financially, I was very well-off in LA. In addition to the Blue Fox charge card, RCA paid my rent, my phone, and shipment of my stuff from New Jersey. Furthermore, I had executive signing authority along with my buddy Tom McKee who was manager of the LA sales office. He and I could sign each others expense reports and that would be sufficient. Since we were not sure how long we'd be away, we kept our house and shipped only absolutely necessary items. Our cars arrived in good order—my 1958 blue Peugeot and Kitty's white 1960 Plymouth Valiant. The Peugeot I bought used in Camden when I dumped my dog of a 1957 Chevy. It was a great car in many ways, but its four-cylinder engine was not equal to the heat and mountains of California; and it was not air-conditioned. The aluminum cylinder head warped and had to be reground several times while I owned it. The Valiant was bought used from a dealer in Collingswood. As we were getting ready to leave California, I sold the Peugeot and had the Valiant shipped east. We kept the six-cylinder Valiant until we traded it for a station wagon in 1965.

While reducing staff, I did what I could, personally, to improve the product and its production methods. Because the product was in trouble, I had frequent visitors from back east, including RCA vice presidents. I took good care of these visiting "firemen," including lunches, dinners, and shows; all of which I charged off to expenses. One of my and their favorite places was the Horn, a Santa Monica night club that featured song and comedy by rising stars. The owner was from New Jersey and had a powerful singing voice. Late each night, he sang "Granada," and it was a socko performance, well worth staying for. Another favorite nightspot was the Pink Pussycat Lounge, a high-class strip joint. I drew the line at taking my visitors to the newly opened Disneyland or to Tijuana, Mexico. They were on their own for those places. Unfortunately, Kitty got stuck taking family visitors to Disneyland, while I pleaded work. I did get there once, however.

Jim Palmer, who was my boss at the time, came out to LA to visit the Lockheed Burbank plant with me. He then asked me to accompany him to Sacramento for a meeting with the air force at their McClellan logistics base. We flew from LAX to SFO (San Francisco), and there we were to catch a plane to Sacramento. At SFO, we learned that we would have a several hour layover and decided to go into the city. We were both excited to visit the storied city by the bay for the first time. I suggested that we go to the Top of the Mark, the famous bar-restaurant on top of the Mark Hopkins hotel. The Mark figured in so many WWII movie departures and, after Fisherman's Wharf, was probably the best-known spot in the city. Jim and I had several libations while enjoying the spectacular view. Since 1962, many taller hotels and office buildings have popped up in San Francisco, but the Mark is still my favorite. Jim and I went on to McClellan and back to LA, but I don't recall that we accomplished much of anything.

Not long after that trip, Jim Palmer left RCA, and Jim Mulligan became my boss. This was quite a shock to me, as I felt that I had the superior claim to replace Palmer as data communications manager. However, I was on the West Coast, and Mulligan was back east under the nose of the higher-ups. I, in effect, had traded my position in the hierarchy for the transitory joys of life in California. While Mulligan and I had been drinking buddies and partiers, along with other managers—Reggie Alexander and Enzo Chierici—back at Pennsauken, we were not really very close. Actually, no one was really close to him. I would miss Jim Palmer as he and his wife, Peg, had always sat with Kitty and I at the Christmas parties. She was a real live wire. Jim Palmer went on to found a very successful company, which he sold out and later became mayor of Moorestown. I last saw him at the Bizmac Engineering reunion held in the early 1980s. On that occasion, we were sitting with the Spielbergs—Arnie and his wife, parents of movie bigwig Stephen Spielberg. Arnie had left RCA about the time I started so I had never worked with him, but knew the name. The Spielbergs had lived in Cherry Hill before moving to California. I believe Jim Palmer now lives at the Jersey shore all year round. After leaving RCA, Jim Mulligan went to Harris Corporation in Melbourne, Florida, and then to the West Coast. I heard an unflattering story about his short tenure at Harris when I made some visits to the company in the early 1980s.

Another visitor from back east was corporate troubleshooter George Kiessling, a Manhattan College engineering graduate who had come to RCA some years prior to my arrival. I had known George as a founding member, with me, of the Delaware Valley Chapter of the Manhattan College Alumni Society. George was very amusing and a welcome guest at my home and office. On one of his visits to LA, he had come out early so that he could make a side trip to Tijuana, Mexico. Instead of leaving his car in the United States as we did, he drove right into Tijuana. As he parked, some kids offered to watch his car for a dollar. He declined their offer, went to a bar and had a few drinks, and then could not find his car. The police told him that his car was well on its way south for resale. He made his way back to Hertz in San Diego and confessed the error of his ways. They just added his missing car to a long list of those that had disappeared in Mexico. We were quite fond of George and his wife, Rosemary, and saw them socially back east.

Kitty and I made several trips back east. We always took the "red-eye" flight, which left LA at 10:00 p.m., primarily because we could fly first class on an overnight flight. As we settled into our seats, we were offered nightcaps of our choosing along with a bottle of wine to hold us until breakfast. As the sun rose in the east, we'd have a champagne breakfast with eggs Benedict. We were usually more than a big groggy getting off the plane in Philadelphia. At one point, the RCA plant in Van Nuys formally shut down EDGE production by throwing the traditional canvas covers over the work benches. This precipitated a panic back east; and I was summoned to attend a conference with Wes Leas, who was now general manager of our business unit, and his staff. I wasn't very helpful to Wes on that occasion, something that I regretted; but the truth was that the EDGE product was a loser from start to finish. Wes left RCA shortly thereafter.

While I was back east, our Coral Gables neighbors tossed a party for Kitty in our apartment. When I got back, our downstairs neighbors, a gay couple, complained to me about the noisy party. They claimed they heard me stomping back and forth and that I was very heavy-footed. They were not mollified even after I explained that I hadn't been in town for the party. The truth was that the construction of the apartment block was very poor—the thinnest possible walls and floors were used. Our next door neighbors were airline hostesses who often arrived late at night. If they opened their medicine cabinet, it would wake me up thinking someone was in our bathroom. If Kitty and I were smooching on our sofa, they'd knock on the wall telling us to behave.

We did a lot of weekend traveling around Southern California, particularly to the old Spanish missions. Our favorites were San Juan Capistrano and San Gabriel. Other popular spots were the Los Angeles Arboretum and the Will Rogers ranch in Pacific Palisades. A favorite eating place was Victor Hugo's in Laguna Beach, about fifty miles south of Santa Monica. It sat on a bluff overlooking the Pacific and some seal rocks. Laguna has a magnificent crescent-shaped beach, and the homes on the hills behind twinkled as the sun set. It was a beautiful sight. We looked at houses at a new development, Laguna Nigel, which overlooked Laguna Bay. We could have bought a beautiful new house there for $55,000 or so. Today they resell for well over one million dollars. Another favorite dining spot was a restaurant practically right in the ocean at Malibu Beach. As you walked in, you were greeted by a talky parrot with a very broad vocabulary. You could ask to be seated in one of the secluded booths by the windows. In addition to watching waves crash against the windows, all kinds of hanky-panky was possible.

Early on, we went to Las Vegas for the Easter holidays. At that time, we had a rental V-8 Chevy that just flew through the desert at ninety-plus miles per hour. It was hard to keep it down. In Vegas, we stayed at the Flamingo, even then one of the really old two-story hotels. It seemed sacrilegious to be at the crap tables at the Sands early on Easter morning and with a girl's choir singing in the background. Our Santa Monica neighbors could not understand why we were traveling so much, even when we explained that we might not be out there for very long, and we had to see and do it all.

Kitty Gets a Job, and We Have More Visitors

Kitty decided to find a job through Kelly Girls, and they came up with a beauty. She worked on the personal staff of Dallas's oilman Tex Feldman. Each day, she drove to his home in Bel Air and dropped off her car for his chauffeur to park for her. Her work was mostly making and answering phone calls, and pulling and updating the meticulous personal records he maintained on each of his business associates and friends. Others on his staff included a retired colonel who served as his aide and a personal secretary. With Kitty on board, the secretary went on vacation, leaving Kitty to work with the colonel to set up Tex's new office suite over on Wilshire Boulevard. When the secretary returned, Kitty's assignment soon ended; but they continued to meet for lunch and gossip about Tex and his wide circle of notables.

My parents and Sister Maureen came out for a visit, and we gave them a taste of California. We took them to Tijuana and to Palm Springs where it was 115 degrees in the shade. We all got out of the hot, hot Valiant for a brief walk around town. At that time, there was no tram to the top of Mt. San Jacinto, a peak that rises right up out of downtown Palm Springs, so there was little to do. I think we picked up something to drink and quickly got back into the car and on our way back to LA. On the way, I took them up into the San Gabriel mountain resorts—Arrowhead and Big Bear, nearly scaring them to death on the narrow, cliff-hugging roads. Later, my parents and my sister went by bus to Las Vegas. On the way back, my father's bag was lost with his camera. He thereafter told everyone that he lost his shirt in Vegas. While they were with us in Santa Monica, we got a surprise visit from Cousin Jackie Sander and a friend. He was in the navy and on leave. My father treated us all to dinner at Trader Vic's. I'm not sure how my family felt about our choice of restaurants, as I doubt they had ever experienced the Polynesian cuisine, which was all the rage in LA. Kitty and I took our guests to Jack's On the Beach, Santa Monica's premiere restaurant. From your table, you could actually see the waves crashing up against the windows. Unfortunately my sister became seasick, and we had to place her with her back to those windows.

By the end of the summer, I had pared the staff down to a handful of people and realized it was time to either return to the east or find a permanent job with RCA in the Los Angeles area. We wanted to have children, and children were not permitted in our apartment complex. The Cairns, close friends and neighbors, had been forced to move out when Jan became pregnant. We'd investigated the housing situation in the LA basin and realized we would not likely be able to find something that would suit us. That meant I'd have to commute over the mountains to the north or down the coast to Orange County. Laguna Beach was great but just too far and too expensive. At the same time, we looked into the possibility of adopting a child, and it did not appear promising. With those negatives weighing on us, we decided that we would go back to Cinnaminson when the job was done.

When the Cairns moved away, we not only lost a convenient bridge couple, but I lost a tennis partner in Jim. At that time, Santa Monica was considered to be the tennis capital of the country and public courts were readily available. As a ladies' dress salesman, Jim lugged his samples over a wide area of California, Nevada, and Arizona. It was a ridiculously huge territory, so Jim was often away during the week; therefore, we played Saturdays when Kitty and I were not off sightseeing. Jim and Jan had also tipped Kitty to the best shopping in the area. In those days, the Bullock's stores were the best around Los Angeles for quality and selection. I even liked to occasionally go with Kitty to the Bullock's in Westwood Village. We liked Bullock's so well that we bought stock in the company; of course, it fell on hard times shortly thereafter. The Westwood Village store became a Macy's and is now an Expo Design Center. San Francisco had a comparable chain, I. Magnin that we also liked very much. They also became Macy's as that brand swept the country.

I had also started sailing with a couple of neighbors or coworkers, I don't remember which. We trailered a "Blue Buoy" day sailer to such exotic spots as Balboa Bay and Redondo Beach. It was a very nice boat to learn on as it was rigged for easy handling by a landlubber such as me. Balboa Bay was a very crowded venue on weekends, with a totally bizarre collection of vessels left over from pirate movies and the like. One day, a Roman galley passed us with a full crew manning the oars. Later there was a popular television show called *The Baileys of Balboa*, but by then, we had moved to Redondo Beach where we were often plagued by lack of wind. I got a serious finger infection from handling the sheets (ropes) after an outing at Redondo Beach. It kept me on land for several weeks.

Our California Grand Tour

In September, Patsy Gross came out for a visit and we, mostly Kitty, took her around to some of our favorite places. We had decided to take our Plymouth Valiant for an extended swing up the coast, through the wine country, and into the Sierra Nevada. Jim Mulligan wanted me to check in with him back east every day while I was on the road. In the days before cell phones, I didn't think that was feasible and said so; he wasn't happy. Patsy joined us for the drive up the Pacific Coast Highway with obligatory photo stops in the beautiful city of Santa Barbara and at Morro Rock. The highway then took us through the beautiful Big Sur area and on to San Francisco, with stops in Carmel and Monterey. I had chosen a hotel for us just a block away from one of the cable car lines. When we got there, we found that we had to almost crawl down and up one of the steepest streets in the city to reach the cars. It was not one of my better ideas. As we had been driving north, we had been in and out of a bank of clouds just off the coast. At first, from our hotel window, we could only see the very tip of the Golden Gate Bridge towers. Then the weather broke and unair-conditioned San Francisco became very warm—up in the eighties. After a boat tour around the city and out toward Alcatraz, we decided to get out of town, and Patsy flew back east.

We set out over the Golden Gate Bridge for the Napa Valley. There we visited two of our favorite vintners—Beringer Brothers and Charles Krug. We toured the grounds and sampled their wines. The Charles Krug winery was absolutely beautiful and on the National Register of Historic Places. Sadly, there were no Krug heirs, so son-in-law Robert Mondavi became the proprietor and gave his name to the great old winery. We stayed the night in Marysville and next day drove up through the Sierras to Squaw Valley, Donner Pass, and Donner Lake. Many of the Donner party of migrants had perished in the pass when trapped in the winter of 1846-47. In 1952, over one hundred years later, a passenger train was snowed in there for almost a week before the passengers were rescued. We next took a drive down to Lake Tahoe and then west to California 49, the route of the gold-seeking Forty-Niners. Heading south on 49, with darkness closing in, we searched in vain for a motel as we drove through a succession of abandoned mining towns. Desperate, we finally pulled into a small settlement with a rather nice-looking motel. Now that we were finally settled for the night, our next concern was dinner.

At the motel, we were directed to a fine restaurant up in the Sierra foothills. We were warned that in driving to it in the dark, we might be tempted to give up and assume we were lost. We were told to keep on going, we'd find it, and we did. I believe the restaurant was called the Water Wheel, and it was opposite the abandoned Kennedy gold mine workings, whose huge water wheels were still standing. The restaurant was run by the Meyers family who had moved to Sacramento from Haddonfield, New Jersey, for health reasons. Mr. Meyers had worked for the Campbell Soup Company and had made frequent trips to their canning plant in Sacramento, that's how they came to choose that location. The whole family worked at the restaurant, which was very fine and attracted clientele all the way from Sacramento. The eldest Meyers daughter, Debbie, had been a top competitive swimmer in the Middle

States, so the name was well-known to me. Since Kitty was a Campbell's alumna, it was like a family reunion. Next day, we went back in daylight to take some pictures of the Kennedy works, but, unfortunately, none of the restaurant. In looking at maps today, it would seem that we must have stayed somewhere in the vicinity of Jackson, a city with good road connections to Sacramento and to the Sierra foothills.

Our next stop down Route 49 was Yosemite National Park. We drove into the village and looked at some of the many waterfalls, mostly dried up at that time of year. From Yosemite, we headed south to Fresno and then into Kings Canyon National Park. There we saw and marveled at the magnificent sequoia trees. Leaving the park, we headed back into the Great Central Valley at Visalia, and it was hot, torrid in fact. We sweltered in our unair-conditioned Plymouth, and by Bakersfield, we had taped newspapers to the side windows to keep out the afternoon sun. Relief came when we finally crossed into Los Angeles County at the famous "Grapevine" over Tejon Pass in the Tehachapi Mountains—then U.S. 99 and subsequently upgraded as I-5. That trip concluded our California adventure, and we started packing for our return to Cinnaminson.

Back to Cinnaminson and Children

Early in October, we got back to our house, which had suffered while we had gone. Some of our furniture developed unsightly cracks, probably from the heat as the house had no air-conditioning, and had to be replaced. I had been reading good things about the all-new Plymouth Belvedere Six, and we bought one. It was a really great car. Gradually, we eased back into the Cinnaminson-Riverton social scene. At work, Jim Mulligan assigned me to take responsibility for the "Edge System Evaluation and Reliability Program." With my negative feelings about EDGE, this assignment did not suit me at all. I went to the new chief engineer Art Beard and asked for another assignment. He claimed there was nothing else presently available and that I should stay with EDGE for the time being. This unhappy situation caused me to seriously look for work outside of RCA. On May 22, 1963, I received a job offer from Digitek Corporation, of Fairless Hills, Pennsylvania. The job was that of a chief engineer at a yearly salary of $16,000—about $70,000 in today's dollars. On May 24, I submitted my resignation from RCA to Jim Mulligan.

We had been disappointed with the adoption process in California, so we immediately applied for adoption through Catholic Welfare of the Diocese of Trenton. While our application was proceeding through their sluggish system, Ed Sullivan approached us with the possibility of a private adoption. One of his patient's and her husband didn't feel that they were ready for children and wanted to place their baby for adoption through Ed. We eagerly accepted this wonderful out-of-the-blue chance to adopt. So shortly after the baby was born on April 16, 1963, in Philadelphia, we picked him up and took him home. On Sunday April 28, we had the baby christened at Sacred Heart, Riverton, as Thomas James. It was a beautiful day, and we had a wonderful house party for family and friends. The godparents were Tom Gaffney and Pat Davis. I had three close friends named Thomas, so the name really resonated with me. We loved Tommy dearly and thought he was wonderful even though he proved to be a difficult baby and youngster—colicky as an infant and then uncooperative as he grew older.

Saturday, April 11, 1964, was a special occasion for me. Manhattan College asked me to represent the school at the Eightieth Anniversary Convocation of the Philadelphia College of Textiles and Science. The college sent me my cap, gown, and engineers' stole for the academic procession in which colleges marched in the order of their inception. Manhattan's date of 1853 put me in a tie with Beaver, one behind St. Joseph's, and two ahead of Penn State. I felt honored to be in the procession, as many of those alongside me were college presidents. Senator Pastore, of Rhode Island, gave the principal address, but I was most impressed by the talk given by Textile's president, Bertrand W. Hayward. He ably articulated the school's mission, which was to serve the business community with technically able graduates. Since then, the East Falls School has changed its name to Philadelphia University. I hope it hasn't entirely lost its focus.

The next big family occasion was the marriage, at St. Margaret's Church, of my sister Maureen to John Crumlish, of Brooklyn. In June of 1964, he was twenty-four, and she was twenty-three. They had met while working at Shamrock Casualty Company, an insurance firm owned by the taxi operators themselves. After the wedding trip, they resided in an apartment in Yonkers until they, too, moved to New Jersey. By the time of Maureen's wedding, my sister Pat and Jim had four children—Terry, Maureen, Sheila, and Denise. The following summer would see the arrival of Stephen, the fifth and last of the Maguire children.

Meanwhile, Kitty and I had completed the legal adoption of Tommy and had gone back to Catholic Charities to reinstate our application for adoption. They turned us down, saying they would not put one of their children in a home with a "risky" privately adopted child. With the help of Joe Carlin, Kitty's father and K of C (Knights of Columbus) Grand Knight, strings were pulled, and voila, our application was suddenly reinstated up in Trenton. In March of 1965, we went to Hopewell and took home a beautiful baby boy who had been born on March 10, my mother's birthday. Though he had been christened by the nuns at Our Lady of Light Chapel, we had a naming ceremony at Sacred Heart, Riverton—Patricia and Jim Maguire were godparents. Thankfully, Christopher Joseph turned out to be an easy baby to care for—or were we becoming more expert?

Before we'd gone to California, Kitty and I had taken weeklong vacations at a number of places, including Cape Cod, Stowe (Vermont), Glenburnie Lodge (Lake George, New York), and the aforementioned trip to Florida. While at Glenburnie in 1960, Kitty learned to water ski and became quite proficient. I found that she was quite a good athlete. Much as I tried, I could not get up on the water skis. Now, with two children, we could see that vacations would have to be nearer and shorter. Kitty's mother, Marie, agreed to look after the children while we took some extended weekends to the wonderful Tides Inn resort in Virginia. The Tides Inn was undoubtedly the nicest and classiest place we had ever experienced. It offered golf, tennis, boating, cocktails, wonderful food, and an exciting after-dinner horse racing game. For some reason, they always treated Kitty and I as newlyweds.

By 1964, we had started going to Ocean City in June, before the season, and had some great places right on the beach. That year a family friend, Fr. James Donnelly, was visiting from Ireland. He and Mother drove down to stay with us for a few days. It had been a cool summer, and there had been a widely publicized outbreak of mosquito-born equine encephalitis. Fr. James liked to sit on the porch, which looked out on the ocean and read his "office." One evening while he was on the porch, the wind shifted to the southwest bringing flocks of mosquitoes from the back bay, and the temperature shot up into the nineties. Fr. James called in and asked what kind of insects were after him. When I said they were "Jersey" mosquitoes, he rushed inside and would rarely step outside thereafter. Fr. James was an uncle of Owen Donnelly and Dominic Ward, future immigrants and family friends.

After Chris's arrival in 1965, we took the backseat out of our "new" Plymouth and replaced it with a portable crib for the kids. This arrangement didn't prove satisfactory, so that summer we bought our first station wagon, a blue Ford Fairlane and sold the Belvedere to Uncle Jim Fletcher. Shortly after, Kitty and I took the wagon on a short vacation trip to the Poconos. We noticed that the car's body leaned ominously on sharp curves. When we got back to Cinnaminson, we took the car back to the dealer, Paul Canton in Delran. They found that, on one side, the bar that ties the car's body to the frame was missing. So much for Ford's final inspection—this was a potentially dangerous condition.

In 1965, Mother had a visit from her next to youngest sister Katherine "Kitty" Quinn who was known as Mrs. Davis, after one of her husbands. Kitty had been a nurse or nurse's aide in England and seemed to have homes in both Ireland and England. John Crumlish and I, and perhaps others, took her to the New York World's Fair. On the way to the Triborough Bridge, we took her for a drive down East 138th Street, our former home turf. The street was very dark and gloomy, and Aunt Kitty couldn't believe that we had ever lived there. That October, the County Tyrone Society held a Seventy-fifth Anniversary Banquet at New York's Hotel Commodore. At our family table, number 25, were Mother and Dad, Kitty and I, Pat and Jim Maguire, Rose and Frank McElroy, Aunt Nell O'Donnell, and Aunt Kitty. For some reason, Maureen and John Crumlish did not attend that affair. However, they did make the October 1964 dinner and dance of the Shamrock Casualty Company. It too was held at the Commodore. On that earlier occasion, we had table 12, with essentially the same cast sans Aunt Kitty. Pete Toal, Dad's boss and longtime friend, did not attend the 1964 dinner-dance and was deceased by the time of the 1965 banquet. His daughter Jane and her husband, Roger Daly, attended both and sat with the Terence (Terry) Toal family.

At some point in the mid-1960s, Dad retired. He sold his fleet of five cabs for about $175,000, about $700,000 or more in today's money. He retained his stock in Shamrock Casualty, an insurance company that the taxi owner's created when other insurers dropped taxicab policies because of high losses. Through Shamrock, the cab operating group became self-insured for part of the risk spectrum. Shamrock also offered auto and other types of casualty insurance to the public. One of Dad's drivers had a spectacular accident on the Brooklyn Bridge, which resulted in the driver and several passengers being burned to death. An accumulation of such potential liabilities in access of assets caused the state insurance agency to take over the assets of Winford Holdings, which owned Shamrock. At the time, it appeared that our Winford stock would be worthless. However, after the dust settled some thirty years later, the assets of Winford were returned to the shareholders, and my sisters and I benefited from a payout of about $15,000 to my mother's estate.

A Change of Jobs

Back in May of 1963, RCA management did not want to accept my resignation and asked if I would like to be transferred to another operation within the company. I had thought about moving to one of the defense or government systems engineering sections but had dismissed that notion because I wanted to stay with computers in some way or other. I decided that I'd like to stay at RCA and transfer to the marketing department. At the time, marketing was coming to the fore as a discipline separate from sales. Another benefit was that marketing was in the relatively new Cherry Hill offices recently vacated by the television business unit. I gladly made the move out of Camden and was assigned to promotion of EDGE (Electronic Data Gathering Equipment). This was obviously not my choice based upon my prior experience in LA. In this assignment, I went to the various sales offices around the country and assisted their efforts by making presentations to their customers and helping prepare proposals. The job involved lots of air travel while lugging around a heavy EDGE demonstration unit.

While there were many interesting trips to places I'd never been, this assignment was not to my liking for several reasons. First of all, EDGE was not a winner, so while my efforts were appreciated by the sales branch managers, they were not able to close many deals. Second, my frequent absences put a burden on Kitty, now that she had young children to care for. Third, working with customers and sales personnel meant lots of after-hours drinking and that played into my tendency to drink to excess. I particularly enjoyed after hours in a wonderful bar next to our Atlanta office. On one occasion, Ted Smith, a senior RCA vice president, happened to be in Atlanta and joined us for a drink or two. It turned out that Ted and I were flying back to Philadelphia that evening on the same flight—he had a first-class ticket, and I had a regular ticket. As we drank and chatted, I assumed that he would eventually suggest that I switch to first class, but it never happened. Was I glad when he was put out to pasture a short time later?

I believe that my tenure in this position ended more or less on November 22, 1963. On that Friday, I was giving an EDGE presentation to a large group at Raytheon, which was located in a Boston suburb. While I was on stage, I noticed that people seemed to be drifting out of the auditorium. When I concluded and went backstage, I learned that President Kennedy had been shot while motoring through Dallas. The RCA team was very upset and thought I should have quickly ended the presentation. How was I to know, I kept saying. One of the Boston salesmen, a native, drove me to the airport while trying to control his tears when we heard that the president had died. It was a very somber flight back to Philadelphia. We spent the next several days glued to our television sets as an unbelievable drama, worthy of Shakespeare, unfolded. The day of the funeral dawned bright and chilly. It was quite a sight to see the world's most famous statesmen marching down Pennsylvania Avenue in the funeral procession. I will always remember two tall men in the front row—Eamon de Valera, of Ireland, and Charles de Gaulle, of France. I think that I held baby Tommy and played with him throughout most of the extended weekend of television coverage. We watched Lee Harvey Oswald get shot by Jack Ruby as it happened. It was an incredible event.

John Macri was a marketing manager with whom I'd worked on several proposals. He had been reassigned to the product planning group and was responsible for data communications products. I learned that he had an opening in his group, and he was willing to take me on. I was moved to his group and hadn't been on board with him for very long when he was assigned elsewhere. I then succeeded John as manager of product planning for the data communications components of RCA's new Spectra 70 computer systems. This was a job right up my alley, as it was a blend of engineering and marketing. I inherited a staff of three or four and added several more during my tenure. Furthermore, I had an immediate boss, W. R. "Bill" Lonergan, whom I greatly respected. Many on my staff became close friends—Art Cherry, Bill Marshall, Pete Patterson, and Joe Tordella—as well as drinking and golfing companions. Just about every day that we were in town, Art and I had the two-martini lunch at the nearby Greenbrier Inn on Racetrack Circle (now gone) on NJ Route 70. Terry, the Greek, owned and operated the Greenbrier where he managed to employ an outstanding crew of waitresses. Terry was prominent in Cherry Hill's business and social life and a founder of the Greek Catholic church located near the RCA facility. On the last day of work before the Christmas break, we usually had lunch at the Greenbrier and moved on to the livelier Charles Lounge on NJ Route 73 in Mount Laurel. As I write this, both places have been torn down.

In addition to everyday duties such as meetings with engineering and sales personnel, my new assignment also offered outstanding travel opportunities to actual and prospective business partners, many overseas. My group was also responsible for RCA's participation in the industry's standardization efforts in the burgeoning field of digital data transmission. Since these standards would affect both business equipment manufacturers and transmission utilities, both sides were represented. Usually the resulting standards were published by the interest groups for each BEMA (Business Equipment Manufacturer's Association) and CCITT (International Telephone and Telegraph Consultative Committee). I personally served on the two committees responsible for defining the bit order and format for transmitting the ASCII (American Standard Code for Information Interchange) over various types of data transmission facilities. My participation is noted in the resulting standards as published by BEMA and its successor agencies. Meetings of my standards groups were usually in excellent locations. There was a memorable session at NOAA (National Oceanic and Atmospheric Administration) in Boulder, Colorado. There I had my first up close view of the Rocky Mountains and of NOAA's Cray supercomputer. Both were impressive. One night, we had dinner in a wonderful steakhouse well up in the foothills alongside a babbling brook.

Another standards committee meeting was held in Houston, where we had a raucous wine-laden dinner followed by a baseball game in the sparkling new Astrodome. From our skybox, courtesy of Delta Airlines, we had a wonderful, seemingly up close view of the field. I remember some heroics on that night by Joe Morgan, second baseman of the Cincinnati Reds. Our IBM delegate had so much to drink we had to almost carry him out of the stadium and back to our hotel, which I believe was the Warwick. I came away very impressed with Houston and its Astrodome. Our committee was scheming for a meeting in Hawaii, but we could never get BEMA to approve it.

RCA had a business relationship with Siemens and that meant trips to their facilities near Munich; and they, in turn, visited us. I became friendly with some of the Siemens staff and had them to dinner at our home. Once we took the afternoon off and took our German guests to the nearby Garden State racetrack. They loved it; but we were far apart on our approaches to new joint ventures. My group also had responsibility for a video terminal product that was designed and built in RCA's Van Nuys, California, plant. That necessitated frequent trips to Van Nuys and visits back East for them. I became very fond of their leader, George Turner, and had him to the house for drinks, dinner, and bumper pool. I enjoyed visiting George at his new home in what's now called Simi Valley.

Some business trips were only as far as Washington or New York. To Washington, I'd often take the PRR's (Pennsylvania Railroad) Congressional from Philadelphia's Thirtieth Street Station. I loved the powerful humming sound the Pennsylvania's GG-1 electric locomotives made as they moved into and out of the station. I can still hear them in my imagination. One winter's night, a fine snow had blown into the locomotives air intakes causing a short circuit, which rendered them unable to move. That morning, I waited hours for a train to Washington. When one finally arrived, it was jammed and the snack car was out of everything, including water. We had to make an extended stay in Wilmington to take on water and supplies. I was very late getting to Washington and had no coffee. On another occasion, I was in New York for a technical meeting with Western Union and other carriers. As I was walking down Broad Street to the meeting place, I was waylaid by an RCA lawyer and told not to attend. It seemed to RCA's legal staff that we might be on dangerous antitrust ground had I gone ahead as planned.

By this time, RCA's Computer Systems Division had adopted a business strategy of being a second source to IBM. Our new Spectra 70 products were to match up with IBM's and have a degree of compatibility at the software level. In support of this strategy, a number of IBM managers were offered good packages to come to RCA. This influx had a drastic effect upon the RCA culture as key people were displaced by IBMers. By 1966, personnel shifts began putting a series of new managers between Bill Lonergan and me—at first Brian Pollard, whom I despised, and later Don Stevens, for whom I had little respect. In addition, we acquired a new executive vice president and general manager, Jim Bradburn, who had held a senior manufacturing position at Burroughs Corporation. To many of us, Jim's hiring was a mistake; we felt we needed broad marketing expertise far more than narrow manufacturing experience. Jim would not be with us for very long. In September of 1967, *The Marketing Magazine* published an article entitled "RCA's Profitless Prosperity at Cherry Hill." The gist was that we were spending big but earning little. It was a shot across the bow and a preview of a possibly unhappy ending for all of us.

Patrick Arrives

One of the perks of working for RCA in the 1960s was business trips to Florida. We had opened a new plant in Palm Beach Gardens at the behest of the John MacArthurs, who owned a lot of property there and had financial ties to the company. He founded the John D. and Catherine T. MacArthur Foundation, from which MacArthur Fellowships and numerous other philanthropic enterprises emanate. At the time of his death in 1978, John was one of the wealthiest men in America. On our 1960s trips to the Florida plant, we always stayed at the Holiday Inn, which was right on Riviera Beach, not far from the MacArthur State Park. The inn and its oceanfront restaurant were great, but it's all gone now as Riviera Beach has become a crime-ridden blot on Palm Beach County. After dinner, many of us went to O'Hara's in downtown Palm Beach for an evening of wine and song. To me, this was one of the world's truly great bars, also long gone. About midnight, the house band always performed their spectacular version of "Little Orphan Annie." Here's a sample of the lyric:

> Who's that little chatter box?
> The one with pretty auburn locks?
> Who can it be?
> It's Little Orphan Annie!

After that stirring performance, we'd usually finish our drinks and head back to the hotel.

For our tenth wedding anniversary in April of 1967, I took Kitty to Florida—a combined business and pleasure trip. We stayed at the Holiday Inn and, naturally, went to O'Haras for "Annie." We enjoyed a fine dinner at the Petite Marmite on Worth Avenue in Palm Beach. I had been there with a crew from work; and it was very expensive, but RCA was paying, so who cared? I knew Kitty would love it, so we celebrated our anniversary there. It became one of our very favorite dining places. During that Florida trip, we visited my godmother Aunt Mary and Uncle John Sander at their new home in nearby Lantana. I believe that Cousin Jackie was away in the navy at the time. It was a small house—two bedrooms and one bath—but the house and the grounds were beautifully maintained by Uncle John. Mother and Uncle Mike Quinn often flew down to stay with their sister and to enjoy the nearby Lantana beach. Some thirty years later, as Aunt Mary's caregiver, I had to quickly dispose of the house before it was condemned and torn down. The meager funds we realized went to her nursing home.

Kitty's pregnancy became a totally unexpected, but joyous, result of our Florida trip. The balmy sea air did the trick after ten years of trying to have a baby. Patrick Edward was born on January 29, 1968, almost exactly nine months afterward. Patrick was delivered at Philadelphia's Pennsylvania Hospital with a full head of red hair. The hospital was great, you could order dinner for guests, and so Kitty and I dined together most of the six nights she was there. Mom-Mom Carlin looked after Tom and Chris while Kitty was in the hospital. To help out when Kitty came home, for two weeks we had the services of Mrs. Ward, a former Carlin neighbor. Mrs. Ward saw to it that Kitty wasn't bothered by the kids, the phone, or the neighbors. She was probably overprotective but served Kitty and I very well. Patrick was baptized at Sacred Heart, Riverton, on February 25, 1968. Kitty's sister Agnes and her husband Jim Fletcher were godparents.

Not long after Patrick's arrival, we realized that now that we were five, we'd have to make some changes. In 1965, we'd bought a station wagon for Tommy and Chris. It was a new blue V6 Ford Fairlane but rather cheaply made, even for the times. However, we did get eight years of use from it. Now we were going to need a bigger house for Patrick. We felt we needed an eat-in kitchen, basement, and two-car garage. We started house-hunting during 1968, but choices became very limited when we decided to stay in Cinnaminson or, possibly, Moorestown. That fall, we saw five-year-old Tommy off to kindergarten. We had prepared him for the big day and expected it to go well, but Tommy did not seem to be ready. It took a bit of doing, but we finally got him off on the bus.

In June of 1967, we took a family vacation with Tommy and Chris at Indian Cave Lodge on Lake Sunapee, New Hampshire. My sisters and their husbands, along with some, but not all, of their children made it an O'Donnell family outing. We each had a cabin some distance from the main building where we went for meals and indoor recreation, including supervised play times for the children. There were lots of outdoor activities centered on the lake and around the pool. Our Tommy—age four—and his cousin Kathleen Crumlish—age two—were inseparable. We also took the children into town for window shopping and on a wonderful outing to the summit of Mount Sunapee, reached by a gondola ski lift. Jim Maguire, John Crumlish, and I got in one or more rounds of golf. We had a lot of fun and agreed to do it again. But for many reasons, including new additions to our respective families, we didn't get back to the lodge until June of 1970. On that occasion, it was only the Crumlish family and ours, and we now had two-year-old Patrick with us. We shared a two or four family lodge, instead of separate cabins, and that worked out well. The weather that June was a bit chilly and curtailed some outdoor activities. The 1970 stay was to be our last in New Hampshire. Sadly, the lodge is no more.

A Swing through Europe

In the spring of 1968, I had Joe Tordella of my staff set up an extended survey visit to Europe for us. It started with a trip to a company in Malmo, Sweden. Joe and I stayed in Copenhagen and took the morning ferry over to Malmo. The ferry offered three classes of service. Naturally, we chose first class and enjoyed breakfast with white table cloths while overlooking the bow of the boat. The Swedes were wonderful people, and they spoke much better English than our German or Italian business associates. What little I got to see in Malmo was very impressive. One night, we went to their best restaurant to experience one of Sweden's classic fish dishes. It turned out to be a salmon that had been buried in Arctic ice for a year, dug up, and served raw. The salmon was not to my taste; however, the restaurant was physically one of the most beautiful I've ever been in. At night, we took the faster hydrofoil boat back to our hotel in Copenhagen. The hotel we had chosen was very nice but too close to the city hall, whose bells tolled the hours day and night.

From Copenhagen, we flew to Munich for another "huddle" with the Siemens engineers who had visited us in Cherry Hill. From Munich, we flew to Milan for a meeting with the Italian firm, Olivetti, a successful manufacturer of electromechanical computer-related equipment. Our mission was to determine if we could form a joint venture to marry their line of business machines to our computers. An RCA vice president, N. Richard Miller, joined us in Milan, presumably to add some "weight" to our delegation. The Olivetti plant was located in the factory town of Ivrea, which is in the foothills of the Italian Alps. Our hosts sent a car into Milan to pick us up. The chauffeur drove like a maniac on the autostrada from Milan to Turin and up into Ivrea. The Italians quickly realized that Dick was a total nonentity and privately asked me why he was there. All I could say was that he was a relative of the Sarnoffs who ran RCA—David, founder and chairman, and his son Robert, current president. I don't know if that impressed or further annoyed the Olivetti folks. For lunch, we went further up into the Alps and dined al fresco on a beautiful spring day. Our hosts didn't buy into our plans, and we parted company amicably.

Back in Milan, Dick Miller went on his way, and I thought I'd seen the last of him; but that was not the case. But that's another story. In Milan, I had a secondary mission. I had been asked to visit the main office of one of the country's largest banks where they had an RCA Spectra 70 computer system. It had been installed by one of our European licensees, either Siemens or Bull. In the basement, under wraps, they had one of our RACE (Rapid Access Computer Equipment) units. They wanted to know if we were making any progress in getting it working on their system. This was very embarrassing because I had no answer. RACE was intended to be RCA's solution to the need for a primary large-scale random access storage device on our systems. While RACE was not one of my products, I had been sent to the Van Nuys plant to monitor and assess their progress in getting the beast working. It was a huge electromechanical monster that shot magnetic cards at high speed along a raceway so that the needed card could be selected by a computer. My experience with RCA mechanical devices told me that it would never work, and I so reported. This served to bolster marketing management's resolve that we find an alternative solution, and we did. We bought IBM's equivalent unit and married it to our computers. I couldn't impart any or all of this to the Italians in Milan, so I simply said that their problem was one of our major concerns, and they would be hearing something shortly. Then Joe Tordella and I speedily left for London.

We were very interested in studying the operations of large-branch banking firms overseas. Historically, laws in the United States prevented banks from having branches outside their home states. That restriction was being lifted, and American banks would soon have extended regional operations. American branch banking experience was limited, but many European banks had branches nationwide. We chose to visit the United Kingdom's Barclay's Bank, which had the most branches of any of the large European banks. What actually transpired at Barclay's wasn't very important, but from its headquarters in London, Dublin was only a short flight away.

An Irish Sojourn

Knowing that I would be in London, Kitty and I had decided to take an extended vacation trip to Ireland (IRL). Mom-Mom Carlin agreed to look after Patrick, and the two older boys would go with us. Kitty's passport, issued April 29, 1968, covered her and the two minor children—Tom and Chris—by name. Now they would each need a separate passport. They took an overnight Aer Lingus flight from Philadelphia to Dublin, with a stop in Boston. Kitty's fellow passengers pitched in to keep the kids amused on the long flight. Kitty was pretty frazzled by the time she checked into our Dublin hotel, the Central. On the other hand, I arrived on a very pleasant afternoon flight from London. I met lots of partiers on the plane and then in the hotel hallways as I searched for our room. Everyone was pleased when Kitty and I were finally reunited. But I was so late that she and the boys had gone to bed. I learned that they had dined in the hotel's fine restaurant; the boys had crepes—pinch-hitting for pancakes—and Kitty had duck. Next day, we visited some of the Dublin tourist spots and then set out for the north in our rented Ford Cortina two-door sedan.

Out trip north took us through Drogheda, then over the border into Northern Ireland (NIR). From Newry in County Down, we went south along Carlingford Lough and then east and north along the coast at the foot of the Mountains of Mourne. We had planned to stay in a hotel in Newcastle, but it turned out to be a rather forbidding place, which obviously catered to Orangemen. We drove on a little further and found a charming inn at the foot of Slieve Donard, the highest peak at 2,796 feet. The Enniskeen was everything one could ask for in hospitality. We had a second-floor corner room with a bathroom every bit as large as the bedroom. The bathtub was big enough for the kids to sleep in, so they did. Next morning while Kitty finished breakfast, I took the kids out on the lawn, and we tossed a Frisbee. We must have been amusing because many of the elderly diners watched us from the large windows that faced the mountains.

The following day took us up to Belfast where we did a quick pass through downtown, making a few photo stops. From Belfast, we drove along Belfast Lough to the beautiful village of Carrickfergus, County Antrim. The weather was nasty, and the old Norman castle was closed, so we were quickly off again. We motored north up the Antrim Coast Road, truly a delight—the Irish Sea to the east and the beautiful glens of Antrim to the west. For scenery, that road is hard to beat. As we rounded the northeastern most part of IRL, we came to the famous Devil's Causeway. As time was running short, we decided to pass it up and made for Coleraine and, from there, south to our hotel in the Dungannon area of County Tyrone, NIR. No sooner had we checked in than relatives and friends started coming to visit us. I ducked into the shower, leaving Kitty stuck with the kids and the visitors—people she had no idea were my relatives.

Someone had steered us to the Altmore House as the place to stay. It was nicely located outside of Cappagh village and on the way to Pomeroy. Originally, it had been the home of the well-to-do Catholic Shields family. A son, James Shields, had been a Union general and, after the Civil War, had served as a senator from three western states. In those days, senators were elected by the state legislatures. The Shields' house had been converted into a modern inn and restaurant. The only downside was the heating system. The weather had turned warm, and there was no way to stop the stored heat from filling our rooms. We threw open our windows and hung out to catch a breeze. The real bonus was that the innkeepers, a very nice young couple, had children the same age as Tom and Chris. They let our kids eat in the kitchen with theirs and looked after them when Kitty and I were visiting in the area. We visited the Joe Quinns at their Cornamaddy home, the O'Donnells on Gort Road in the Brantry District, and the nearby Tommy Donnellys, longtime family friends. Tommy Donnelly and his wife, Nora, visited us some years ago; and we took them to the historic sites in Philadelphia. Tommy, who was so helpful to the O'Donnell family in NIR and to the O'Donnell visitors from the USA, died in 1997 at age sixty-six.

We also spent a couple of evenings at parish events run by Fr. James Donnelly at the Church of St. Columcille in Carrickmore. This was the church of my great-uncle Canon Francis Donnelly, who is buried in the churchyard. Fr. James was a brother of Tommy Donnelly and eventually left the priesthood and went to England. There he became quite well-to-do with substantial real estate holdings. However, he recently returned to NIR where he is in a nursing home, suffering from the ravages of Alzheimer's disease.

After several busy days and nights in the Dungannon area, we headed west toward County Sligo. Along the way, we stopped in the beautiful town of Enniskillen in County Fermanagh. The town sits on the waterway that connects two large lakes, Upper and Lower Lough Erne; consequently, it has wonderful water views in all directions. We visited a large and handsome stone army barracks, no longer in use, which sat along the waterfront. Our next stop was at Drumcliffe in Sligo. There we visited the old, circa 1200 AD, Norman church in whose graveyard the poet William Butler Yeats is buried. We arrived on a rather bleak and chilly day, which was just perfect for the setting. There was not a soul to be seen in any direction, and the church doors swung in the wind. The entrance to the lane that led to the church was guarded by a large Celtic cross on the right and a well-preserved round tower on the left. Ireland is dotted with the ruins of churches, monasteries, and round towers. The latter were built as a defense against the Viking raiders that beset Ireland up to 1000 AD. The tower at Drumcliffe was unusually well preserved.

From Drumcliffe, we motored into Mayo and through the county town of Castlebar. There we ran into our first pathetic beggar women. We were so shocked at the sight that we passed by without giving them anything. We chose to drive to Galway by way of the Connemara Peninsula—for Ireland, usually arid and desolate. The Connemara Motor Inn became our overnight home as we experienced our first light rain in a week. Next day, in the rain, we explored Galway City and the Salthill beaches on the bay. The city and surrounding area is well worth an extended stay, but we had to push on. I particularly enjoyed the area around the salmon weir on the River Corrib. We had a delightful lunch at Galway's Great Northern hotel, one of a chain in Ireland. All four of us had fish, and it was the best I've had before or since—even the kids enjoyed theirs.

From Galway, we set out for Ennis, the county town of Clare, and the weather turned sunny and warm. On the way to Ennis, we stopped at a ruined abbey near the Galway town of Kinvara. By then, the kids were objecting to our frequent photo stops, saying, "Not another ruin." Across from the ruins was a shop that sold ice cream, a rare treat for the kiddies. The town of Kinvara is where our son Patrick met Janet Pederson many years later. Bathroom breaks, or lack thereof, were a problem traveling in Ireland; there were no public facilities. We had a real problem with Chris who was barely potty-trained. From Kinvara, we drove to the Clare Atlantic coast and the famous Cliffs of Moher, truly one of Ireland's most spectacular scenes. In Ennis, we stopped for refreshment at the Old Ground Hotel, Clare's finest. We landed right in the middle of a marvelous wedding reception but were able to get a garden table from which we were able to watch the guests as they wandered past our table. I was particularly impressed by the many priests, all in formal frock coats. I'm thinking they were Church of Ireland, rather than Catholic, priests, but I could be wrong. It's just that I've never seen Catholic priests in frock coats.

That night, we stayed at the Shannon Shamrock Inn near the airport and next to Dromoland Castle, a tourist trap. The inn was big, but all on one floor. When we found that our room was a long walk from the hotel entrance, we started climbing in and out of our window to the car we'd park nearby. Next morning, we climbed out of the window with our luggage and left for the airport without paying our hotel bill. I wonder if anyone saw us. Anyway, from the airport, I called the hotel and made arrangements to pay by check after returning to Cinnaminson. I don't remember anything about our return flight, so it's likely everything went well.

A Death in the Family

The balance of 1968 was one of good news and very bad news. The good news was that my favorite football team, the New York Jets, had risen to the top of the AFL (American Football League) eastern division. I had gotten interested in pro football in 1964 when I read that someone named Joe Namath had been given $400,000 to play for the Jets. This was an unheard amount of money for a football player. The early Namath years coincided with Vince Lombardi's great Green Bay packers teams with Paul Hornung, Jim Taylor, and Bart Starr. The Packers were exciting and so were the Jets with Namath at the helm. I went to several Jet games with my brother-in-law John Crumlish. In one of them, the Jets beat Kansas City on six field goals and a pass interception by Johnny Sample. The great Jet offense did not score a point. The wags in the stands said that Broadway Joe was too hung over to pass accurately. The Jets went on to beat Oakland for the AFL championship. Earlier in the season, the Jets had lost to Oakland in the famous "Heidi" game. We were all at Mom-Mom Carlin's for dinner that Sunday and enjoyed what seemed to be any easy Jet victory. At 7:00 p.m., with a few minutes remaining in the game, the NBC network switched to a broadcast of a television production of the Heidi novel. In the final minutes, Oakland scored several touchdowns and won the game. Much of the country was outraged at NBC, and ever since televised games have been shown to completion.

For the January Super Bowl, the NFL (National Football League) Colts were favored by seventeen points. In the office, I took many bets as everybody, but Art Cherry and I thought the Colts would win. On game day, I sat glued to my television set, and while drinking a pitcher of martinis, I exulted in the Jets relatively easy win. I picked up an easy $50 from my office mates. The following weekend, we started serious house-hunting. Our Cinnaminson neighbors the Kellers had read about a new Scarborough development in Delran and suggested we take a look. Scarborough had built several very nice communities and that sounded very good to us. What didn't sound good was the poor school situation in Delran—no middle school, no high school, and so on. But we figured that by the time our kids needed them, the schools would be there as had happened in Cinnaminson. Consequently, in January of 1969, we went to Tenby Chase, tromped through the snow, picked out a lot, and put a down payment on a two-story colonial on what was to be Westover Court. Our Yardley model had four bedrooms, two and a half baths, basement, walk-in attic, mudroom, and two-car garage. The lot was heavily wooded, which we thought we'd like—our Cinnaminson lot had been bare of trees and remained so after ten years. Best of all, it would be on a cul-de-sac, a safe location for our young family. Occupancy was set for mid-July.

During 1967, Dad's health was obviously failing; and he had begun treatment for emphysema by Dr. Paul Breitenberger. His condition was undoubtedly due in part to his smoking—Camels were his cigarette of choice. I think there was a large environmental contribution to his condition. After all, he had spent many years driving vehicles in New York City at a time where automobile exhaust was full of pathogens. From the late 1940s, he worked indoors but in a taxicab garage that was filled with automobile exhaust fumes during shift changes. His treatment included stays at St. Joseph's in Yonkers. During one such stay, he was visited by Uncle John Sander and Aunt Mary who had driven up from Florida to see him. While in New York, John had a fatal heart attack and died in St. Elizabeth's Hospital in August. We had an unexpected funeral on our hands, which ended at a crematorium in Upstate New York. It was a rainy day, just perfect for the occasion. John' ashes were sent to Sweden for burial. Mary and Cousin Jackie returned to Florida and never came north again.

Dad died at home on February 9, 1969. He was only seventy-two years of age. His death occurred during a snowstorm that shut down the New Jersey Turnpike for several days. Kitty and I finally got to the city in time for the wake at Williams Funeral Home on North Broadway. On February 13, after the funeral at St. Margaret's, Dad was buried at Gate of Heaven Cemetery, Valhalla, New York. This New York Archdiocesan cemetery holds many famous Catholics, mostly Irish, including Babe Ruth, the Yankee slugger. While I was unable to get to New York, brother-in-law John Crumlish ably handled all of the funeral arrangements; not the least of which was finding parking places amid the great heaps of snow that covered the northwest Bronx. This was John's finest hour.

Off to Europe Again

In April of 1969, I made a trip to Germany to attend the huge trade fair held each year in Hannover. Also attending the fair was our new general manager, Jim Bradburn. In addition to attending the fair, Jim and

I were to meet with some Siemens executives. Jim and I had dinner the night before meeting the Siemens group, and I tried to prep him because I knew that he had little knowledge of the Siemens issues. Because of the interest in this huge trade fair, all accommodations were made through a central organization. As a high-ranking executive, Jim was assigned to the Intercontinental—one of Hannover's finest—while I was assigned to a pension in the student quarter.

The year of 1969 saw student demonstrations all over Europe and the United States. They started out as antiwar marches, but quickly became antiestablishment protests. As I was leaving my pension to find a cab to take me to my dinner meeting with Jim, I heard a commotion in the street. When I went outside, I looked down Kurt-Schumacher-Strasse and saw an army of protestors heading toward a stopped streetcar. Chanting, "Seig Heil, Seig Heil!" they rocked the streetcar to overturn it. The people in the front row of the march were wearing Prussian army helmets and looked very ominous, like something you'd see in the films of the Nazi takeover of Germany in the 1930s. I ran down the street away from the action and tried to find a cab among the many vehicles that were fleeing the oncoming marchers. I raced across the intersection and got a cab just as he was turning away. At dinner that night with Jim, he wondered why he was hearing so many sirens around the city. Typical of Jim, he had no idea what was going on.

The next day, I toured the fair's electronic exhibits, which were in a separate building. I don't remember that I saw anything special, but the booths and the overall atmosphere struck me as very professional. That evening, Jim and I met at the Siemens booth to spend the evening with the Siemens folks. After some preliminary talks, they took us off to dinner in a company Mercedes limousine—I was really riding high. During a dinner break, one of the Siemens executives pulled me aside, saying Jim didn't seem to know much about the computer business and the RCA-Siemens relationship. I could only say that he had just come aboard and was not up to speed. I had one free evening in Hannover and got a taste of a "hot" German nightspot. Because I was staying in a pension, I had no bath or shower in my room, only a sink. To take a shower, I had to call the desk and get assigned a bathroom. It was very awkward, so I didn't take many showers.

Since I was flying to Dublin from Germany, my flight was out of Frankfurt. I wanted to try the vaunted German railroads and went to the main station and bought a ticket to ride. My train arrived and I went to my assigned seat to enjoy the daylight ride. There I sat and sat, until I heard an announcement—in German, of course—that ended with the word "kaput." I couldn't believe it, my first ride in a European train, and it was kaput. I got up and went to the bar car to get a beer. The car was jam-packed, and the beer was also kaput—all gone. So much for German railroads.

I spent a long weekend visiting the folks in Northern Ireland and made my way back to Cinnaminson. The next week, I was back in my Cherry Hill office when people started to ask me where was Jim Bradburn. I told them I had no idea, I'd last seen him in Hannover, and he didn't say anything about his travel plans. After many days, his secretary located him in a Swiss hospital. Unknown to most of us, he was an avid skier and had gone to Switzerland for a weekend on the slopes. He had fallen and gotten a serious leg break, putting him in the hospital. Once we learned that he was an avid skier, we understood why he had determined to move the computer division to Marlboro, Massachusetts, close to the New England ski slopes. No one but Jim was happy at that prospect, and people started to jump ship. I started to again think seriously about life after RCA.

We Move to Tenby Chase

Through the spring of 1969, we made frequent visits to the Delran construction site to see our house rise, and rise it did. We hadn't realized that it would have a steep north-facing driveway that would plague us for some twenty-nine years. As summer approached, the family prepared for relocation to our new home at 174 Westover Court. Kitty had joined the Pheasant Run Swim Club so that she and the kids could enjoy the pool and start making new friends closer to Tenby Chase.

As our July moving day approached, we had day after day of rain; and it became doubtful that they'd be able to pave our driveway, front walk, and sidewalk. If not, our move would be delayed as we wouldn't be able to approach the house through the mud. The move did come off as scheduled, with one hitch. We had marked certain items to go into the basement. The moving men took the first batch down and came back looking kind of questioning. I thought nothing of it until I heard them splashing down there. I went to see—there were several inches of water on the floor. Because of the incessant rain, the builder hadn't been able to grade the property so the mounds of dirt were funneling water into the basement via the window wells. Everything had to be taken out of the basement and stashed in the garage. Even after final grading, surface water would continue to be a problem because of our hillside location.

After the dust—or in this case, mud—settled, our new home cost us $37,500. Kitty and I put up the $7,500, and Mother gave us a mortgage for the balance. She did the same for my sisters. We had made no money on our August sale of our Cinnaminson house. We had paid about $22,500 and sold—ten years later—for $25,100, less real estate commission and points totaling $2,750. We did, however, come away with $7,600 of recovered equity. There had been little interest in the house while it was on the market with John McDermott of the Walter D. Lamon Realty firm, and we were happy to be rid of it at last. We had lots of family visits after our move, and in between, we shopped for furniture for our nearly empty new home.

Over the years, Mother provided a massive feast whenever Kitty and I came to dinner, especially when she hosted family Thanksgiving and Christmas. In addition to roast turkey with stuffing and gravy, she usually had mashed potatoes, creamed onions, green peas, carrots, string beans, candied yams, and her favorite, mashed sweet potatoes with melted marshmallows on top. Riverdale had a good Jewish bakery, so we always had delicious rye bread and tasty hard rolls, with lots of butter. For dessert Mother usually made apple and mince pies, and the former was always served with ice cream. Naturally, we had cocktails before dinner, wine during dinner, and after-dinner liqueurs. Almost every Sunday, Kitty's mother, Marie, had a similarly huge family dinner in Moorestown, except she had roast beef instead of turkey. Marie's excellent desserts included strawberry shortcake with real whipped cream and mince pie laced with bourbon whiskey for added flavor. Marie Carlin continued to turn out splendid meals until she was almost eighty years of age. As I write this, I wonder how we survived those wonderful meals.

My mother's large-scale cooking pretty much ended after my father's death and her subsequent sale of the house in Riverdale later that year. The sudden sale took us by surprise, and we hustled up to New York to help clear the house for closing. There were lots of things that people could have used, but we really had no way to transport large items, so most stuff went out on the curb for trash pickup. It amused me to find that while we were inside, the stuff we'd just put outside had already disappeared. Years ago, Mother had acquired a used set of Encyclopedia Americana published in 1926. They were beautiful books but not very useful forty years later in the space age. Mother reluctantly let me put them on the curb, and they too quickly disappeared. From the house, Mother moved to an apartment block in Riverdale where she lived for several years before moving to Lincroft, New Jersey. We later wondered whatever happened to Mother's piano. She said since she couldn't move it, she took an ax, chopped it up, and put it out for trash pickup. Patience was not one of Mother's virtues.

In the late summer of 1969, I attended the first NY Jets versus NY Giants game ever played. Though a preseason home game for one the teams, for some reason it was switched to the Yale Bowl, New Haven, Connecticut. We Jets fans assembled at the usual place, the Hilltop, a saloon in north Yonkers. There we got our usual school bus for the trip up I-95 to New Haven. The more prosperous Giants fans made the trip from a nearby saloon in their air-conditioned coach. Even though it was a Sunday, construction made the trip horrendously long. We were indeed fortunate to get to the stadium for the kickoff. There was a lot of mostly good-natured kidding between the supporters of the two teams. The Jets won the game quite handily. After that game, the Jets fans upgraded to an air-conditioned coach for the fall season. After the Yale Bowl outing, I stopped going to the games, so I never got to ride in the new coach.

Final Days at RCA

In the late 1960s and in 1970, I spent a lot of time on one of our more successful products, something we called a VDT—Video Data Terminal. Art Cherry of my staff was our principal contact with the engineering and production team at the Van Nuys plant. However, there were occasions when I had to become involved in order to back him up. Jim Bradburn decided to make a visit to Van Nuys, and I had to be there to brief him, a rather thankless task. I got to know the Van Nuys team rather well, and George Turner was a guest at our new home. By then, I'd gotten a bumper pool table for the basement where George and I played while sipping our martinis. Not long after, Jim Bradburn was swept away in a rising tide of IBMers. L. E. "Ed" Donegan became our new vice president and general manager, and he moved in a lot of his IBM cronies. My function was soon replaced by a multi-discipline committee—the IBMers loved committees. My committee was chaired by Orville Wright, late of IBM and subsequently to be president of MCI. Neither of the famous Wright Brothers ever married—perhaps they were too busy suing other air pioneers—so I guess, this Orville was neither a descendant nor a relative.

After a series of pointless meetings, I told Orville that I'd taken minutes and would he like me to publish them. He said it was OK by him, but it didn't seem that it mattered to him at all, one way or another

In the late summer of 1969, baby Patrick came down with a mysterious rash. He was initially treated with antibiotics for roseola, but the rash or inflammation became so severe that Dr. Mitrotz had us take him to Cooper Hospital, in Camden. There, a team of doctors was confounded by his condition. Scarlet fever was mentioned as one possibility but quickly gave way to others. He continued on antibiotics and was given oatmeal baths to sooth his sore skin. Consequently, a dermatologist became one of his principal caregivers. I don't know if Patrick was in pain or not, I certainly hope not. He was so very glad to see us on the day we came to take him home. After five days, he had become a favorite of the pediatric staff, and they gave him a nice sendoff. Years later, I spoke to the attending dermatologist and asked if they had ever figured out what ailed Patrick. He remembered the case and said no, they didn't. Not long after Patrick's stay, they had a few more cases that year and treated them the same way. Their best guess was that the inflammation was triggered by a virus or a strep infection.

In February of 1970, I was invited to attend the Quota Club (Marketing Achievement Club) blast given each year to honor the salesmen and sales managers who had met or exceeded their sales goals. Others who contributed to a successful year were invited, along with RCA executives and relatives of the Sarnoff family. The 1970 sessions were held in San Francisco's Masonic Temple, a really fabulous venue. The sessions went on for some three days and were far and away the most elaborate ever held by RCA. I was blown away by the professional video presentations and speeches. I was assigned to a room in the Fairmont Hotel atop Nob Hill. My first entry to my room was after dark, and I was stunned by the view of the city that poured through the huge windows on two sides of my corner room. Coincidentally, my roommate was Ron Ludwig; I didn't know him then but later learned he was a fellow resident of Tenby Chase. Ron had friends and relatives in the Bay Area, so I never saw very much of him. However, I did see someone I didn't want to see. One day as I got off the elevator at the Fairmount lobby, there waiting to greet me was Dick Miller, the RCA vice president that I'd last seen in Italy. He thought we'd just have a wonderful time together in the Bay City. Actually, since neither of us had much to do with the gung ho sales crew, we were both fish out of water and lacked for company. In off hours,

Dick and I did hit some of the tourist spots, particularly along the Pacific shoreline. The salespeople naturally took off for the area's famous golf courses. Back in Cherry Hill, everyone wondered what this extravaganza cost RCA for video production, auditorium, room, and board. Guesses started at $500,000.

I recall a business trip to the Bay Area in the winter of 1971. It was very cold and rainy, and no fun to be out and around town. I did visit the new tourist venue called Ghirardelli Square, which had recently opened in some abandoned loft buildings downtown. It was one of the first of its kind, and I was quite impressed. What is really memorable about the trip was my visit to Jack London Square in Oakland. I loved his stories and enjoyed sitting at a bar, which looked out upon the very docks from which he sailed in 1893.

Back at the office, RCA's president, Robert Sarnoff, had surrounded himself with former auto industry executives. They had convinced him that RCA must drive for a 10 percent share of the computer market or exit the business. The current working plan was to grow from the current 3-4 percent share and achieve the 10 percent share by second-sourcing IBM. That's why we were hiring so many ex-IBMers. To grow RCA's share of market to anywhere close to 10 percent would require a massive investment that many on RCA's board of directors opposed, particularly the executives of the other RCA divisions that also needed cash. These same directors were also unhappy with such extravagances as the recent 1970 Quota Club meeting. Under founder David Sarnoff, the RCA culture had always been conservative and cost-conscious. Outside viewers of the computer industry, lumped RCA with the other companies trying to make headway against the IBM monolith. The scenario was described in the press as IBM versus the seven dwarfs.

In March of 1970, I was asked to participate in a Time Sharing Applications Symposium to be held at the Engineers Club in Philadelphia. Multiuser computer time-sharing was a new and hot topic at the time. The symposium was cosponsored by several local chapters of the major engineering and computer organizations. I was assigned to the Terminals and Displays panel. Today, as I look at the program, I observe that all seven members of my panel were from companies that no longer exist.

In May, there was a huge shakeup in my department, and I was shunted off to a meaningless management position. Later that month, I received a letter from our general manager inviting me to a small group meeting with him. Ed Donegan loved the concept of bypassing his immediate staff to get input from their subordinates. I don't remember much about the meeting, but everyone seemed pleased with the notion. Although the formal announcement hadn't been made as yet, we knew that the computer division would shortly leave the Camden area. Some functions had already moved to a temporary facility in Framingham, Massachusetts, and others to our existing plant in Florida. By October of 1970, my department had grown enormously to a staff of seventy-six and was now called Product and Programming Planning. Nonentity Don Stevens was still manager and my immediate boss. That month's departmental organization chart noted that certain groups would be going to Massachusetts or to Florida. It was clear to the rest of us that we were expected to go to Massachusetts when our new Marlboro facility became ready.

Some of those asked to relocate declined to do so. When their groups moved out, they were assigned to temporary offices in Cherry Hill. There they finished up whatever project documentation was required while looking for jobs elsewhere. Since I had no plans to relocate and hadn't been asked to, I often met with them to discuss job opportunities. It turned out that many of them were involved in starting new businesses in the area. The overall computer job market dried up in the recession of 1970-71, but I was ready to leave RCA as I had lost faith in the company's future. On November 4, Ed Donegan called an "all hands" meeting for the movie theater located in the Cherry Hill Mall. There, he formally announced what we all knew, that the division would close up shop in the Camden area and move to Massachusetts and some to Florida. Many of those who were scheduled to move to Marlboro were upset that he didn't offer a cost of living adjustment, housing, and other costs being much higher there.

A few weeks after the meeting in the theater, Ed Donegan asked me to come see him for a one-on-one session at his Cherry Hill office. By that time, product planning was in a building in Cinnaminson. I don't recall how the meeting started, but at one point he said that his people had to be committed and loyal and asked how I felt about those attributes. I told him I was certainly committed, having spent over fourteen years with the company. Actually this was a lie; I was ready to jump ship at any moment. As for loyalty, I said that he'd have to earn it. With that, he said I was finished and walked me out the front door, taking my badge on the way. A few days later, they let me back into my Cinnaminson office to clean out my desk and say my farewells to those friends and staffers who were left in the area. Actually, I did have a nice farewell party and received a fine golf cart as a gift. I had been playing golf with my staffers since the company's turmoil allowed us to come late and leave early. The bottom line was that I was out of work with a baby due in a matter of days and a big new house to furnish.

Postscript

In September 1971, ten months after I left, RCA's board shut down the computer division. Most of those left from the ten-thousand-employee peak in 1970 were laid off. One friend of mine closed on his Massachusetts house on a Friday morning and got his layoff notice when he got back to the office. It was a disaster for many people I knew, who were now stranded in Massachusetts. The Florida people fared better and most remained in Florida when the Palm Beach Gardens plant closed. The RCA installed base of computer systems was transferred to Sperry-Rand-Univac, based in Blue Bell, Pennsylvania. The two to three thousand employees needed to support those systems were retained by Univac for some time thereafter. They fared better than most. Coincidentally, my brother-in-law Dick Davis had left the RCA plant in Moorestown and taken a job with Univac sometime back. Initially, he had commuted from his home in Moorestown, but wear and tear led him and Pat to buy a house in Blue Bell around the time we bought ours in Tenby Chase. However, Dick had nothing to do with the RCA systems while at Univac.

RCA's decision to take on IBM led to the first large monetary write-off in American corporate history—about five hundred million dollars. RCA had lost its way as an electronics leader and had become a conglomerate, all the rage at the time. The new RCA family included Hertz car rentals, Coronet carpets, and Banquet frozen foods, among others. The company still had the television manufacturing and broadcasting operations, as well as the defense electronics operations in Camden and Moorestown. RCA's top management had allowed Ed Donegan and his crew of IBMers to lead the company by the nose to the brink of disaster. Robert Sarnoff was eased out of the presidency in favor of a tough RCA insider Anthony Conrad. With that move, Ed Donegan's days were numbered. I'm told that Ed went on to wreak havoc at other computer/software firms. Fifteen years later, RCA disappeared entirely when it was acquired by GE (General Electric Company) in a "fire" sale.

Kitty Carlin—glamour photo taken when at Dunbarton College.

Kitty Carlin at her desk at RCA, Moorestown, summer of 1956.

My first new car, a 1957 Chevrolet.

I'm at work on my first computer project at RCA, the Transcribing Card Punch. The mechanical punch unit never worked, so the project was a failure.

Kitty and I were married in Moorestown, New Jersey, on April 27, 1957.

With our parents on our wedding day. On left are Marie and Joe Carlin, and on the right are Peg and Jim O'Donnell.

Our first home was an upstairs two-bedroom apartment on East Coulter Avenue in Collingswood.

Our complete wedding party. *From left to right:* Jim Fletcher, Bob Foran, Bob Tomasulo, and best man Tom Gaffney. *From right to left:* Patsy Gross, Pat O'Donnell, Peggy Byrne, and maid of honor Pat Carlin. Joanne Carlin was flower girl.

On wedding trip, Kitty at Fort Fincastle near Nassau, the Bahamas.

Our rental MG sports car at Fort Fincastle. I had always wanted to drive one.

Kitty and I enjoy a break from her sister's wedding festivities.

Pat Carlin married Dick Davis, of Albany, New York, in June of 1957, also in Moorestown. The bride's sisters Kitty (in hat) and Joanne were her principal assistants.

My sister Pat was married to my Manhattan classmate Jim Maguire on October 11, 1958, at St. Margaret's Church in Riverdale, New York. I was in the wedding party. Pat's bridesmaids from left to right were Anna Ward, her sister Maureen, Peggy Lennon, and her cousin Eileen Quinn.

In the summer of 1959, we moved into our first house in Cinnaminson. It was on Kathleen Avenue in a new Max Odlen development called Riverton Estates. We had the house painted black, but quickly tired of it and had it repainted a soft green. Our 1957 Chevy is in the driveway alongside Jim Maguire's Ford.

In 1959, I got rid of the Chevy and bought this 1958 Peugeot, also with manual transmission. It had many fine qualities but some serious drawbacks as well as we found out when we drove it to Miami in 1961.

The electronics cabinet for RCA's Transcribing Card Punch as it was ready to go out the door, circa 1958. By then, I was lead engineer on the project. The unit was programmed by means of a wiring panel located on the left, below the control panel.

The Elm Street house on the left was the Carlin home in 1906 when the death of Kitty's grandfather Thomas Carlin caused the family to move into Moorestown from an outlying farm. It has been substantially reworked since then.

The Carlin family with Kitty's grandmother Mary Anne Callahan, 1906. Kitty's father Joseph, age six, was the only boy. Clockwise from lower left corner, the aunts are Margaret, May, Bessie (baby), Catharine, Alice, Agnes, and Etta.

Marie Catharine Regina Gegan was married to Joseph James Carlin (center) in Philadelphia on October 10, 1928. In attendance were Mary Shober and John Fallon.

		Column 1
1ST TURN SWITCH RIGHT TO CLOSE CURTAINS 2ND MARK YOUR BALLOT AND LEAVE MARKS SHOWING →	WARNING—YOUR MARKS MUST BE SHOWING FOR VOTE TO REGISTER 3RD TURN SWITCH LEFT	**DEMOCRATIC**
1	Member of House of Representatives (Fourth Congressional District) (VOTE FOR ONE)	Regular Democratic Organization FRANK **THOMPSON, JR.** ☐
2	Sheriff (VOTE FOR ONE)	Organization FRANCIS P. **BRENNAN** ☐
3		
4	Member of Board of Chosen Freeholders (VOTE FOR ONE)	Organization ROBERT F. **NASH** ☐
5	Coroner (VOTE FOR ONE)	Organization PETER W. **SMITH** ☐
6	Member of Township Committee (VOTE FOR ONE)	Organization CHARLES W. **FRENCH** ☐
7	Member of County Committee (Male) (VOTE FOR ONE)	Organization PAUL J. **WALLACE** ☐
8	Member of County Committee (Female) (VOTE FOR ONE)	Organization CATHERINE C. **O'DONNELL** ☐
9		

Primary ballot when Kitty was elected to the county committee, April 1961.

After promotion to engineering management in 1960, I took over the development of several fully transistorized new products. Shown here is RCA's DaSpan, a paper tape transmitter-receiver.

Here I am in March of 1962, dining at Rome's popular restaurant Da Meo Patacca. On the right is my RCA colleague Don Cianto. Our host was an executive of ItalCable who brought his wife along, though she didn't speak any English. I tried to converse with her in broken Spanish.

Santa Monica welcomed us in the spring of 1962. The statue stands on Ocean Avenue at the foot of Wilshire Boulevard. We stayed at the nearby Miramar Hotel for several weeks.

We eventually moved into the top-right front apartment at the Coral Gables on San Vicente Boulevard in Santa Monica. It was two blocks from Ocean Avenue, which overlooked the Pacific Coast Highway.

Kitty at California's Cajon Pass on the road to Las Vegas, 1962.

Kitty on the Colorado River just north of Boulder (now Hoover) Dam, 1962.

My secretary Marie May made this sketch in the summer of 1962.

Mother and Kitty are at the poolside in Coral Gables, Santa Monica, 1962.

Mother and Dad at San Gabriel Mission, 1962.

Kitty and I with Mother and Dad at one of the California missions, 1962.

The family visits Tijuana, Mexico, 1962. That's my sister
Maureen on the right.

We make a pit stop in torrid Palm Springs, 1962. That's Kitty's
Plymouth Valiant complete with the then-popular tires with
white sidewalls.

Kitty and Patsy Gross check out some art on display at the Victor Hugo Inn in Laguna Beach. It was one of our very favorite restaurants as it overlooked the beautiful crescent beach.

We loved the new houses going up at Laguna Nigel but felt that $55,000 was too much for us in 1962. They are reselling now for upward of a million.

Kitty and Patsy join the festivities at Disneyland. Kitty had to take a number of visitors to the new theme park.

This is a picture of Santa Barbara's city hall, which took us as we started our trip up the coast to San Francisco. It is one of my all-time favorite snapshots. I love the way I framed the evergreen and mountains beyond.

Kitty and I are taking a break at Morro Rock on the California coast, 1962.

In San Francisco, Kitty and Patsy struggle up the horrendous hill from the cable car line to our hotel, 1962.

Kitty and I stand atop San Francisco's Twin Peaks, 1962.

This is Donner Summit in the Sierras, with Donner Lake below. The pass and the lake are named in memory of the Donner party, many of whom perished in the 1800s trying to get over the pass when winter came upon them.

Kitty holds our new son Thomas James on christening day, April 1963.

Kitty's sister Marietta and her husband, Dr. Ed Sullivan, on the day of Tommy's christening.

Tommy's great-aunts. *From left to right:* Nell O'Donnell, May Carlin, Etta Carlin, Peg Carlin, and Agnes Carlin King.

Tommy and I, summer of 1963.

Jackie Sander, a first cousin, in his navy uniform, around 1963. Jackie died in Florida in 1995 at the age of fifty-six.

My sister Maureen married John Crumlish of Brooklyn at St. Margaret's Church, Riverdale, in June of 1964. Her sister Pat was maid of honor. The leftmost bridesmaid was Kitty O'Donnell and the rightmost, Eileen O'Connor, also of Riverdale. The other bridesmaid is a sister of the groom.

Members of Maureen's wedding party: Kitty O'Donnell and best man Kevin Crumlish, John's younger brother, now a retired NYC policeman.

The Crumlishes join Kitty and me for a libation or two at the 1964 New York World's Fair.

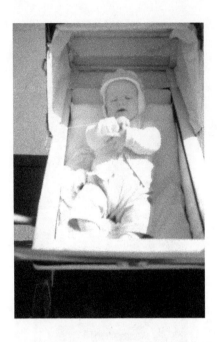

Chris was born on March 10, 1965. Here he is at the
Philadelphia Zoo that summer.

Kathleen Avenue Christmas with Tom and Chris, 1965.

Tom (left) and Chris, 1966.

**Kitty and I took a tenth anniversary work-play trip to
Florida in April of 1967. Here, Kitty is taking a break in
one of the many beautiful courtyards along Worth Avenue
in Palm Beach.**

Kitty gets splashed on the beach at the Holiday Inn, Riviera
Beach, 1967.

We visited my aunt Mary and uncle John Sander at their new
home in Lantana, Florida, 1967.

We spent two family vacations—June 1967 and 1970—here
at Indian Cave Lodge at Lake Sunapee, New Hampshire.
We stayed in cabins rather than in the main building
shown here.

The pool at Indian Cave Lodge was probably too cold for
Chris, 1967.

Tom and Chris at Riverton Yacht Club, 1967.

Patrick arrived on January 29, 1968

The boys gather in our cozy den on Kathleen Avenue, spring of 1968.

The family takes a break at the Wolfe Tone memorial in downtown Dublin, Ireland, 1968.

The Altmore House was our home while we stayed in the Dungannon-Pomeroy area of Northern Ireland, 1968. On the right, our boys can be seen playing with the sons of the innkeepers.

Altmore Chapel where Mother went to mass with her family. The chapel was heavily damaged during the sectarian strife of the 1970s and 1980s. It has since been restored.

My aunts Cassie and Maggie enjoyed our visit to the O'Donnell homestead at Gort, the Brantry, Dungannon. The girls on the right are my first cousins Geraldine (in red) and Claire (far right). The other girl is a neighbor. The cousins are the children of my uncle Pat O'Donnell and aunt Mary.

Uncle Pat O'Donnell gives Chris a ride on his tractor.

Tom and Chris at the grave of Irish poet W. B. Yeats at Drumcliffe, County Sligo, 1968.

Tom and Chris dodge the rain in Galway City, 1968.

The kids were tired of looking at ruins, so we took an ice cream break near Kinvara, County Galway, 1968. Years later, our Patrick met his wife-to-be in a Kinvara pub.

Our last overnight was at this inn near Shannon airport. We got used to going in and out of our room window, so we accidentally left without paying. That's Dromoland Castle in the background.

Family dining at Kathleen Avenue—now we are five, 1969.

We check out our house while under construction, spring 1969.

Patrick does a final tooth-brushing on Kathleen Avenue, 1969.

Our friends Mickie and Hank Smith. I was godfather to one of their sons. I worked with Hank for a while in the 1970s. He was a rabid right-winger and John Bircher, but a nice guy.

Our new house at 174 Westover Court, Tenby Chase, Delran, fall of 1969. In the driveway are my 1966 Ford Mustang and Kitty's 1965 Ford Fairlane wagon. The house had gray siding with blue trim.

First cousins Sean Quinn (left) and Gerry Donnelly at old the Quinn homestead, Cornamaddy, 1969.

First cousins Patrick Quinn (center) and Barry Quinn (right), with their spouses—Dolores (seated) and Frances (standing).

Patrick (center) with Tom on his right and Chris on his left. A star is born, summer of 1969.

Chris (left) and Tom enjoyed a trip to the Hudson River whenever we visited Mother and Dad in Riverdale, circa 1968. That's the cliffs of the New Jersey palisades in the background.

Portrait of Patrick, 1970.

Patrick and I try the pool at Indian Cave Lodge, June 1970.

With Jimmy on christening day, in December of 1970.

PART 5

Troubled Years (1971-1983)

We Become Six

Before I had a chance to worry very much about my jobless situation, our beautiful new baby boy arrived. James Michael was born on Thanksgiving Day, November 26, 1970, at Memorial Hospital (now Virtua) in Mount Holly. We were now a family of six. Kitty's mother, Mom-Mom Carlin, looked after the other boys while I was at the hospital. After leaving the hospital, I returned home, relieved Mom-Mom, gathered up the boys, and took them to the Carlin family's holiday dinner at the Sullivan home in Springfield, Pennsylvania. The boys knew something big was afoot and behaved very well throughout that long day and evening. Next day, my mother arrived to help with the boys so I could spend time at the hospital with Kitty. After five or six days, Kitty came home, and Mother left for New York. I guess the boys were too much for her. As always, Marie Carlin was a great help to us while Kitty recuperated. On December 20, Jimmy was christened at our new parish church, St, Peter's in Riverside. The godparents were my sister Maureen and her husband John Crumlish. Within a few years, we would have our own parish in Delran—Holy Name—on Conrow Road.

When I left RCA, I withdrew my $8,000-plus contribution to the pension plan, figuring that I would never again work for the company. That sum, plus some severance pay, would have to tide us over until I found a job. In the new year of 1971, I started to seriously job hunt, and at the same time, I applied for unemployment compensation. In those days, collecting an unemployment check was like running over hot coals; they didn't make it easy. On one of my frequent trips to the unemployment office in Burlington City, I met my former colleague Joe Tordella. He had left RCA a few years before to take a job as a copilot with TWA—Trans World Airways. Joe had been a multiengine pilot in the U.S. Air Force prior to joining the Friden Calculator Company from which I had hired him. The recession had hit the airlines hard. and many of the new pilot hires were being laid off. Joe went on to medical school and became a doctor. He never seriously practiced medicine but became involved in international business ventures that had a medical facet. Kitty and I visited Joe and his wife at their lovely home on Centennial Lake on several occasions.

With a recession under way, many other firms reevaluated their commitment to the high-investment computer business. By 1971, many decided to leave or had already left the computer industry—Philco, General Electric, and Honeywell were among them. It became apparent that my prospects of finding employment in computers or electronics in general were not good. Furthermore, relocating out of the area was not an attractive option. We enjoyed being close to family, and Kitty and the boys enjoyed swimming and tanning at the nearby Pheasant Run Club. What was I to do?

Our friend Hank Smith was always into some money-making scheme or other. He suggested that I join him in land sales for an International Paper Company development in the Poconos. I went to the property with Hank and learned what was involved; basically, the company would give me leads that I would follow-up. My job was to induce the prospect(s) to make the trip to the property where they would be met by "closers." I would receive a small commission for getting the prospect to the property and a larger commission if they actually bought a lot. I had some misgivings about the whole setup but decided to work the first set of leads I was given. It turned out that most of the prospects were older people, who had requested literature and had only a mild interest in buying anything. I found myself bullying them into making the trip to the Poconos. The sales calls were not a pleasant experience, and I soon tired of the land sales game. As the Poconos developed, there were many instances of fraud and poor site planning. I was glad that I had no part in it.

After I dropped out of Pocono land sales, Hank Smith suggested I try local real estate sales. He arranged for me to meet with Don D'Amato, a local real estate broker. Don helped me to get my real estate license and put me to work in his office, which was just off U.S. Route 130 on Cinnaminson Avenue. Most of Don's business had been in small houses located in the river towns from Palmyra to Burlington. My focus was on the more lucrative houses on the west side of U.S. 130, most particularly resales in my home community of Tenby Chase. It had become a popular stopping place for young executives, who were frequently transferred. By the summer of 1971, the new Tenby Chase Swim and Tennis Club was up and running, making the community very attractive for active families. Kitty and I dropped our Pheasant Run membership and joined the new club. We had a bit of a problem, redeeming our Pheasant Run bond, but eventually succeeded with the help of their Dan Seymour.

I steered all financially qualified prospects to Tenby Chase, regardless of what they said they wanted. I felt that the community would sell itself. Two of my early Tenby Chase sales were to Lucy and Tom Dzakowic and to Carol and Jim Lake; both families became good friends. Working with Lucy was a real challenge because Tom was working in Mexico during her house-hunting. I knew that she needed to be in Tenby Chase, but I had to bully her into buying the house on York Road. Lucy and Tom quickly settled into Tenby Chase, and their children became strong swimmers for the local team.

In aid of my new sales career, I sadly traded in my 1966 Mustang two-door coupe for a used 1968 four-door Delmont V-8 Oldsmobile I bought from Danny Mento at Burlington Motors. Though the car turned out to be a lemon, I regarded Danny as a friend. He was an excellent trumpet player, and we often went to hear his band at local night spots. I kept the Olds for little more than a year and traded it in for a new V-8 Ford Torino. This was an excellent car that I kept until I got a company car in 1976. However, over time with D'Amato Realty, I found that I was more or less running the sales force while Don pursued other activities. I spoke to Don about this situation and asked for a share of the company. He declined and I resolved to move on.

On to Fox & Lazo and Southern Parts

While I was working with Don, Mom-Mom Carlin had mentioned from time to time that a cousin managed the Fox & Lazo real estate office in Moorestown. I knew that Fox had a good reputation in the local real estate market and decided to pay a visit to their office. When I first met Jim McGlone, he already knew of my work in the local area—I had sold some of their listings—and was eager to bring me aboard. Jim's wife, Sylvia, was Mom-Mom's first cousin, once removed. I started with Fox in the fall of 1971 and adopted Tenby Chase as my home base. My principal strategy was to go door-to-door to meet people and ask for an opportunity to make a listing presentation. On one occasion, I surprised soon-to-be close friend Liz Oberschmidt when she came to the door without her wig. I quickly built a strong Fox presence in Tenby Chase and knocked out Cherry Hill's Gallery of Homes, which had been making inroads in the community. I was soon recognized within Fox and Lazo as a real comer. Later that year, I was invited to join the Fox group who were taking the famous Dale Carnegie course in effective speaking and human relations. My class of forty included seven Fox salespeople. This was an intense and enjoyable experience, and I received my signed and sealed diploma on March 8, 1972.

By 1973, I was flying high and became Fox's all-time highest recorder of home listings, with forty-seven. It was time-consuming to make proposals and sign up that many principals and keep them all happy with Fox & Lazo. Whether at the office or at home, I was constantly called to the phone. With the arrival of son Patrick in 1968, I had given up smoking my beloved Lucky Strikes. By the middle of 1973, my life had become so hectic that I started smoking again, after five years of abstinence. This time, I chose the presumably milder and filtered Salems. My alcoholic intake also increased and in time, would become a serious problem. My inflation-adjusted after-tax income for 1973 was about $100,000, the highest of my entire working life. As a sign of my arrival at the upper tier of real estate sales, I was accepted into the New Jersey Board of Realtors "Million Dollar Club." Kitty and I attended their annual awards banquet held at Zaberer's restaurant near Atlantic City. I really enjoyed meeting socially many of the salespeople with whom I competed throughout the year.

I took a nice break that summer in the form of a family motor trip, south to the Great Smoky Mountains. Along the way, we visited historic Harper's Ferry and toured the Luray Caverns, then made an overnight stop on the Blue Ridge parkway in Virginia where we stayed in a rustic cabin. The boys were excited to see deer on the road as it became dark. The next day, we detoured to Charlottesville for a tour of Jefferson's home at Monticello. There was quite a crowd and a long wait to get into the house, so the boys ran about the grounds. We stayed in a local motel and pulled in just as a big thunderstorm hit. We were all getting into the room when lightning struck nearby. Poor Jimmy was still outside at the time, and when we got to him, he had turned white from fright—the poor kid. The following night, we stayed in an Indian-run motel on the Cherokee Reservation in North Carolina. The motel was very poorly run but was adjacent to a lovely trout stream where the boys tried to fish using poles supplied by the motel. In the evening, we browsed around a trading post in town that offered the usual rustic junk. Next day, we went into the mountains and hoped to see Clingman's Dome, the highest peak in the national park. Naturally, clouds had moved in, and we couldn't see much of anything at the higher elevations. As we left the park heading west, we saw the bright lights of Gatlinburg, Tennessee. It reminded me of Times Square or Coney Island. I was not impressed.

Quickly passing through Gatlinburg, we came to the town of Sevierville and the Great Smoky Mountain Fish Farm. For a small fee, the farm offered fishing gear and a chance to catch a trout. There were lots of them visible in a big pool, so it looked easy to catch one. The boys got their fishing poles and bait and took up appropriate positions around the pool. Lucky Patrick was the only one to catch a fish for which he got—I don't remember what, if anything—and he didn't get to keep the fish. We were anxious to move on because the boys, Tommy in particular, wanted to visit Boonesboro—it was the heyday of television shows featuring Daniel Boone and Davy Crockett. On the way, we stopped at Norris Dam on the Tennessee River, part of the Roosevelt Era's vast TVA (Tennessee Valley Authority) project. When we got to Boonesboro, however, we were all disappointed. It was just a spot on the muddy Kentucky River with a small marker noting that it was the site of Daniel's cabin. The side trip to Boonesboro meant another overnight stop on the homeward leg, this time at Parkersburg, West Virginia. It was quite a trip and covered a lot of ground, but Jimmy wasn't yet three and probably remembers none of it.

In December of 1973, we lost another family member. Aunt Nell O'Donnell, age sixty-six, was found dead on the stairs of her walk-up apartment building in the Inwood section of upper Manhattan. Her purse was missing, so it may well have been taken from her in an assault. Nell Daly was from Cork City, Ireland, and was the estranged wife of my godfather Frank O'Donnell. Nell often came to Sunday dinner with Mother and was always a lot of fun. She was buried at Gate of Heaven Cemetery near my father's plot. Uncle Frank died in 1955 at fifty-five years of age. He died ugly from the consequences of alcoholism, an addiction he couldn't shake. In later years, Dad wouldn't have anything to do with Frank and asked us to shun him as well. Mother did surreptitiously help Frank when he'd turn up at the house, looking for a meal or a place to stay. He was a terribly sad case.

Back at work, I went into 1974 on a high note—I was taken into the inner circle at Fox and Lazo. At company pep rallies and at training sessions, I was always asked to speak on the art and science of securing real estate listings and working the home market. I very much enjoyed working with the key people in the Moorestown office—Jim McGlone, Howie Dvorin, Selo Vitale, and Joyce Cannone. I also enjoyed working with people at the other Fox offices, some of whom were ex-RCAers like me. Management mandated a Tuesday company-wide tour of all new listings, and we sales associates often felt it was a waste of our time. However, it was informative and an occasion of fun, usually at the management's expense. All of this would change when Fox decided to open a new sales office in Burlington Township and asked me to select a site and become broker-manager. If the office were successful, I would earn a percentage of the office net. However, on my own time, I would have to take the course for a broker's license and set up the new office. I also had to attend management meetings, including the quarterly meetings held at the Fox home in Ocean City, New Jersey. While these new duties were challenging and interesting, they were without pay. I still had to list and sell houses on my own account to earn any income. Consequently, 1974 was to be bad financially—my real net income was to fall some 25 percent for the year.

To get away from it all, in November, Kitty and I took the family to Disney World by train—Tom was eleven and Jimmy, four. We stayed on-site at the beautiful Polynesian Village and had a wonderful time. On the weekdays, we had no waits to enjoy the various attractions in the Magic Kingdom. I've always enjoyed amusement park rides and was happy to take the kids on the rides, particularly the scary ones like the Haunted House. Tom and Patrick enjoyed riding the monorail on their own, while Chris and I sailed on the lagoon, and Kitty sunbathed while Jimmy napped. For evening meals, the older boys went to the coffee shop while Kitty and I dined in splendor, and Jimmy slept under our table. The kids really enjoyed the train trip. On the way north, a man named Bill entertained them for hours in the otherwise empty lounge car.

In 1975, I got the new Fox and Lazo office up and running at the new Cooperstown Plaza on Sunset Road in Burlington. The company insisted that I quickly bring on board some twenty sales associates. We particularly needed people with experience in the large Willingboro market. While I was looking for good local people, the company was sending me people that, in some cases, had been dumped from other Fox offices. I soon realized that I was going to have to make do with much less than an optimum talent mix. I worked with the younger, better prospects but left the rest mostly to their own devices. After all, I had to list and sell myself to make ends meet at home until the office started to pay off, but it never did. The combination of marginal staff and minimal supervision led to some nasty scandals and lawsuits. One of my associates had no home and just squatted in whatever vacant house we had available. He was finally caught when the police were called by a suspicious neighbor. One of my associates listed and sold a house with what he said was a first-floor powder room. At closing, the buyers learned that the toilet was just sitting on the floor, there was no plumbing.

By the summer of 1975, I was becoming increasingly unhappy with my staff, my reduced income, and the demands placed upon me by the firm. With my high-office costs—rent, telephone, and furniture—and low-office production, I wasn't earning any managerial override as were most of the other company managers. My resentment fed my growing addiction to my favorite poison, the very dry Beefeater martini. By the fall, I decided that I had to get out from under and make a major change. I could stay in real estate and try to open my own office or, find a job in industry. I decided that I wasn't really interested in staying with the very stressful but not very stimulating real estate business. I would try to find another job; and as a first step, I resigned as broker-manager. This came as a real shock to the Fox & Lazo management, as no one had ever done anything like that before—I was always a pioneer. I returned to the Moorestown office as a sales associate and started an all-out job search. We had such reduced income that year that we were unable to pay our mortgage. Fortunately, Mother didn't need the money and soon canceled the mortgage.

I recently read an obituary notice for G. William Fox who died from the ravages of Huntington's disease at age sixty-six. Bill was the younger of two grandsons of the founder of Fox Realty, later Fox & Lazo. While I was at Fox, Bill managed the residential real estate arm of the firm. He was a very effective manager, probably the best I ever worked for. He built Fox, through mergers and acquisitions, into one of the region's largest real estate firms. In 2003, Bill retired from active management of Prudential Fox Roach and died in 2007, at age sixty-six, of Huntington's disease.

On to AMF Electrosystems

I had never really launched an all-out job campaign, so I read lots of material on how to write a good resume. I remembered that when I was doing the hiring, the resumes that had come to me were awful, usually photocopies that were the worse for wear. I decide to follow the prescription for an accomplishment-driven resume, but one that was professionally printed on blue paper. I soon got a call from Bill Gaskill, sales manager for AMF Electrosystems, a division of the old-line firm, American Machine and Foundry (AMF). I met with Bill at his home in Moorestown and found that he and many others at AMF were ex-RCAers. Bill said that he had gotten lots of resumes, but my blue one always ended on top of the pile. I was hired to be northeast regional sales manager, responsible for sales and service in North Jersey, New York State, and all of New England. I would have an office in the AMF office suite on Madison Avenue in New York City, but I would usually be traveling throughout my very large territory.

When I joined AMF in December of 1975, the company had moved on from its origins in food-handling machinery and become a conglomerate, all the rage at the time. Among its components were head sporting equipment, Harley-Davidson motorcycles, sailboats, bicycles, and bowling equipment. My division sold point-of-sales systems—essentially programmable computers—to food service establishments. We had a major inroad with the Burger King chain, and their outlets represented the bulk of our business. The division was based in Vandalia, Ohio, and I made several trips there to meet the executives and see the engineering facilities. From the outset, I was not impressed and would soon learn that the equipment was just as primitive as the organization.

I was given a Ford Granada as a company car and sold my Ford Torino. What a mistake—the Granada was a piece of junk and crippled me on long trips. Fortunately, I had made my early longer trips in the Torino. One such I particularly remember, I drove my Torino straight-through to Delran from Cape Cod to beat out an advancing snowstorm, and I did. The job quickly came down to working in New York City with the Horn and Hardart management—they had Burger Kings around Manhattan—and the Pudgie's Pizza chain, which was based in Elmira, New York. Our equipment was totally unsuited to the harsh conditions in busy Manhattan eateries—it was failure prone and difficult to reload after a crash. To keep Horn and Hardart happy, I had to buy gifts for their management, and they didn't want junky AMF stuff. Dealing with Pudgie's was much more pleasant, so I tended to spend more time with them as they opened new stores. However, this meant that I was making lots of back-breaking trips to Elmira in my crummy Ford Granada.

Pudgie's management was interested in opening stores in Pennsylvania, the territory of my colleague Jim Thomas. Jim suggested that he and I become partners and open a Pudgie's in the city of York. To explore this idea, I packed up the family, and we drove out to York. The proposed store site was on the west side of town and seemed to be OK. We then went to a real estate office to look at houses. The housing selection was very limited and surprisingly expensive. Because of the family's heavy involvement with our Tenby Chase club, we hoped for a similar facility in the York area. We were disappointed to learn that nothing like that was available. The only swimming available locally was at the indoor YMCA, and it was in the city, a long way from suitable housing. We did tour several properties and had a near panic attack when Patrick went headfirst down a flight of unprotected stairs in a badly lighted attic. Fortunately, Patrick was not badly hurt but got a good scare as we all did. It didn't take much more to convince us that York was a "no go" situation. Furthermore, I couldn't see myself working in a fast-food place as I'd have to in getting started. Jim Thomas was very disappointed that I didn't want to join him as a Pudgie's owner, and I don't know if he ever pursued it further on his own.

While I was at my New York office, I usually had to deal with our field service men and their problems. It was difficult for them to get good replacement parts, and they usually had to drive to an airport to pick them up. Sometimes they were mugged and their equipment stolen, even out of the lockers they used. I did my best to arrange things to minimize theft losses and reduce the time lost in travel. I soon had my own personal service experience. A new independent Burger King was opening on 125th Street in the heart of black Harlem. I was asked to go to the store to be sure that our equipment went in and was working properly. On-site, I found that the six cash drawers we supplied were too small to fit on the built-in stainless steel brackets below the counters. I looked at the drawings and found that we'd given the cash drawer dimensions for an older and larger type drawer. The problem was our fault, and the store was set to open the next day. I walked outside and noticed carpenters working in an empty store a few doors down. I asked if there was a nearby lumber yard and was directed to one a few blocks away. It was an awfully hot July day as I set out in my dress suit for the lumber yard. There, an elderly mill worker quickly cut six pieces of heavy plywood to my dimensions, and voila, I had what I needed to fix the problem. But, how was I to get it back to the store? Easy, they found me a trucker heading my way, and I was soon back at the store with my shelves. Jim, the AMF hero, saved the day, and the store opened on time.

On days that I was planning to stay in the city, I took the 7:30 a.m. Trailways bus from Mount Laurel to the Port Authority terminal on the west side of Manhattan. From there, I walked crosstown to Madison Avenue and down to my office at Thirty-ninth Street, usually arriving about 9:30 a.m. If I needed my car, I put it in one of the nearby garages, fortunately at company expense. If my boss came with me, he and I would have lunch and immediately set off for home. I quickly learned that my bosses were not knocking themselves out for dear old AMF.

One summer day, I had an appointment with the Bedford-Stuyvesant investment group in Brooklyn. They were planning to open one or more Burger Kings in that black and blighted part of the city. A glance at the subway map showed a station near my destination on what had always been a prominent street in Brooklyn. Since I would be carrying a heavy bag of stuff, I figured that I would get a cab from the subway stop to the Bed-Stuy offices. When I got to the street from the subway, I was shocked to find that there was no traffic on the street. In fact, the buildings had been torn down for blocks in all directions. Realizing I was in trouble, I started walking toward a far-off street that seemed to have some traffic. As I walked, I noticed a few young men looking in my direction. Now I had always felt that, because of our family involvement in the legal taxicab business, I would never take a "gypsy" (illegal) cab. Now I was desperate, and when a beat-up gypsy pulled up, I was glad to get in and be on my way. The people at Bed-Stuy could not believe that I had taken the risk that I had and called a radio cab for me when I was leaving.

Problems at Horn and Hardart became so severe that an emergency meeting of division brass was held in Manhattan with their executives. Afterward, the AMF team met separately in a nearby restaurant. I understood the problem much better than anyone else at AMF and made several suggestions for improvement. I then offered to go to Vandalia and take over the engineering team. My offer was totally ignored, and I realized that I should probably be moving on. AMF headquarters were at White Plains, New York. While in Vandalia, I had met AMF's marketing vice president, and he asked me to visit him when next in the area. I went to see him and found his office festooned with pictures of all of AMF's sexy products, but not ours. This visit confirmed my suspicion that AMF Electrosystems was not important to the company and would soon be spun off or folded. It was best to get out before that happened.

Through the summer of 1976, we had a heavily promoted sales contest with substantial prizes. It was kicked off with an all-hands meeting in Vandalia, during which we were given a glossy folder describing the contest terms and prizes. Because of my Pudgie's account, I came in second and was entitled to an AMF Sunfish sailboat. When I asked the marketing management when I should expect the boat, I was stonewalled and eventually told that "management" had not budgeted the prizes. I called the chief financial officer who was of no help. This issue was still on my mind when I left the company in September and started with L. D. Lowery. I finally decided to call the president of AMF Electrosystems and reached his secretary. I explained my problem to her, and she said she would look into it and get back to me. Before long, she called and asked for my shipping address as the boat would soon be mine; and it was. The lesson: never underestimate the power of a secretary. I still have the boat, but AMF is long gone from the American business scene. I believe that AMF Bowling Systems is all that remains of the firm I knew in the 1970s. The only thing I miss about my AMF days was the excellent delicatessen next to our building. Every day that I was in the city, I bought a ham and cheese on rye sandwich, and I've never had better.

Tenby Chase Swim and Tennis Club

By the summer of 1975, the family became fully involved with competitive swimming. Patrick had joined the Tenby Chase swim team, the Sharks, and was the anchor man on their very successful seven—to nine-year-old boy's relay teams. In fact, for the first time, we had to delay going to the shore until August so that Patrick could swim for the Sharks in the tri-county championships. Patrick's early relay teams held the club records for many years. Tri-county swimming involved some twenty-plus private swim clubs in the adjacent counties of Burlington, Camden, and Gloucester, so travel was not an issue. However, our swimmers—boys and girls—were encouraged to swim all year round. Patrick, at various times, swam in the off-season for Willingboro and for Jersey Wahoos at their indoor facilities. Competitive winter swimming meant travel throughout the Mid-Atlantic States—to Dover, Delaware, and Harrisburg, Pennsylvania, among others.

With Patrick so involved in swimming, I was invited to join the board of the Tenby Chase Swim Club. Mike Thornton and I went on the board together for terms of three years each. Also on the board at that time were Tony Sciano, Jack Alfieri, and Jack Anderson. Over time, I became the club's swim director, and Mike became the tennis director. My major task was to interview and hire our swim coach and his/her assistants. We'd had a very popular and successful coach who'd moved on to his home club and found it difficult to keep and retain a head coach. During the swim season, we had very good parental support for the team. Parents served as timers, referees, announcer, and manned a refreshment stand for home dual meets. Sadly, during my tenure, the team gradually descended to lower divisions—from C to D and on down, eventually to F. Essentially, our strongest swimmers moved away or grew too old to participate.

Many individuals played important roles in supporting the team during the Saturday dual meets. I have a recurring vision of John Dennen, who often served as a timer or referee, cheering Patrick on as he swims the anchor lap. As I cast my mind's eye on the 1970s swim deck, I see Mike Thornton and Joanna Wormley serving as timers and Bob Christie as swim referee. I started out as a timer and finally became the meet announcer, a position I enjoyed for many years. In those days, the diving competition was held during a break in the swimming events. Many people were bored by the diving, and there were very few good divers around. Anyway, for lack of a qualified dive referee, I was often forced to serve in that capacity. It was the referee's job to determine that the diver did the dive that was announced and, if not, to disqualify the dive. The referee also requested and read out the scores the five judges gave the dive. I had taken the diving judging course, so I had some rudimentary knowledge, but that's all. Naturally, on my very first day as referee, I had to disqualify Brendan Bossett, my neighbor's son. In fact, he was the first diver of the day and was very nervous—that was two of us. I was astonished at his first dive; it didn't look like anything I'd seen in my diving book. Then I realized that Brendan had just fallen off the board; he hadn't done a dive at all. Still, it wasn't an easy DQ for me. I guess my fame became such that I was pressed to serve as diving referee at some of our away meets. Serving as diving referee at home was not nearly as bad as having to do it on the road in front of a hostile crowd. After my time, the divers had a separate meet one evening during the week.

Running in parallel with our swim program was our tennis program run by Mike Thornton. For the first time, the club arranged for a professional to give group tennis lessons. Kitty and I found them to be hugely beneficial and enjoyable. The pro, named Lynn Fulmer, was just terrific. She lost the use of her right arm and had to switch to her left, but you'd never know it. We held weekend round robin events for the men and during the week for the women and tournaments for both. The patio alongside the courts became a popular spot to sit and enjoy a libation while watching the play. One year, my partner, Maryann Mancini, and I advanced to the final of the mixed doubles tournament. The final was rained out week after week and was finally canceled as winter drew closer. Furthermore, Maryann, who was pregnant at the time, could no longer play. The 1970s were the peak years of American tennis, and its popularity started to wane badly through the 1980s. By the 1990s, most of the club's tennis players had become rather well-to-do and switched to golf. Mitch Sheairs, a lefty, and I continued to play as a team into the 1990s. We were in an indoor contract, first at Delran's Millside Racquet Club, then at Burlington's Crossroads Racquet Club. From time to time, we were challenged by other players to come to their courts for a match. We usually did well, mostly because of Mitch's wicked spin serves and drop shots. Patrick took lessons at Millside Racquet Club and played on their age-group teams. He was a good player but found soccer and swimming more to his liking.

During my Tenby Chase board tenure, we accomplished quite a bit. Until my time, club members used to take care of the grounds, paint the buildings, and the like. This was feasible since many of the early board members had local jobs and got home well before evening. The newer board members such as Mike and I were often out of town or home late during the week. On weekends, we had to take care of our own homes and grounds and prepare for the competitive swim season. Consequently, we decided to hire people to do the routine maintenance jobs around the club. One of the first people I hired was Tom Wade for lawn care. Tom worked for the club for many years and took care of my lawn as well until quite recently. Many of the older members were unhappy with this turn of events; they felt that they had done the work themselves to save the club's money and now we were squandering it. To keep peace in the family, we decided to paint the clubhouse ourselves and did so. Other accomplishments during my term included resurfacing the tennis courts and construction of a pavilion to provide shelter during summer rainstorms.

In addition to the active swimming and tennis programs, we had some really great cookouts and socials through the years. Looking back on those times, I realize that there was an excess of alcohol, which contributed to some unseemly rowdiness. After one of the parties, I drove Liz and Leo Oberschmidt home to their house in the far corner of Fenwick Court. As I backed out of their driveway, I hit the electric transformer box and knocked it awry on its concrete pad. I guess someone reported it, and the electric company came out to put it right. Of course, when asked, Liz and Leo had no idea who had done it. Over the years, we had many laughs as many times I came close to again knocking the box off its perch. Men's volleyball games were a big Sunday treat. They were a lot of fun at first, but over the years, they became so rough that I stopped playing. Regretfully, I did play volleyball one more time. It was at a Pennsylvania wedding of one of our neighbor's sons. He was a West Pointer, and many of his buddies were in the game. I received a football-style hit that made my hip black-and-blue for a long time. After that, I knew my volleyball days were far behind me.

The Tenby Chase Swim Club was located in a hollow that seemed to be a target for any passing storm. One day, while sitting on the tennis patio, I saw a whirlwind—mini tornado—heading toward the clubhouse and sucking up towels as it bore down on us. It hit the patio and sucked up the beach umbrellas, lifted them over the twelve-foot fence, and dumped them on the near court, narrowly missing the players. On another occasion, a severe storm hit the club during swim practice. The kids ran for their lives, as the wind blew down the tennis practice wall as well as a huge tree that tore down the back fence and fell over the creek alongside the club. The tree made a neat bridge over the creek until the township had it removed months later. At the height of that storm, Patrick found himself besieged by young girls asking him to save them. Eventually parents arrived by car to rescue the kids. Fortunately, while many were scared, no one was seriously injured.

The destruction of the aforementioned tennis wall annoyed me very much. I had written a specification for the wall, which required that reinforcing rods be used to connect the wall's concrete blocks. I received several bids and was waiting for a final bid when I had to leave for vacation. While I was gone, the board received a final low bid and went ahead and had the wall built without anyone checking the work. At a membership meeting, during which the destruction of the wall was discussed, a former board member rose to speak. He said he saw the wall going up and that no reinforcing bars had been used. I asked if he had reported this to anyone on the board, and he said that he hadn't—dumb and dumber. Fortunately, our insurance company paid for a new and properly constructed wall, which still stands. Speaking of storms, I was in the old Delran K-Mart one evening when a tornado passed overhead. It sounded as though a locomotive was tearing through the store, and all the false ceiling tiles were sucked up one after another. It seems as though the area no longer gets the fierce storms that we had in the 1970s and early 1980s. There will be more on this point later.

As with everything in life, there came a time when our boys moved on to other things, Kitty went back to work, and we rarely used the club. One year, we realized that we had been there to play tennis six times while paying $240 annual dues. At $40 per session, it was expensive tennis. So after many wonderful years, we resigned from the club, and our bond was repaid the following year. So ended a significant chapter in our family history.

On to L. D. Lowery and Andrew Wyeth

When I realized I would have to leave AMF, I thought about what I wanted to do and how I'd go about doing it. It seemed to me that selling electronic components and/or equipment might be a good way to exploit my sales and engineering background. Out of the blue came an opportunity to join an electronic component sales firm, and I seized it. L. D. Lowery was a manufacturer's representation firm based in Broomall, Pennsylvania. They handled some of the best known manufacturers of electronic components—these firms were known as principals. Unfortunately, I was to learn that, with the exception of Harris Semiconductor, most of their lines were dated. Newer firms were developing the products that were needed in a field rapidly adopting microelectronic technology. However, their old-line products did generate enough income to keep body and soul together for a time. A real plus was my company car, a red full-size Chevrolet Caprice. It was one of the best cars I ever had.

I started with Lowery in September of 1976 and was assigned a territory, which included northern Delaware, Pennsylvania's Delaware County, Philadelphia, and southern New Jersey. I would receive a base income plus a bonus based upon sales in my territory. At first glance, this seemed reasonable, but I soon learned otherwise. In my territory, I was only responsible for direct sales to manufacturers; whereas my electronic firms were all low-volume parts users who bought from electronic distributors, not direct. This meant that I got no credit for RCA's purchase of our products, which they bought entirely from distributors. Still, I had to spend a lot of time at RCA taking care of engineering requests for literature, samples, and dealing with product problems. As a result, I got to know the purchasing people at RCA and enjoyed seeing them and dropping off modest gifts. I suspected that the distributors were dropping off much more substantial gifts than I was. While I soon tired of the mundane nature of the "rep" business, it did have some rewarding moments, usually when I was able to help a firm solve a technical problem or when one of our principals came to town for a visit with target prospects I'd selected.

One interesting aspect of the job was visits to the plants of our principals. An early visit was to the Indiana plant of Robinson-Nugent, a manufacturer of sockets for transistors and integrated circuits. There were lots of issues with such sockets—cost, materials, wear, and reliability. Consequently, wherever possible, equipment manufacturers were phasing them out. We made a winter-time visit to Clarostat in Durham, New Hampshire. Not long before our arrival, a major fire engulfed the town center, and the area was covered in ice from the fire hoses. Clarostat, a manufacturer of potentiometers and rheostats, was a real throwback. It was located in an old mill building, which had ten-foot-long icicles hanging from the roof. During our meeting, a huge icicle fell onto our window air-conditioner, ripping it out of the window and sending it crashing to the ground. After that, we stayed well clear of the building. One evening, we went to dinner at a marvelous restaurant in the New England Center of the University of New Hampshire. During dinner, I was able to look out of the large picture windows and see moonlit pristine snow-covered fields and pine trees. It was truly a winter wonderland.

One of our major principals was the electronics division of Corning Glass Works. I made several trips to their film resistor plant at Bradford, Pennsylvania. One of them was a three-day extravaganza for the entire Lowery team that included a tennis and golf outing at the Bradford Country Club. I was very interested in the resistor-manufacturing process, from drawing the glass rods to depositing thereon a resistive film that was automatically trimmed to the desired tolerance and a protective coating applied, with value marking. On another occasion, I went to the division headquarters in Corning, New York, with an RCA purchasing agent. RCA wanted Corning to take back some $100,000-plus product, which had become surplus when a project was canceled. The two companies had a close relationship because Corning supplied the coated glass bottles for RCA's television picture tubes. Furthermore, Corning had used RCA computer systems, the very ones that had been wiped out in the great Chemung River flood in the early 1970s. However, Corning would not take back the product, saying something to the effect that they were sellers, not buyers, of electronic components.

The Lowery team also toured the Corning tantalum capacitor plant in Biddeford, Maine. This too was very interesting from the technical perspective. On this occasion, we stayed at a hotel in nearby Portland and got to see the famous Portland Head lighthouse. Our hosts took us to dinner at a wonderful seafood restaurant on the Portland waterfront. While there, we ate every lobster they had and took over the bandstand when the band departed. The police were called to evict us from the restaurant and see us back to our hotel. It was quite an outing. The Corning folks were not pleased at our behavior or at the excessive cost of the evening.

I learned that Lowery's president Bill Sylvester and vice president Mike Dermott were avid tennis players, and both were well over six feet tall. A salesman who came aboard after me was also an avid tennis player. In addition, Joe McDonald was an experienced manufacturer's representative, a top-notch salesman, and a good steady drinker. Accordingly, a match was arranged with Joe and me playing Bill and Mike. We played at an indoor tennis facility near Broomall. Well, it turned out that my partner was a very good player, indeed. Our two huge opponents loved to rush the net, allowing us to lob over their heads to send them in retreat. With some help from me, my partner easily beat them, and they were forced to buy drinks and dinner. They must have concluded that our victory was a fluke because they again challenged us. The result was pretty much the same, and we never heard about tennis again. As I said, Joe was an amazing drinker. Since he also lived in New Jersey, I rode over to Broomall and back with him on several occasions. He liked to drive with a glass of Scotch ready to hand, claiming it calmed him while behind the wheel. I thought it was a wonderful idea and thought I would emulate him when next I drove. You have to wonder what we were thinking.

While with Lowery, I traveled extensively in the area that surrounded the hamlet of Chadds Ford, Pennsylvania. As I was approaching Chadds Ford one wintry day, I noticed a sign for the Brandywine River Museum. Being a museum rat, I quickly pulled into a driveway that led up to what seemed like a barn. Actually it had been a gristmill, now it was a museum that housed many works of the Wyeth family of Chadds Ford—N. C. Wyeth, his son Andrew Wyeth, and his grandson Jamie. As I strolled through the galleries, I caught a glimpse of the partially frozen Brandywine River that runs alongside the building. It was a beautiful scene, just perfect for a Wyeth landscape. I enjoyed the museum for its informality and its wonderful collection of Wyeth works.

Some ten years later—mid-1980s—I read that a vast collection of new paintings by Andrew Wyeth had come to light. From 1971 to 1985, Wyeth had created over 240 individual works of neighbor Helga Testorf without telling a single person, not even his wife. He stated that he would not have been able to have finished the project with everyone looking at it. At that time, the collection could be seen at the Brandywine museum. I don't know how I was able to do it, but I got over to see them before they were sold and taken away. It was truly a spectacular collection, and I felt privileged to have seen the pictures before they became really well-known and scattered about. I was turned off a bit by "improvements" at the museum—it was bigger, more costly, and more formal. It's still a great museum, and Kitty and I have stopped in from time to time.

Wyeth may have prospered in Chadds Ford, but my Lowery career died in Florida. I didn't like component sales and really wasn't suited for it by temperament. It didn't pay very well either, but it did offer many occasions for drinking and I made the most of them. Here's how it happened. We made a company-wide excursion to our new principal,

Harris Semiconductor. We toured their new plant near Melbourne, Florida, and participated in several product training sessions that I enjoyed very much. On the last day, we had a cocktail party and dinner. I drank way too much and passed out during dinner, making a fool of myself. The Harris people were very annoyed and told Lowery management that they didn't want me to represent them. I wasn't doing very well anyway, so Bill Sylvester took the opportunity to let me go in September of 1978. So once again, I was jobless and not by choice. Bill was very fair and returned my two-year share of the pension fund, with interest. However, I had to give up my car. That left us with one car, Kitty's dark green AMC Hornet wagon with the Gucci interior. We would have to make do with that.

The Soccer Wars

Sometime in the mid-1970s, I read a notice in our local newspaper that Delran was going to have open soccer and needed coaches. Previously, the town had only age-group traveling teams, which accommodated only a few elite players. Figuring that there were not many fathers who knew much about soccer, I turned out for the inaugural meeting. There I met a number of men who had played soccer in their home countries. They were willing to help but were not available to actually coach the many teams we expected to have. That first summer, the parents signed up their children for soccer in large numbers. Teams were established by age group, and the players, mostly unknowns, were assigned by a draw.

That first fall, I was assigned to work with Chuck Haller who was coaching an age-group team, which included our Christopher and his son, Gary. Chuck was very knowledgeable and great to work with. The following year, Chuck's team formed the basis of the mininovice traveling team, and he decided to stay with them while I was assigned to an older age-group team that included our Tom. Our team, the Panthers, did not do very well that first season, but I learned a lot. In subsequent years, my team developed quite a rivalry with a team coached by Phil Betancourt. He and his wife, Mary, played in our duplicate bridge group. Phil was an art history professor at Temple University, and Mary taught at Moorestown High School. As our soccer rivalry developed, we started to violate the spirit of the league by recruiting players. This came about unintentionally, at least on my part, when players asked if their friends who were not then on any team could join our team. I said sure and found that I had acquired some really good players. As my team started sweeping the age group, Phil noticed that I had some new players and did some recruiting of his own. For a couple of years, our teams dominated the age group. One year, my team won it all and another year we tied for the age-group championship.

Phil's son, John, was very tall and the star center forward on his team. He was hated by my team, and my guys did their best to knock him out of the game. I had several stars on my team: Mark Teasdale, Bob Weaver, Mike Jewell, Mark Bohn, and Mike Vermes. His younger brother, Peter, became a professional and played on the U.S. Olympic soccer team. The main Delran soccer complex is named for Peter Vermes. John Betancourt became a published writer whose forte was science fiction. Phil and Mary moved to Moorestown so that John could finish high school there.

Somewhere along the way, I became involved with soccer politics and found myself township commissioner for boy's soccer. This turned out to be a real can of worms. I was beset by parents, coaches, and players, as well as my superiors on the township recreation committee. I had to recruit coaches and manage the draft of players so as to keep the teams as even as possible. I'm sure that there were those who were thwarting my efforts one way or another. A real problem was getting referees for our games. High school players were supposed to fill this need but often failed to show up. I had to referee many games while coaching my team. This was a really unpleasant experience, but it was better for the kids than not playing at all. One happy experience was a game where the players faced off against the parents. For some reason, I was not a starter on the parents' team; but when I got in, I scored the only goal of the game. I took a pass from my left side and pushed it into the goal. I've forgotten who set it up for me. It was the only time I ever set foot on a soccer field as a player, and I won the game.

After the exciting soccer years in the top division with Phil, I dropped down to a lower age group to coach Patrick's team. In the years I worked with his age group, I had three decent players—Patrick, Mike Oberschmidt, and Billy Andrews—but not all at the same time. One year, I had Mike on offense and Patrick on defense, and the following year Billy replaced Mike. We just didn't have enough talent, but I had good support from their fathers—Leo Oberschmidt and Jim Andrews. Of these teams, only Patrick went on to play high school soccer. After they moved on, I coached for one year the upcoming team, which included my son Jim. The team was very poor, and I was tired of putting in so many hours on soccer-practices, games, and meetings. I wasn't doing well for my team or for myself—I was finding that soccer was cutting into my drinking time. So after seven years of coaching, I retired entirely from all involvement with Delran soccer.

A sad footnote to my soccer years was the mysterious death of Gary Haller, son of my first coach. Gary died at sixteen, I believe, while doing some foolish experiment with a plastic bag over his head. Chuck and Sis Haller had been longtime soccer enthusiasts and major supporters of Delran soccer. They were members of the Tenby Chase club, and Chuck worked for RCA in Camden; we had a lot in common. They were so devastated by the death of their only child that they moved away.

On a happy note, in October of 1977 Kitty's "baby" sister Joanne was married to Phillip Jeffrey Hoops. Phil was originally from Alabama, but until recently had been living in the Los Angeles area. After graduating from the University of Bridgeport or "UB" as it was called, Joanne had been working for a publishing company. I recall that they were married by Judge McGann, a family friend, in the dining room of the Carlin home on South Church Street. A wedding reception followed at Kenney's suburban house. Joanne and Phil rented in the Collingswood area until they bought a split-level house in Haddon Township. They would have four sons while living there.

The Early Grim Years

After leaving Lowery, I once again trekked up to Burlington City and applied for unemployment compensation. Nothing had changed, you still had to stand in one line after another and report in every week to prove that you were unemployed. It was made difficult and humiliating so as to discourage applicants. I persevered and had just started to receive my checks when I landed a job. Actually, my new resume had drawn quite a bit of interest, and I quickly had several interviews. Naturally, as always, I had doctored my work experience to accentuate the positives and eliminate the negatives. But what drew many inquiries was my experience in swimming and soccer, which I had featured in the resume. Several of my prospective bosses had children who were just getting into those sports, and they enjoyed talking with me about them. I soon had several job offers, but they would have required relocation, so I passed on them. I just didn't see how I could start relocating at that stage of life—nearing fifty years of age and with four kids enjoying the good life in Tenby Chase.

My wait paid off when I got an interview with a local firm, GRM (formerly General Research and Manufacturing), a subsidiary of

Inductotherm Industries. Mike Yurko, GRM's president, had also noticed my swimming experience—he had a daughter who had been a competitive swimmer. I interviewed with Mike and with Tom McGarrity, his sales manager. I accepted their offer of a sales position. The base salary was low, but there was a bonus program based upon company profits; and I got a company car, which solved my immediate transportation problem. My job was to secure electronic manufacturing jobs where we would subcontract for a prime such as RCA or GE. Somehow Mike was convinced that there were lots of those business opportunities out there. There were a few, but they were hard to come by, and I had to travel extensively to dig them out. Often GRM's facility and equipment were not deemed satisfactory by prospects. We were located in old buildings at Burlington County Airpark in Lumberton. They were pretty shabby, but a convenient drive for me, so I was OK with them. We had no computers, nor did we have any microprocessor know-how. So I began a campaign to upgrade our capability so that I could present a stronger picture to my prospects. There was a lot of resistance at first, but I got help from an unexpected place.

When I joined GRM, I was asked to handle an existing customer, namely Inductotherm Corporation, the senior firm of our parental conglomerate. Inductotherm made induction furnaces, which were used for melting metals by newer and more nimble steel companies worldwide. GRM manufactured the electronic controllers for the Inductotherm furnaces. There were lots of problems with the controllers, but no one was sure whether their design or our manufacturing was at fault. I met with their engineering staff to try to get to the bottom of the problem. It was clear to me that the controller needed a redesign using microprocessor technology. At the time, their business was off, and the Inductotherm management did not want to spend any dollars for a risky, to them, redesign. I couldn't push too hard because we didn't have the know-how at GRM to take on the job.

Henry Rowan was chief executive and major stockholder of Inductotherm Industries, a conglomerate that owned six or eight companies in the metals and electronics businesses. With the exception of Inductotherm Corporation, Henry left the companies to operate pretty much under their preacquisition management. That was the case with GRM. Henry did take a personal interest in the management of Inductotherm and became very annoyed with expensive field failures of our controllers. Mike Yurko and I met with him on several occasions to discuss the problem, but Mike did not want me to bring up any of my "radical" ideas. I just sat there and spoke when spoken to. I had been pushing Mike to invest in a programmable automatic electronic circuit board tester from either Hewlett-Packard or General Radio—each would cost about $200,000. Mike didn't want to seriously consider such an expenditure, partly because he didn't think Henry Rowan would approve it. Finally, Henry called a meeting and badly beat on Mike and me to improve our product quality. Not knowing what else to say, Mike mentioned that we were considering spending over $200,000 for a piece of test equipment. He expected Henry to pooh-pooh on such a large expenditure, probably the largest Mike had ever considered. Instead, Henry said $200,000 was a drop in the bucket compared to their field repair costs. Not long after that meeting, a Hewlett-Packard tester arrived at GRM. Years later, Mike told me that the tester was saving a lot of money for GRM and improving quality. I said it was one of his best ideas, and he looked at me to see if I was pulling his leg. I asked why he chose Hewlett-Packard when I was pushing General Radio. He said he owned stock in Hewlett-Packard. That's how decisions were made at GRM.

While I don't recall seeing Henry after the "tester" meeting, his influence was all around us. He owned a large industrial tract in Westampton Township and several of his companies had their facilities there. He also had a private air strip for his company planes. He told Mike that he didn't like the GRM company name, he didn't want letters, and he wanted a meaningful name, even if it were Grim. My boss Tom and I lobbied for something other than Grim to no avail. Mike felt he had to go along with Henry's notion and put the big red Inductotherm over the *i* in Grim. Mike had other options but made the first of many bad decisions I was to experience. From the sales and marketing point of view, Grim Corporation's new name became a bad joke. Bypassing Tom McGarrity and I, Mike had taken a subcontract to assemble and test computer keyboards for IBM. Consequently, we needed a lot more space, and Mike set up the huge IBM operation in one of Henry Rowan's vacant buildings near Inductotherm on Indel Avenue. Sales and marketing also moved to Indel Avenue and into a nicely refurbished office suite. By then, we had added several additional salespeople, and I was promoted to industrial sales manager. We also acquired our own secretary who tried hard to please everyone, but failed with Mike—he hated her.

Life on Indel Avenue was good when I was in town. Tom McGarrity and I usually had lunch at a nearby restaurant on Route 38. Our production manager would meet us there, and we'd have a few drinks and a lot of laughs usually at Mike's expense. For several years, prior to the ill-fated

IBM subcontract, we earned good bonuses and prospered company-wide. In one of those years, I got Mike to agree to fund a lavish Christmas party at an area restaurant, a company first. I arranged it and acted as emcee as we honored various employees for jobs well done. I also made a slide presentation that showed what we'd done in the past and what we expected to do in the future. Everyone loved the party and the presentation, even Mike. But once he saw the bills, he decided not to have one the following year. Mike soon learned that trying to make money working for IBM was tough. Tom and I had been opposed to the whole idea, but Mike felt it was sour grapes as he had gotten the job himself. After a year or two of losses, Mike scrapped the IBM job, closed the Indel facility, and moved sales and marketing back to Lumberton. On Indel Avenue, we were located close to Rancocas Creek and suffered a number of severe electrical storms. During one storm, lightning hit the vacant building next to ours and totally destroyed its electric service panel and wiring. It seems to me that serious storms always moved eastward along the creek.

While we were on Indel Avenue, a small computer showed up in an otherwise empty room next to our office suite. I learned that it was part of a package of two Apple IIs and an Apple III that someone had sold to Mike Yurko for inventory control. Typical of Mike, he just plunged into the computer age without any notion of what he was getting into. He never asked for input from me, the only person in the company with a computer background. No one was using the nearby Apple II, so I adopted it. I quickly found that Mike had acquired no application software. It did have a Basic language interpreter, which I used to create various report templates. Once Mike realized that he'd bought a pig in a poke, he got rid of all the computers, even my Apple II. Whoever he was dealing with next sold him a big and expensive Data General minicomputer, which was installed in a room in the Lumberton main office. Again, there appeared to be no application software, and since it was password protected, I couldn't find out what was installed on it. While I was with Grim, I never saw any output from the computer, nor did I ever see anyone using it. Much as I enjoyed working with the Apple II, I never thought of buying one.

Other Family Doings

While Patrick, Kitty, and I were heavily involved with swimming and tennis, Chris was doing his thing, mostly in our garage and basement. From an early age he showed his mechanical skills—at age four, he helped me to build an enclosed workshop in the basement. He wasn't happy with his first skateboard, so he built his own out of fiberglass. He then started to make them for sale to other demanding skaters. Our neighbors often called on Chris to fix lawn mowers and bicycles. Chris and some of his friends started to trap muskrats in the nearby creeks. This meant that he had to get up and out before school to tend to his traps. Sometimes he'd be able to sell the complete rat, but often he had to skin it himself. We were not happy to have dead rats and rat skins in our garage. With the proceeds from trapping and other jobs, Chris bought his first moped. Kitty and I figured that he couldn't get into much trouble on a little bitty moped. However as time went by, we noticed that his moped was morphing into a killer bike. I soon realized that he was probably doing the same for his buddies and that Delran would soon have a plague of souped-up mopeds. I feared that we would soon hear from the police.

It didn't seem that Chris was in Delran High School for very long when cars became an issue. I don't recall how it came about, but Chris soon had his driver's license and wanted to buy a car. As was the case with the moped, Chris ran through a succession of cars, each more powerful than its predecessor. At the same time, he acquired lots of parts that he needed to repair his car and those of his friends. It seemed that our garage was turning into an auto chop and repair shop. Little Jimmy was a keen observer of the doings in our garage and in turn, would fall heir to the family auto repair heritage. The father of one of their friends owned a nearby auto junkyard that gave our boys easy access to lots of low-cost parts. I can recall that one weekend, Jimmy put three junkyard transmissions into his car until he found one that worked. I was amazed at the boys' skill with cars but very unhappy with the spilled oil and junk in our garage.

Meanwhile, Tommy had dropped out of high school and in 1980 earned his New Jersey High School Equivalent Diploma, known colloquially as a GED. Always a roamer, he hitched rides out west and ended up near Aspen, Colorado. There he worked in Snowmass Village, primarily a winter ski resort. On a business trip to Colorado Springs via Denver in the summer of 1982, I made a detour to visit Tom and stayed a night or two at the Snowmass Village Inn. It was my first experience of the high mountain passes in the Rockies, and it was most exciting. While allowed to stay at the inn's crew quarters, he wasn't able to find work in the off season. Unhappy with his job prospects, Tom came back to Delran and joined the U.S. Army. He went on active duty for three years in November of 1982.

Tom was assigned to the signal corps and took basic training at Fort Gordon, Georgia. From there, he went for advanced training with an artillery unit at Fort Sill, Oklahoma. His next stop was to an overseas artillery unit stationed in the mountains near Venice, Italy. There he was adopted by an Italian family and enjoyed many Sunday meals with them. He was selected to train so that he could represent his unit in the annual four days of marches from Nijmegen to Arnhem—see the film *A Bridge Too Far*. This July event commemorates the World War II liberation of the Netherlands. Tommy completed the march and received a presidential citation. We wanted him to remain in the army, but he took an honorable discharge in October of 1985. On his way back to the States, he visited family and friends in Northern Ireland. His visit occurred during the height of sectarian strife, and he had several run-ins with British troops and police. His U.S. Army ID usually sufficed to defuse confrontations.

Mother's high-rise apartment in Riverdale suited her very well. The boys liked to visit her because I would take them down to the Hudson River to see the boats and the trains on the New York Central's main line to the north and west; and they liked to ride the elevator. While living there, she participated in a Riverdale Neighborhood House sewing project to honor the national bicentennial in 1976. She contributed a panel to a quilt that was displayed in New York's Metropolitan Museum of Art. The museum sold a color print of the quilt, which I have—somewhere. However, she was in a very cold and icy spot come winter. While her building had an elevator, outside she had to negotiate a flight of steps to get down to the shops on Riverdale Avenue. Those steps faced east and got little direct sunlight, so ice tended to linger there and on the west side of Riverdale Avenue as well. After she had fallen a few times, my sisters and I resolved to have her move to New Jersey. A new subsidized senior citizen housing project was going up in Lincroft near my sister Patricia. So in 1978, Mother moved in with Pat while she awaited approval to move into the new Luftman Towers. To qualify for the towers Mother had to shed most of her remaining financial assets by making gifts to me and my sisters. It took a few years to do so in a way which did not incur adverse tax consequences. Finally, Mother moved into the towers in 1980. Until illness overcame her, she enjoyed life in her third-floor apartment. The towers had many amenities that suited her—cards, sewing, shops within an easy walk, and bus trips to Atlantic City.

Back to tennis—briefly. One day I read in the *Philadelphia Inquirer* that the USTA (United States Tennis Association) was holding the 1981

Girl's National Grass Court Championships at the Philadelphia Cricket Club. I'd never seen grass court tennis, so Patrick and I set out to see the matches. With considerable difficulty, we found the club, parked, and walked up to the nearest court with a match under way. We sat on the ground at the end of the court to enjoy the match and the wonderful ambiance. When the girl at the far end of the court came to our end, she started talking to us. She said she was from California and had never seen, much less played on, a grass court. It turned out that her name was Andrea Leand, and she went on to win the tournament and become a touring tennis professional. We strolled about the grounds and saw an enormous expanse of grass courts. They were arranged one after another, with no sidelines. To limit wear on the grass, the odd-numbered courts were used on day one and the even on day two, and so on. It was all like the setting for the movie version of *The Great Gatsby*. A few days later, we went back for the semifinals. On that day, we had to pay to park and see the matches but were able to view the center court matches from the clubhouse veranda, just as they did in the movie.

In the 1980s, Philadelphia hosted a major men's tennis tournament each year at the Spectrum, an arena at the sports complex near the old navy yard. Patrick and I were avid fans of John McEnroe, and so we went to see his matches for several years in a row. One year, we arrived at 1:00 p.m. for his singles match, which went to five sets and over four hours. John and his partner, Peter Fleming, were scheduled to play a doubles match that evening. We decided to stick around for that as well. John was given a brief respite between matches, and we got something to eat to hold us until we got home for dinner. When the doubles match got under way, John and Peter were just awful. We couldn't bear to watch, so we left early. Later that evening I learned that John and Peter had won the doubles match in straight sets, but all three sets went to tie breakers. They won 7-6, 7-6, and 7-6. That probably took another two or three hours. I don't recall how it all played out, but I hope they gave John a day off after his grueling six to eight hours on court.

During the 1970s, presidential politics had soured for me. While a registered Republican, I had pretty consistently voted for Democratic presidential candidates. In state elections, however, I often voted for the Socialist Labor candidate, Bernie Doganiero. He never won, but I liked his audacity in taking on the party hacks. One year I believe, I voted for a socialist presidential candidate, but I can't remember his name. In the 1972 presidential election, I did vote for Richard Nixon. He had done some good stuff and had gone to China, finally "getting real" and opening a dialog with that awakening giant. Thereafter, I missed a string of presidential elections because of business travel. I remember that Kitty and I were on a combined business-pleasure trip to Connecticut on Jimmy Carter's victorious election day. We were very annoyed because all the bars and restaurants were closed until the polls closed. I don't remember voting in any of the presidential elections during the 1980s, but I did vote for Bill Clinton in 1992 and 1996.

In May of 1982, Mother's sister Alice Quinn, a.k.a. Sister Jane Frances Quinn, had been professed as a Dominican nun for fifty years. A large celebration was to be held at the motherhouse, St. Mary of the Springs, Columbus, Ohio, and all of her relatives were invited. Although Mother wasn't feeling too well, she was determined to make the trip, so she flew out in the care of granddaughter Kathleen Crumlish. Others who made the trip to Columbus included Kitty and I, with our Patrick and Jimmy; my sister Pat and Jim Maguire; my sister Maureen Crumlish, with daughters Kathleen, Trish, and Dianne; my cousin Patrick Quinn; and my cousin Barry Quinn and his wife, Dolores. There may have been others, but they don't appear in any of the pictures.

It was a wonderful occasion, and we had a great time. We stayed in the magnificent new convent and retreat house where we had a great party after the nuns and young people went to bed. Our aunt was a seamstress who made beautiful nun dolls, which were sold to raise money. Her workshop overlooked the beautiful campus of the college, formerly St. Mary of the Springs, a girl's college, but now Ohio Dominican College, open to men and women. The southeast corner of the campus abutted a bad part of Columbus, but you'd never know it; the campus seemed truly idyllic. On the trip home, many of us stopped for a group dinner somewhere in eastern Ohio. It was a wonderful family experience, one of the best.

Kitty Goes Back to Work, and We Go to Quebec

With my earnings in a slump in the late 1970s, Kitty decided to seek part-time employment. She signed up with Manpower Temporary Services and had some short-term local assignments. In May of 1979, Manpower asked her to participate in a back-to-work seminar they sponsored at the Echelon Mall. Guest speakers included the late Jim O'Brien, who was an anchorman at ABC's channel six Action News. Kitty served on a panel with two Manpower executives. The housewives' back-to-work movement was really taking off, and Kitty became a role model. That fall, Kitty got a brand-new Ford V-8 LTD wagon, the biggest then available. It had power on everything, but everything soon started going bad. One of the rear window panels even blew out during a severe thunderstorm. It was to be the family's last of a long series of station wagons.

Popular as Kitty was at Manpower, they were not able to find many of the limited-hours, short-term assignments she wanted. Accordingly, she also signed on with J and J Temporaries. There she also became a model for the back-to-work movement. The *Burlington County Times* of May 29, 1982, had a feature article titled "Keeping pace with technology" and subtitled "Agency offers 'temporaries' advanced training." The picture accompanying the article shows Kitty at a word processor as Jack Malady, J and J's owner, looks on. The gist of the article was that J and J was offering advanced training for its "temps" and that Kitty was one of their first "guinea pigs." Speaking of Kitty, the article goes on to say: "'It was the easiest, safest, most pleasant way for me to make my reentry,' smiles the attractive mother who has since enjoyed six years in the employ of J and J Temporaries of Cherry Hill." Kitty is quoted several more times as she extols the wonderful training she received. The article was mostly fluff, but the underlying fact was that Kitty did indeed seem to be a pioneer in the women's back-to-work movement.

While "temping," Kitty always liked to have her summers off. In her free summer of 1982, we decided to take a motor trip to Quebec City, a place I'd always wanted to visit. With Tom in Colorado and Chris visiting in Wyoming, we only had Patrick and Jimmy along with us. On the drive north, we stopped at the historic and scenic sites around New York's Lake George, including Fort Ticonderoga. Once in Canada, we bypassed Montreal and went directly to Quebec. The city was built by the French as a fortress on a bluff overlooking the Saint Lawrence River. Towering over all is the famous Chateau Frontenac Hotel. It was rather upscale for us, but Kitty and I had always wanted to stay there, so we did. It is an ornate, huge, and spooky place that the kids enjoyed exploring. At dinner in the chateau, we had a table by a window that had a gorgeous view out over the lower city and the river. Next day, we took a tour of the city and found it to be truly beautiful—I'd love to go back and spend more time there. But we were soon off to Bar Harbor and Acadia National Park in Maine. It was a very long and lonely trip along the route of the famous ill-fated march on Quebec by Benedict Arnold in 1775. We stayed in Bar Harbor and toured the park, then went on to Portland Head lighthouse and on down to coast to home. Of course, we made the obligatory stops at L. L. Bean and other shops in Maine. I don't remember that we bought anything.

Back on the home front, Kitty's involvement with temporary agencies soon ended when she signed on as a "temp" with the Campbell Soup Company at their world headquarters in Camden, New Jersey. At Campbell's, she took a number of temporary assignments during the school year, but took summers off to be at home with the children and to enjoy the sun around the pool at the swim club. However, as she became more involved with the people at Campbell's, she became pressured to work full-time, especially as the company was planning to curtail the use of even its own "temps." She finally took a full-time position with the packaging group but with an opportunity to extend her normal vacation allowance. Kitty remained a full-time employee of Campbell Soup until her retirement in 1996. She even came away with a small pension.

A big family event in 1979 was the second marriage of Kitty's Uncle Vincent Francis Gegan. Skip had been married for forty years to Edith Burroughs, a real Southern belle with a son from a previous marriage. They always came to Carlin's for Easter; otherwise, we never saw them. Vincent enjoyed a few drinks with the family, but Edith would say "Vincent" in a rather stern and formal manner. Skip was a Wharton School (University of Pennsylvania) graduate and worked at the BLS (Bureau of Labor Statistics) in Washington, DC. He and Edith had a small but comfortable home in the northwestern part of the city. Not long after Edith's death, at age seventy-one, Skip married Dorothy Ann O'Connor, thirty years his junior. After their church wedding, our convoy made a wild and exciting ride through Washington to the reception at the officer's club at Bethesda Naval Hospital. It was a wonderful occasion; and we all liked the new bride who, I believe, had three daughters by a previous marriage. Sadly Dorothy Ann died in 1989 of cancer; she was fifty years of age.

More Grim Years

While I handled industrial sales to OEMs (original equipment manufacturers), Tom McGarrity handled government sales, primarily to the FAA (Federal Aviation Administration), our most important customer. Every year, a communications show was held in Washington, and Grim secured a booth. Tom and I decided to put on a better show one year and got approval from President Mike Yurko to have a booth professionally designed and built for us. Since it was my idea, I had to interview firms and get cost estimates. I was able to strike a reasonable balance between cost and design, and so we had a brand-new look at the show that year. It was a nice change from the utterly stodgy Grim look of prior years and was well-received, even by Mike. Tom and I shared a room at a large and grand hotel in downtown Washington. One evening, Mike came down, and we took our FAA contacts to dinner at a really fine restaurant at the Mayflower Hotel. I believe the restaurant was called the Bicycle Shop or the Bombay Shop, something with an Asian Indian name and décor. It was beautifully decorated in a Rudyard Kipling style and had great food. It was a wonderful evening.

Except for the annual show in Washington, Tom and I didn't usually travel together. However, the state of New Jersey sponsored a two-day export seminar at the old Robert Treat Hotel in Newark. Mike wanted Tom and I to go and submitted our names. We were not crazy about going to Newark, but it turned out to be a nice experience. During our free time, we got to walk around downtown and do some sightseeing. There were some beautiful churches, and the hotel itself was very interesting. We didn't gain a whole lot from the formal sessions, but it was good to know that the state was behind us. I later received a letter from New Jersey Congressman Peter Rodino of Watergate fame thanking me for my participation.

During a meeting of the Grim brass, I brought up the idea of using independent sales firms—known as reps—to represent us around the eastern United States. It was decided that we would try this, and I was assigned the task of finding and recruiting reps. I don't know how I did it, but I was able to locate a number of firms that were interested. I then traveled to their territories for face-to-face meetings prior to inviting them to visit our facilities. Not all prospective reps were willing to make the trip at their own expense, but that did not disqualify them. I was able to bring on board several reps that signed contracts and started to work for us. I was pleased when one of the reps secured a small piece of new business, with the possibility of more. When we booked the business, the rep firm expected the agreed-upon commission. Mike balked, saying that he had to cut his price to the point that he was unable to pay the commission. Naturally, that ended our relationship with that firm. When I learned about this, I thought it was typical of the small-mindedness of Mike and his vice president, Joe Fattorini. I was doing my best to drag the company forward, but they fought me every inch of the way. Since my personal reputation was now at stake, I called the other rep firms and told them that we had decided to cancel independent sales representation. So ended that ill-fated episode.

I made several trips to Melbourne, Florida, where I was developing a very promising relationship with the government systems division of Harris Corporation. A nice feature of a call on Harris was that they arranged hotel accommodations for you on the top floor of the Holiday Inn on Melbourne Beach. The hotel, the beach, and the area restaurants were wonderful, so I had Kitty accompany me on one or two trips. When Mike learned that I was making productive trips to Harris via Orlando, he asked to go on the next trip. I assumed that he wanted to help me to secure a foothold at Harris. This was not to be the case at all. He had bought a house in one of the new housing tracts in the Orlando area and wanted to check on it. Thereafter, when traveling with Mike, we made only perfunctory visits to Harris and rushed back to Orlando so he could work on his house. This seriously cramped my style—I couldn't stay on the beach and enjoy the amenities of the Melbourne area, and I wasn't developing any new contacts at Harris. One day, I prevailed upon Mike to visit the new upscale housing tract called Bay Hills. We went through the samples, and I asked many questions to the salespeople. I was very interested but realized that relocating to Florida, and to Bay Hills in particular, was not in the cards at that point. Mike became very upset, thinking that I was about to make a deposit on a house, one that would be so much finer than his low-end place. I will say that Mike was a good host—he cooked a nice breakfast for us every morning. Despite my best efforts, I don't think we ever closed a deal with Harris.

California Follies and I Get a Computer and Go Back to School

I had another ill-fated venture, this time in California. Somehow, we had developed a possible business situation at Vandenberg Air Force Base, north of Santa Barbara. I made arrangements for my associate Eileen Muzina and me to fly to Los Angeles and drive to Santa Barbara where we'd stay at the Santa Inez Inn. It seemed that we'd have no problem driving from there to the base for the meeting at 10:00 a.m. the following morning. It took almost thirteen hours to get from home to the inn, and we were exhausted. I called home and spoke to one of the boys who wondered where we had gone. I explained that we'd flown to Los Angeles, rented a car, drove north to Santa Barbara, and then on to the inn. He still found it hard to believe that it could take thirteen hours to get anywhere within the Lower 48. I didn't mention that we detoured to Santa Monica, so I could check out our old neighborhood. It was disappointing to see that high-rises now lined Ocean Avenue overlooking the Pacific Coast Highway. I couldn't get into our old apartment block as it now had a security gate, a sure sign of the times. At the inn, Eileen and I had a nice dinner and turned in for an early departure to the base.

Next morning as we prepared to leave for Vandenberg, we learned that many area roads had been washed out by heavy rains the previous week. We set out with our road map and soon ran into a road closure. Without a cell phone or an OnStar to help us, we took whatever roads seemed to go in the right direction and ran into road block after road block. By dumb luck, we arrived at Vandenberg at noon, two hours late for our meeting. It had taken over three hours to go some twenty-five miles from the inn. After lunch, we were able to see a few of the people we came to visit, but the key people were tied up and were annoyed that we hadn't shown up for the long-planned morning session. Unhappily Eileen and I returned to the inn to hang out until it was time to go to the Santa Barbara airport for an evening flight to San Diego. Fortunately, the inn had an "honor bar" that enabled Eileen and I to make first-rate martinis and relax in the empty lounge. There we learned that the inn occupied the former ranch of actor Ronald Coleman. The management had put together and set out a book of letters from happy and unhappy former patrons. Reading them easily sent us into fits of laughter, propelled no doubt by the martinis and our exhausted state. Eventually we pulled ourselves together and drove to the airport.

The evening flight to San Diego turned out to be just awful, partly because of the layover in Los Angeles. Worst of all was that we arrived in San Diego about midnight and could not find a restaurant that was still serving. On a guess, we took a taxi to the hot spots in the center of the city and found a big restaurant that was still open. Inexplicably, they had a long line—for dining at 1:00 a.m. Pleading starvation—we hadn't eaten for over twelve hours—we were allowed to go to the front of the line and, finally, get something to eat. Next day, we called on Logicon Corporation, an up-and-coming software house that seemed to be a good fit for our hardware facilities. The visit to Logicon was life-changing for me. The causal atmosphere—bicycles and cutoff jeans inside and tennis courts outside—was like a siren's call. I resolved to "go native" when I got back east. But the capper was that each engineer seemed to have one or more computers on his desk. Wow! We never got any business from Logicon, and it was soon gobbled up by a larger firm.

After my visit to Logicon, I resolved to dress casually except on special occasions and to get my own computer. I had read that IBM was planning to offer something new for them, a personal computer; but one that was aimed at the business market. This held more appeal for me than the Apple II, which seemed to be more of a hobbyist's computer. The IBM PC was introduced in the summer of 1981 and became an immediate sensation. It was backed by a clever television advertising campaign that featured the Charlie Chaplin creation—the Little Tramp. By reading the technical magazines, I learned that the initial machine only offered 65,536 (64k) bytes of memory for programs and active files and that this was really not adequate. Since Grim was not going to give me a computer, I decided to buy an IBM PC for myself. As soon as the upgraded model—256K versus 64K of memory—became available, I bought my first PC from a computer store in Cinnaminson. It was for my fifty-first birthday on April 13, 1983. I paid $4,500 for the computer, an Epson printer, and the IBM Disk Operating System (DOS). The latter came in a nice box with disks and two-ring binders of technical information that I wish I'd saved for posterity. I separately bought a Peachtree Software application suite that included word processor and spreadsheet programs. I thought I was set for life, what could be better than this. However, I soon found out that I would need a lot more hardware and software. In 1988, I added a 20-megabyte (million bytes) hard drive, probably the most significant hardware advance for PCs and a forerunner of things to come. The Peachtree software package, with upgrades, served me for many years, that is, until Microsoft moved from DOS and BASIC to a full suite of applications, and came to dominate the world of PC software.

With all bedrooms occupied, I didn't have any good place to set up my big clunky computer other than on our good dining room table. There wasn't anything that interested Kitty or the boys on the computer, so I had it all to myself. I was determined to become expert in all aspects of the PC, especially its programming. I started to write my own programs using the built-in BASIC interpreter. I soon found that its limitations were irksome and found the answer when the first version of Turbo Pascal appeared late in 1983. It was an integrated program development tool (an IDE) for the existing Pascal programming language. It compressed Word processor, language processor, and program generator in a single package, a relatively new concept. Best of all, on its debut, Turbo Pascal retailed for $49.99. I bought a copy and used it, with upgrades, successfully at home and later at work for many years. The programs created with Turbo Pascal were much more compact than BASIC programs, ran much faster, and were easier to modify. Night after night, I sat at my computer, smoking away and making a mess of our beautiful dining room. My fascination with my PC and my dedication to learning computer programming were shortly to have a big payoff.

In the summer of 1983, my sister Patricia flew to Ireland and met her daughter Maureen in Dublin. Maureen—or Mo, as we call her—had been backpacking in Europe with friends after graduation from Douglas, now a part of Rutgers. Pat and Mo spent four days in Dublin and then ten days in the Dungannon area of Northern Ireland. There they stayed in Gort with Josephine Casey, a friend of Pat's and a neighbor of the O'Donnell clan. Later, Josephine's daughter became a nurse and came to the States with her boyfriend for an extended stay. While here, she helped out with care of my mother while she was recovering from an operation for colon cancer.

The fall of 1983 was also significant in that Christopher started at Drexel University. He had expressed interest in mechanical engineering, and I thought I would help him by getting information on suitable schools. Because I felt that I had missed out on the full spectrum of student life by going to a local school, I steered him to more distant schools. I thought it would be good for us to be apart for a while, so my first choice was the University of Tulsa. Chris wouldn't hear of it, he wanted to go to Drexel and live on campus. Since he had done very well in AP (advanced placement) physics, they offered a decent financial package that included a Pell grant and generous student loans. We helped Chris pack and move his stuff into the freshman dormitory, wished him well, and set out for home. We soon learned that he would spend two quarters at school and in spring begin the first of his four co-op (cooperative work-study) assignments. Sadly, Drexel's financial inducements were to be scaled back in subsequent years.

The Grim Finale

Over time and out of necessity, I had to become more involved with engineering and plant operations. On a lead, I had secured a nice piece of business with Checkpoint Systems wherein we manufactured and tested the circuit boards that were the heart of their antitheft systems. The reason we got the business was because I promised that we would built some reel-to-reel handlers for their reels of radio frequency merchandise tags. These tape handlers enabled them to test a reel of tags by counting and marking (optional) the bad tags. Our engineers had no experience with mechanical design, so our chief draftsman and I designed the tape handlers to work with Checkpoint's tag testing electronics. I then worked on-site with Checkpoint's engineer to test and refine the setup. The final design was a success and enabled Checkpoint to determine the percentage of good tags on their reels. Thus, they were able to price tag reels based upon the percentage of good tags since even bad tags were usable—prospective shoplifters could not tell the difference. Checkpoint became our biggest customer until they opened their own manufacturing plant in Florida a few years later.

I became heavily involved with one of our plant operations out of necessity. One of my important prospects came for a plant tour and closely inspected our wave-soldering process for circuit boards. He determined that our process did not conform to industry and military standards. A major problem was that we were not using an approved solvent and were not disposing of the waste in an approved manner. When I investigated, I found that we were buying solvent in unmarked barrels and putting the waste into those barrels for disposal by some guy that came by in an unmarked pickup truck. I investigated and determined how we could bring our process in conformance and convinced Mike Yurko and Joe Fattorini, our VP and purchasing agent, to do so. We were soon using the approved DuPont solvent and had the waste taken away by a licensed contractor.

During this whole episode, I had high praise for Pedro, who ran our wave-soldering operation. He was mighty helpful in bringing the whole process into conformance, but he soon left Grim. I found out that he asked for a raise and was turned down, so he quit. He was retired from the U. S. Army, and while the job wasn't critical to him, his pride was hurt. It was typical of Mike Yurko that he didn't always value the contributions of individuals. He seemed more concerned about his bonus than rewarding the outstanding efforts of others. To me that's the downside of Grim's bonus system. While Grim had good outcomes from my increased involvement with engineering and manufacturing, Mike wasn't happy because it was me, not him or his ragtag engineering staff that won the laurel. Generally, to get things done, you had to convince Mike that the original idea had been his. It was not a formula for success.

With raises and bonuses, my take home pay steadily improved on into the early1980s. I know that Henry Rowan wasn't happy with Grim as we didn't seem to be a significant player in the booming world of computers and electronics. This probably accounts for Mike Yurko's desperate search for new worlds to conquer. He hired a lead microprocessor engineer, Jim Parise, and a supporting engineer, Manny Fard. For there first venture, Mike wanted them to produce a programmable electronic sign such as were appearing everywhere at the time. As newly appointed marketing manager, I decided to look into the prospects for such a product. I made a detailed study that showed that we could not compete with the cheap signs coming in from overseas. Mike was not happy that I shot down his idea. We never were able to come up with a product or service that suited the peculiar organization of Grim and Mike's desire to minimize investment. I never really understood what Jim and Manny were doing as my microprocessor know-how was minimal at the time. But Manny and I did play tennis together after work and that was a plus for both of us. Manny was an Iranian who had gone to engineering school in the United States, where he had played competitive tennis. He was much too young and good for me. Long after our Grim days, Manny would pick me out of a crowd to work with him.

I was coming on five years with Grim, and there seemed to be little or no possibility of further advancement. I had read in the company literature that all Inductotherm components offered company-reimbursed opportunities for advanced degrees. I thought it was a good time for me to get a masters in business administration (MBA) and that it might help me to move elsewhere within the Inductotherm empire. I studied and took the exam required for admission to graduate business programs at Rutgers and elsewhere. Apparently, I placed in the upper third of those who applied to Rutgers and was accepted into their MBA program at the Camden campus. I then approached Mike Yurko to get his support for my plans. He admitted that the company had such a program but would not approve my participation. When asked why not, he said he wouldn't do so because I would leave the company once I got my degree. I told him that it seemed to me that he was forcing me out of the company and that I would proceed without company funding. I guess that was the last straw for Mike, because not long after I started at Rutgers in the fall of 1983, he said he was letting me go. Fortunately, Tom McGarrity was aware of this budding break and had made some contacts on my behalf. Not long after my last day at Grim, I was hired by the RCA Service

Company at Cherry Hill. So this time, there was no need to head to the unemployment office in Burlington City.

Here's a historical footnote to my Grim experience. Henry Rowan of Inductotherm became very widely known in 1992, when he gave $100 million to Glassboro State College to establish an engineering school. The entire school was then renamed Rowan University in his honor. He was a graduate of MIT and started his business in his garage. According to Wikipedia, there are now some fifteen thousand induction furnaces in use around the world. Burlington County is buying the former Rowan estate for $4.5 million to preserve it for public use. That sum will be paid to the National Philanthropic Trust, to which Rowan and his second wife had donated the property. The 88-acre Westampton estate on Centerton Road reportedly has 1,800 feet of Rancocas Creek frontage, a 3,400-square-foot historic home, and outbuildings that the county plans to improve for public use. As far as I know, Henry, at age eighty-four, is living in an extended care facility.

Kitty and the boys early in 1971.

Kitty with Chris and Patrick alongside our doughty 1965 Ford
Fairlane wagon.

Kitty with the boys on Black Rock Summit on Skyline Drive in Virginia, 1973.

The boys on Boonesboro, Kentucky, historic site marker, 1973.

With support from Tom and Chris, Patrick goes off to his first day of school, September 1973.

Jim and Pat with Santa, 1973.

We are with Jimmy on top of Split Rock in the Poconos, 1974.

A leasing agent hands me the signed paperwork for my real estate office at Cooperstown Plaza on Sunset Road, Burlington Township, 1975.

I show one of my sales associates how to install a sale sign, 1975.

Tom graduates from Delran Middle School, 1977.

Patrick, age nine, and Brendan Bossuyt with Tenby Chase swim coaches, 1977. Head coach Ken Kennedy is on the right, and his assistant is Danny Pancamo. The swim team reached its peak during Ken's tenure.

The Astros soccer team and coaches, 1977. Jimmy, age seven, is standing next to me. This may have been my last year as a soccer coach.

The boys with Mindy, the world's best dog, 1978.

Chris "enjoys" a typical horrible Tenby Chase winter, 1978.
Parked cars would slide down the hill when it was icy and end
up in the street. Fortunately it was a cul-de-sac.

Chris raises the Sunfish sail, Sea Isle City, New Jersey, 1979. We rented the house with dock for several August vacations. The Oberschmidt boys—Leo and Michael—are standing on the dock with Patrick and others.

Our motorboat at Sea Isle City, summer of 1979. Patrick is up front, and Chris is at the controls. The boat was handy for pulling the Sunfish out of the lagoon and into open water.

Patrick swims the breast stroke for the Tenby Chase relay team.

The record-setting Tenby Chase relay team, summer of 1980. Patrick (at left) usually swam the freestyle anchor lap. The others (*left to right*): Matt Jerd (butterfly), Ed Socci (breaststroke), and Jeff Ogan (backstroke).

We didn't want Jimmy playing peewee football, but here he is dressed for a game, fall of 1980.

The group at the fiftieth anniversary celebration for my aunt Alice, Sister Jane Frances Quinn. It was held at the St. Mary of the Springs's motherhouse in 1981. *From the left:* Kitty, me, my cousin Patrick Quinn, Sister, our Patrick, and my niece Kathleen Crumlish.

Kitty was featured in a newspaper article extolling the training program offered by J and J Temporary Services, May 1982. Company officials look over Kitty's shoulder as she learns how to use a Wang word processor.

To run a swim meet required a lot of help from the parents, summer of 1982. That's me at the announcer's table at the far right rear of the picture.

Kitty and I are in our Sunday best, spring of 1983.

My boss Tom McGarrity (right) and I manned the GRM "booth" for the 1980 AFCEA (Armed Forces Communications & Electronics Assn.) show in Washington. Our booth was so pathetic that I had a professionally designed and built booth ready for the 1981 show. It was very well received.

Dinner at Carlins, circa 1983. Mom-Mom Carlin at right and
Aunt Etta Carlin in center.

Mother pays a visit in the early 1980s.

Tommy's official army portrait, circa 1983.

The Carlin home at 121 N Church Street, Moorestown. During Sunday dinners many of the nineteen Carlin grandchildren had an opportunity to play in the loft of the carriage house at the rear of the property. After dinner, the men sat on the front porch and watched the shore traffic go by. The home was sold to a law firm in 1984.

PART 6

Years of Big Changes (1984-1994)

Back to Cherry Hill

When I went back to RCA on January 3, 1984, I had more or less gone through a back door into the RCA Service Company (RCASCo). I wanted to go back to engineering in Camden and had submitted a resume to them through a contact I had in engineering management. Typical of the slow pace of the personnel—now human resources—people, I had already accepted the service company offer of the position of senior engineer when Camden finally contacted me to set up an interview. The service company was originally established to service RCA products such as console radios and television sets and subsequently, RCA's computer systems. They later branched out to provide government services, initially in support of RCA military radios and radars, and later for any kinds of services needed by the government. By 1983, they had gotten into the telephone business, supplying complete turnkey private phone systems for businesses. They had set up a telecommunications services department to bid on large government contracts in that rapidly advancing sector. It was this latter group that I joined. The really great thing about the job was that it was in the RCA Cherry Hill campus, always my favorite place of work.

My boss, whom I will call John, was vice president and manager of this new high-powered—for the service company—engineering group. He was a florid Irishman who was an ex-military and had worked with my former Grim boss Tom McGarrity at Philco. My first assignment was to manage the EISN—Experimental Integrated Services Network—contract that Camden had taken to supply engineering services to the army at Fort Monmouth in New Jersey. On my team, I had an excellent technician, whom I will call Ray. Fortunately for me, Ray was up-to-date on the latest telephone equipment and instrumentation—it was almost twenty years since I'd worked in telephony. Ray and I made frequent trips to Fort Monmouth to monitor project progress and get new assignments from the project engineer. We only got paid for the tasks requested and approved by the project engineer, so Ray and I worked hard to think up new tasks for ourselves. On the way back from the fort, Ray and I always had a nice lunch and a few brews at the Ground Round in Eatontown.

The overall EISN project was to integrate satellites into the national telephone systems so as to improve post-nuclear survivability. It was a very sophisticated concept and involved a high-powered technical team from major universities, including Lincoln Labs at MIT (Massachusetts Institute of Technology) and the University of Arizona. Tucson. The satellite antennas and related terrestrial telephone equipment were located at several sites, including Fort Monmouth and Fort Huachuca, near Tucson. Ray and I got to make some excellent trips to Fort Huachuca, one of the most interesting places I've ever been. The beautiful original barracks and parade grounds have been preserved as designed by General "Black Jack" Pershing prior to World War 1. Though the fort is located in the desert, it has a most ingenious water works. I enjoyed staying at the Thunderbird Inn outside of the fort in the town of Sierra Vista. One evening, Ray and I drove south out of town and stopped at an Italian restaurant done in the local desert motif. It turned out that the owner was born in Italy and had worked in Philadelphia for many years. His goal was to make enough money to go back to Philly and open a restaurant there. He treated Ray and I like family, and we had a wonderful time with him. In addition to visiting Fort Huachuca while in Tucson, I always visited the team members at the beautiful university.

Another wonderful site I got to visit was the NOAA (National Oceanic and Atmospheric Administration) Research Center at Boulder, Colorado. Their facility was right up against the Front Range of the Rocky Mountains. We got to visit their huge computer facility with its Cray supercomputer. One night, we had dinner in a wonderful steakhouse well up in the Rockies alongside a babbling brook. However, the army only approved my trip to NOAA; poor Ray had to stay home. However, our most important trip was to MIT's Lincoln Labs at Cambridge, Massachusetts. The lab's team was responsible for the special software needed for speech processing and switching between terrestrial and satellite transmission facilities. They were developing their software using the C programming language. Until that time, I was doing my personal software projects using the Pascal programming language. While at MIT, I saw that C, developed by Bell Telephone Laboratories for the Bell System, would be important to my future advancement. Consequently, I bought a C compiler—language processor—for my IBM PC and made a concerted effort to become proficient in the language. It was to pay off handsomely.

No one at RCA was happy with the terms of our "as needed" contract. Furthermore, the army seemed ill-suited at the contracting agency for this very advanced project. John, our VP, set up a meeting with the army at Fort Monmouth to see what could be done to get a more meaningful contract, one that would allow us to assign to it full-time personnel. When Ray and I got to the fort, we found that John had brought along his good-looking young secretary, whom I will call Linda. During the meeting, it became clear that Linda had no idea what it was all about. The lead army engineer suggested that he give her a tour of the site after lunch. John said that he had to leave immediately. I said no problem; Linda could ride back with Ray and me. At that, John exploded, saying that she would ride back with him, right now. Not many days later, John disappeared. It turned out that Linda had reported him for sexual harassment, and he had been summarily fired. I suspect that the situation had been brewing for some time and that management was really quite happy to get rid of John for reasons you will see shortly. I think John had a serious drinking problem as well.

While not working on EISN, I assisted the telecommunications engineering group by joining in their proposal efforts. The one that broke the camel's back was for the army's Dugway Proving Ground in Utah. They wanted a fiber optic system to upgrade communications on that huge base. We understood that we would be competing with AT&T, whose lightwave system we considered old-fashioned. While at Grim, I'd taken a company-sponsored course on fiber optics technology and marketing, and that background was helpful on the Dugway proposal. Our team proposed a really high-tech solution that was easily upgradeable to provide more channels than were needed in the first round. I was personally responsible for the choice of the fiber optic equipment manufacturer and the specific terminal equipment that we would use. We then put together our proposal and held a final review with top management. It did not go well. Management saw that the high cost of purchased equipment far outweighed the possible return from RCA-provided manpower. From their body shop point of view that made the project highly risky and convinced them that our engineering operation was not consistent with their business model. Because of all the work we'd done on the proposal, we were allowed to submit the bid. We lost it to AT&T, the army's choice all along. They said our proposal was excellent, but . . .

I had been preparing the way for transfer to Camden for some time, and the outcome of the Dugway effort fueled my desire to move. As I said earlier, Camden had fobbed the EISN contract off to RCASCo. The secure voice business unit in Camden that actually held the contract was headed by Don Parker, and his engineering manager was Bob Chan. I had known Don back in the early 1960s when he headed the advanced technology unit in Camden. I had made a point of keeping Don and Bob up-to-date on EISN, and Don occasionally came to Cherry Hill for my management presentations. In private, I convinced Don that RCA should not extend the contract on the present terms and just let the contract expire. I then requested transfer to engineering within his business unit. Don went along with my ideas and arranged for my transfer with RCASCo management. I'm sure they were happy to get me off their payroll and onto his. The transfer was consummated sometime in the late summer of 1984.

Family Matters

While I was getting settled at RCA, our extended family was undergoing great changes. Our son Patrick had started at Holy Cross High School, Delran, in the fall of 1982. He initially became involved in the swimming and soccer programs. In time, he decided to focus totally on soccer and eventually made the varsity. I used to sneak out of work early to attend the games. Soccer was a fall sport; and the great weather, the turning leaves, and everything about the games was exhilarating. I did miss a few of the faraway games but pretty much saw all of the others. It was a great experience.

Also in the good news category, in 1984, Chris completed his first year of mechanical engineering at Drexel and got a co-op assignment at RCA in Moorestown. However, he didn't like the assignment because they didn't let him do anything significant or challenging. Consequently, at the end of the assignment, he asked not to go back there. Also, late in 1984 Kitty's younger sister Joanne and husband, Phil Hoops, had their first child, a boy they called Jeffrey Carlin.

But the really bad news was the death of Kitty's parents early in 1984. Her father, Joseph Carlin, known as Pop-Pop to the children, died of congestive heart failure at home on January 30 at age eighty-three. He was diabetic, and his health had been failing for several years. His wake and funeral were attended by Knights of Columbus in full regalia. The funeral was held at the family church, Our Lady of Good Counsel in Moorestown. The burial was at the parish's Mount Carmel Cemetery. All the grandchildren were in attendance. Joe was preceded in death by his eldest sister, Mary Brigid, in 1980 at age ninety. She was known to us as Aunt May, a bright and remarkable woman who had worked for sixty years at the West Jersey Title Company in Camden. Marie Carlin, Kitty's mother, suffered from late-stage diabetes and died at Zurbrugg Hospital in Riverside from a heart attack on February 9, only ten days after her husband. Mom-Mom was only seventy-nine years of age. I remember visiting her in the hospital and watching her heart monitor show a slowly fading heartbeat.

The illness and death of Kitty's parents ended a long and wonderful run of Sunday dinners at Carlin's. They usually had up to fifteen grandchildren and ten adults for dinners most Sundays, not to mention frequent guests. Pre- and postdinner adult beverages ran the gamut from martinis and manhattans to Drambuies and Gran Marniers. We thought nothing of drinking heartily and driving home, even to Pennsylvania. Over the years, the children loved getting together for Sundays at Mom-Mom's where they played in the attic and up in the loft in the carriage house-garage. Later, Joanne would add four posthumous grandchildren, for a total of nineteen.

The Carlin house on Church Street was quickly sold to a law firm that operated out of the house next door. It was with a real sense of loss that the family gathered to empty the house. The girls, the grandchildren, and their friends took what they wanted, leaving plenty behind in the house as well as in the garage. We got the biggest dumpster that would fit in the driveway and started pitching stuff into it from three stories of windows. For some reason, Joe Carlin had saved many old automobile tires and we had to carry them from the garage and lift them into the dumpster. It was difficult leaving so much behind, including things that our children had used and played with. But none of us had places big enough to take more than a few of the smaller items. So ended an era.

On a pleasant note, in January of 1984, Kitty was recognized by the Campbell Soup Company as "Soup Month Souper Star." The accolade was given during National Soup Month—January—and the celebration of the eightieth birthday of the "Campbell Kids." Some years later, Kitty earned another Campbell's "award" as "Packaging's Best Dresser and Shopper!"

Back to Camden

On the appointed day, I showed up at the office of Bob Chan who was engineering manager for the voice/data systems business unit. I knew of Bob during my first tour of duty at RCA but had never met him. He may have heard of me as well. As we discussed my assignment, he strongly recommended against my going into a software management position, saying those young guys would run rings around me. That notion was to prove dead wrong and really betrayed Bob's lack of software savvy. I agreed to take a "hardware" engineering position he recommended, and it turned out to be winner for all concerned. While I waited for my Secret clearance, I settled into a cubicle at the riverside end of building 10, not far from the very place that I had occupied in 1956. Murray Rosenblatt was manager of the engineering team on a final proposal effort for something called FSVS (Future Secure Voice System). He was located inside the secure "tank" at the other end of the second floor, so I required an escort to go to his office until I got my clearance.

Murray dumped on my desk a pile of documents describing the requirements for two FSVS components, something called a Secure Gateway, and the other was a cryptographic Key Distribution System. He also gave me the relevant sections of the preliminary proposal RCA had made to the government, in this case the NSA (National Security Agency). I had always worked in the realm of commercial electronics and had to learn the curious ways of government contracting. I quickly learned that it was a bad form to ever mention the supersecret NSA; they were simply called the government. Murray's desires were not always clearly stated, but I got the impression that I was to come up with a final proposal for the two components mentioned above. To help, he assigned to me Barry Lada, a senior hardware engineer.

Although neither Barry nor I had ever worked in the field of cryptographic or secure systems, we did have some experience with voice and data systems. It turned out that we were ideally suited to the task at hand. We analyzed the requirements and determined that the preliminary designs of both components were poor and did not reflect the state of the art. Barry felt that the front end of the gateway should consist of an array of microprocessors, some two hundred in all. That array would be supervised by a single minicomputer. I could see that Barry's design would be superior to that proposed by the original systems engineer Günter Ludwig. When Günter heard we had scrapped his approach, he complained to anyone who would listen that microprocessors were no good and would never do the job. Thankfully, nobody paid much attention to him.

While Barry finished his design of the gateway, I worked on the key distribution system, a set of networked computer systems. I found that Günter's design called for far more equipment than traffic analysis indicated. I was able to pare down the design and put together a more comprehensible system. Barry and I then put together our cost estimates for manpower and material and turned them over to the "bean counters" who worked up the final cost to the agency. With management's focus on other more cost-sensitive aspects of the overall project, our design and costs were never challenged and went directly into the final bid package.

By now, I had gotten my Secret clearance and had moved into the "tank," the project's secure area behind a single cipher-locked door. I was advised that I would be a member of the proposal team that would go to NSA at Fort Meade to make the final "pitch." At that time, I learned that the key distribution component had already been awarded to GTE (General Telephone and Electronics) and that I would simply pitch the gateway. I was very disappointed that part of my elegant design effort was wasted, but I was relieved of a great deal of complexity, and I could focus on the gateway. On the big day, our large team drove south to the fort and one by one pitched our pieces of the overall system. I made a confident and elegant presentation of the gateway near the end of the all-day show. The engineering types were then excused, and our management met with theirs to come to terms. On the way home, I asked my boss Murray whatever had been decided about the gateway. He seemed surprised that I asked and said, "You sold it to them. You've got your piece of the system." Barry and I were in business to design and build what would be called the MGS (Manual Gateway System) or gateway.

One of the first things I'd noticed when I got back to Camden was the presence of women in engineering positions; there were none when I left in 1963. The female engineers were mostly in software positions so that, at first, I had no occasion to work with them. When I did, I was uniformly unimpressed. Also, upon my return, I found that the clerical pool consisted of administrative aides, not clerk-typists or secretaries. It turned out that this was a position created to get around the union's hold on the older positions. I was always fussy about my letters and reports and found myself asking my aide to redo my work multiple times because of misspellings and typographical errors. Finally, one day, she said that I should ask Lisa Ireland to do my work. I asked why, and she said that Lisa had gone to Catholic school, as though that explained everything I needed to know. I was more than happy to have the lovely Lisa doing my work. In time, this problem went away as we got computer terminals and then personal computers, that enabled us to do our own reports.

My First Project

With the job in hand early in 1985, I went to Murray and then Bob Chan to get the engineers I needed to staff the project. Typical of RCA, even though we'd promised NSA that we had a crew of technically qualified engineers on staff, everyone was otherwise assigned—it was during the height of the Reagan defense buildup. Therefore, it became my

additional task to go into hiring mode. Because of the shoddy support I
got from the people in human relations, I had to personally go down and
look through resumes every day. I brought a number of prospects in for
interviews and wined and dined those I wanted over lunch at Victoria's
Station in Cherry Hill. The ones I didn't care much about, I took to the
prescribed lunch at the company cafeteria. I was fortunate to find Ron
DiUbaldi for the interface microprocessor work and Craig Martin to give
him a hand. Craig had worked his way through Trenton State College after
serving in the air force and had a young family. Although his engineering
credentials were weak by RCA standards, I thought he was an excellent
young man and took a chance on him. Barry, Ron, and Craig were the
core of my team as we undertook this really challenging project. I would
add other engineers for specific parts of the system, and eventually, I had
a crew of seven, plus mechanical engineers for packaging and equipment
layout. It turned out to be a highly skilled and dedicated team. At the end
of the project in 1988, everyone wanted my guys for other projects.

While my engineers did their thing, I found that I had to be involved
in an amazing, at least to me, number of administrative tasks. The first
major task was to decompose the whole multiyear project into small steps,
ones that could be completed in a month. These small steps were called
inchstones. Since we were on a "cost plus" contract, we needed to report
accomplishments every month in order to be paid. To me, it was new and
bizarre, but I got help from the administrative people, particularly from
the loud, pushy, and amusing Eli Gevirtz, deputy program manager. Each
month I then had to report at the monthly program review whether or
not I had accomplished the inchstone and if not, why not. I also made
several trips to the NSA to help the project's management evaluate and
cost estimate scope changes, usually a way for us to get more money.

I soon ran into another startup problem—the lack of equipment for
microprocessor design and test. RCA's budget for capital equipment was
very thin, and it took a year or more to get anything out of the company.
I quickly learned that the trick was to buy or lease what was needed with
contract funds. The government was not happy about this, feeling that
RCA should have invested in the necessary equipment in order to be
a player in the new world of microprocessors. Of course, the field was
evolving very quickly, and one never knew what would be needed until the
project was under way. If the equipment was for general purpose, it could
be purchased and shipped to the government at the end of the contract.
However, most of what was needed was very specialized so that all of

the engineering teams on the FSVS project leased an enormous amount of expensive equipment, particularly Hewlett-Packard ICEs (in-circuit emulators). Since we kept the leased equipment for two or three years, we paid for it many times over. But that's how the game was played.

In addition to project responsibilities, I also had personnel duties that were often unpleasant. Many times I had to interview job candidates for open slots on other projects in order to help management make hiring decisions. Over time I found that if I turned a candidate down, he didn't get hired. We liked to hire engineers from Drexel Institute of Technology—now Drexel University—because they already had hands-on experience from their co-op assignments, often enough at RCA. I remember being asked to interview a young woman who had a dual major, electrical engineering from Penn (the University of Pennsylvania) and liberal arts from Oberlin, a prestigious school in Ohio. She was fun to talk to, so I took her to my preferred lunch spot, Victoria Station. While there, I realized that there was no way that she could fit in with the rough and tumble ways of engineering life at RCA, so I had to turn her down. We had found that even women from Drexel had difficulty adjusting to our ways.

Another onerous duty was evaluating our departmental engineers every year on a "global" basis. This came about because the engineer's union—ASPEP (American Society of Professional Engineering Personnel)—had negotiated a system, whereby raises were pegged exactly to the rating given the engineer for the performance review period. Thus, if an engineer was given a rating of 680 out of 760, his raise was a specific amount in dollars. The performance review period varied, it was twelve months for a senior engineer and fifteen months for consulting engineers. Staff engineers and unit managers were not in the bargaining unit and were usually reviewed every fifteen months. In evaluating engineers, the departmental managers meet to do a global ranking of all of the engineers so that the result conformed to a normal distribution—some high, some low, and most in the middle. While the rankings were supposed to be global, the system forced me to drop rating points from one of my guys to reward another; essentially, a "zero sum" game for my small group. It was a poor system because most engineers who lost points filed grievances and got their points back because we were not able to justify the decrease. The grievance process was fairly speedy, but still took time on the part of the engineer, the union, the responsible manager—me, for example—and human resources. The latter had no idea how the system worked and just wanted the file closed, so they gave back the points. At grievance meetings, I always said that the engineer did a fine job and everyone—except the complainant—avoided asking why his points had been reduced. Bob Chan, our departmental engineering manager, always gave me hell if he had to give out extra points on my account. He wanted his bell curve to be just perfect.

Speaking of Bob Chan reminds me that we had a new FSVS program engineering manager, Phil Liefer. Phil succeeded Murray Rosenblatt who was moved to the program office. The idea was to forestall Murray's tendency to add new and improved features, thereby delaying completion of the job and exceeding the budget. New to RCA, Phil had five unit managers reporting to him, each of whose assignments was technically complex and on a tight schedule. It seemed that Phil had been brought to RCA and assigned to the FSVS project without the blessing of Bob Chan. Consequently, Bob treated Phil very badly, so badly that Phil always asked me to go with him when Bob summoned him to his office. Phil had observed that Bob did not abuse him when I was present. I enjoyed working with Phil, but his days were numbered; Bob fired him as soon as he could and replaced him with Bill Lawrence, one of his own men. In consoling Phil, I explained that RCA was a very clannish place and tended to reject outsiders, especially in midmanagement positions. Phil wasn't consoled, but I think it was good for him to be out from under the lash of Bob Chan.

I Get My MBA and Become an AA

I reached a personal milestone in 1986, when I graduated from Rutgers with my MBA—master of business administration. When I got back to RCA, my continuing educational expenses were assumed by the company, a very nice perk, indeed. The three years I spent nights at Rutgers were interesting and rewarding in several ways. First of all, I had never been in school with women; and they were close to a majority in the school of business. After a semester's final classes, we usually went out to celebrate with a few drinks and a few dances. On one project team, I had a woman who had been born in Russia, although you wouldn't know it right off. She was a great gal but not much of a contributor to the project. I had said to her that I didn't think premium vodka was better than the less expensive—really cheap—brands. For carrying her through the project, she gave me a bottle of Stoli—Stolichnaya—a famous Russian brand. As she suggested, I put it in the freezer and took a nip now and then and quickly finished it off. It went down very easily and convinced me that she was right.

I enjoyed most of the classes, particularly those that had a mathematical component. My favorites were the accounting courses, and I took the optional cost accounting, which made me, in essence, an accounting major. Most Rutgers graduate students avoided cost accounting and any course that involved mathematics. Although I was interested in finance and statistics, the instructors were so poor that I just did the minimum to earn my As. We had really fine professors for economics and international business, so those courses were very good. In fact, managerial economics was my first graduate course, so it became my first final. I had never studied with another student, but one of my classmates asked if we could work together. He came up to the house and asked how I thought we should prepare. I suggested that we agree on the ten most likely topics and study only those. My scheme worked to perfection. Of the ten topics we studied, nine appeared on the final and my partner and I got our As. There was no real need for me to get an MBA; I just did it for the fun of it. I took my degree in absentia, almost exactly thirty years after my bachelor's degree.

Over time and with an unhappy Kitty on my hands, I had to face the reality that I was an alcoholic. One drink would set me off on a path to oblivion, and I organized my daily schedule around opportunities to imbibe. I had convinced myself that all would be well if I ever got back to an engineering position, preferably at RCA. Well, I was back and responsible for a dream project, but I was drinking more than ever. On school nights, I rushed home and had a quick dinner and then off to Rutgers, with no time for my usual martinis. One night, I took a near empty bottle of scotch with me and finished it off in the car on the way to class. When I got to the Rutgers parking lot, I couldn't find a trash can for my empty. It struck me as awful that they didn't have a trash can for serious drinkers such as me to discard our empties. I had to leave the bottle standing near a light post, out of the way. Homework kept me up late many nights, so with everyone in bed, I was able to sneak drinks and smoke into the wee hours.

In March, while sitting in the kitchen one evening, I reached a personal crisis. For some reason, Chris was there with me and saw that I was distraught. He asked if he could help me somehow. I said yes, I'm an alcoholic; get me to AA (Alcoholics Anonymous). He made some calls and found a nearby meeting for the following night. The Delran chapter met in the Methodist church on Conrow Road. When I started going there, I found that it was an absolutely perfect group for me. In the group were priests, lawyers, and other professionals with whom I could relate. I remained with the Delran chapter for many years and became chairman when we moved to the basement of the Holy Name rectory. Smoking was the cause of our ouster by the Methodists; and shortly after we arrived in his basement, Father McGovern said we could no longer smoke there either. This became a real problem as drinkers were usually smokers and had to duck outside to light up, disrupting the meeting. Fortunately, I had given up cigarettes by then, but that's another story.

Because we were encouraged to attend meetings every night and help out elsewhere, I started to travel to more distant meetings. Often times, guests were asked to give a talk at those meetings, and I worked up my spiel so that it went over very well. I made some visits to detention centers for young offenders and really clicked with them. But my favorite place was the big Moorestown chapter that met in the basement of the beautiful Episcopal Church. Actually, at Moorestown, several meetings went on simultaneously, including a women's session and one for Al-Anon (for spouses and children of alcoholics). The main room was reserved for the large men's chapter, more than twice as large as Delran's. This group included some very well-known local sports stars that became hooked on alcohol as their careers were ending. But the best thing of all at Moorestown was the men's beginner meeting, which I called boot camp for alcoholics. I often chaired that meeting and adopted a drill sergeant—no nonsense—approach that served very well. What worked so well for years at Moorestown fell on bad times because of pressures to accommodate new twelve-step programs such as NA (Narcotics Anonymous). The men's sessions were cut back, and I stopped going there after several really good years.

I concluded my association with AA when I served out my term as Delran's chairman. I felt secure enough in my sobriety to stop going to meetings. However, we are obliged to stay on to help others, and I did so for several years. The courts were dumping drug addicts on the aforementioned NA in lieu of jail. Many addicts didn't find the NA meetings very helpful and started coming to our meetings where they didn't fit in. They were just not mature enough for the AA approach. They tended to wander in and out for smokes while they waited for the meeting to end so that I would sign their parole cards. AA was for people who quit drinking and wanted to stay sober. The drug addicts who came to us were, for the most part, still using. It was not a happy situation, and I didn't want to deal with it any longer. By that time, most of the people I had known at the outset of my AA experience were dead, back drinking and would soon be dead, or were safely off enjoying a life of sobriety. They were no longer around, and I missed them.

We Become GE, and the Project Ends

Early in 1986, rumors of a merger with General Electric Company (GE) were confirmed in a lengthy handout from Robert Frederick, RCA's president and chief executive officer. In June, the "merger" was formally concluded, and the process of integration was started. All of the defense business units were combined and placed under John Rittenhouse of RCA. It soon became apparent that the "merger" was a fiction. In fact, GE had taken over RCA, and Bob Frederick made a pile of money out of the deal. We RCAers were then subjected to continuous brain-washing—the GE way was the best. RCA managers were welcomed into GE's Elfun Society, which offered some perks, including preferential treatment by the Electric Insurance Company. It turned out that while they were welcoming us on one hand, they were selling off RCA's assets on the other—gone were RCA records and RCA televisions. They quickly eliminated the 4 percent company contribution to the manager's 401K plans; essentially, giving us a pay cut. In time, they eliminated the unit manager position entirely but asked us to do the same job. At that point, I chose to be an individual contributor, that is, not a team leader. But that was in the future, long after successful completion of the FSVS project.

All the takeover turmoil and management changes did not deter my team from completing the gateway on schedule and within budget. In many ways, we exceeded the requirements. We were the first to employ fiber optics in a product. In so doing, we eliminated the messy cables we would have needed for our many input-output data ports. Barry Lada had proposed to me that he write a monitor program in the C programming language for our controlling minicomputer. It was to allow an operator to view activity on the ports and record activity levels for various reporting periods. I gave him the go-ahead and during our final demonstration to NSA, the monitor proved to be a big hit; and Barry got some well-earned praise.

In the latter phases of the project, we hit a roadblock when we needed a speech generator. This was an analog audio component, and all of my engineers were digital. At about that time, RCA/GE was shutting a division, which freed up a number of capable audio engineers. We acquired a senior engineer named Tom who was able to make a successful design for us to integrate into our system. But I recall being very annoyed with Tom because he failed to complete some elements of the job. His attitude was that he had never done anything like what I expected of him. My attitude was that my senior engineers were expected to do whatever it was and were doing it without complaint. His foot-dragging soured our relationship and led to my dropping his rating and him filing a grievance. It was the first time I had such a disagreement with anyone in any of my groups.

As the gateway wound down, I had to reduce my team. First to go was Barry. The desperate need within the overall Camden operation was for software people, and there was a great need in another building. I convinced Barry that this would be a great opportunity for him to grow, and with misgivings, he went along. He was very successful in his new assignment but thereafter left the company to start up his own business. I placed Craig Martin in a software position as well, and years later, he told me that he was very glad that he had taken my advice. Ron DiUbaldi stayed on the FSVS project after I left and remained with RCA for several years thereafter. He and I often played tennis after work until he moved to a job in Pennsylvania, closer to his home. As the rest of my team scattered, I issued a final project report in January of 1988. However, by then the FSVS project had taken a new direction, and the gateway was no longer needed. It was consigned to storage at NSA. I was to learn that this kind of unhappy ending was a common occurrence in government contract work.

While on the FSVS gateway project, I received a notice from human resources that I could buy my way back into the pension plan and get my continuous service date adjusted. To do this, I needed to come up with $21,000 in cash. I talked to Murray Rosenblatt, my rabbi at RCA, and asked what he thought of the proposition. He said in no uncertain terms that it was the deal of a lifetime and that I should go for it. Fortunately, we had gotten our share of the Carlin estate, and I had the cash to invest and did so. In January of 1987, I received a notification that I was back in the pension plan with a continuous service date of July 30, 1969. This would prove to be the best investment we ever made. Murray became unhappy with his administrative position and retired around 1989. He was a very good engineer and had exceptional mathematical ability. Management's problem with him was that the design never ended while he was on the job. He and his wife live in a Moorestown condo, and after retirement, he started teaching calculus at the Moorestown Friends School. I'm sure he did very well in that capacity, and may well be doing so even now as he is still listed in the phone book

Chris Graduates and Goes to Work

In June of 1988, we had our first college graduate when Christopher received his mechanical engineering degree from Drexel. Actually he didn't receive his degree on graduation day due to misadventure, as follows. In his senior year, Chris was on the team whose project was to design and build an amphibious car. The car would then be entered in the annual competition among engineering schools. Each school was given the same engine—a flat Briggs and Stratton. Chris devoted many hours to the project and volunteered to take the test drive down a ramp and into the Schuylkill River. Voila, it floated! The team took the car to Canada for the trials, and to Chris's annoyance, someone else was elected to be the driver in the competition. In spite of poor site reconnaissance, they did well—second overall and first in design for their unique composite body. The following year, Chris was asked to come back and help that year's team and go to the trials at the University of West Virginia. Because his project team had focused so much on the car itself, they had failed to submit their final report on time. Once it was completed, they received their degrees.

College senior year in my time was primarily devoted to interviewing and finding a job. I was fortunate that 1956 was an excellent year for graduating electrical engineers, so I only interviewed four places and had three offers. The other firm I had told that I was not interested since it was out on Long Island. Christopher frustrated me no end because he made no effort to find a job, even after five years at Drexel. He hadn't even prepared a resume and had no interviews; the car project seemed to be his all-consuming interest.

Tom Meade had been in the Delran schools with Chris but had moved to Pennsauken before their high school years. Chris and Tom stayed in touch throughout their school years and afterward, up to the present. Tom's parents were divorced, and his father had moved to Landers, Wyoming. One summer, Chris and Tom drove out west to spend several weeks with Meade Sr. Tom had graduated from Penn State a year before Chris, also with a degree in mechanical engineering. He was working for a Camden firm, Joseph Oats Company, fabricators of huge tanks for chemical plants. Tom wanted to leave Oats in order to resume a business he had run while at State College where he built furniture suites for college dormitory rooms. When a job opened up at Oats, Tom recommended that they consider Christopher who was interviewed and hired immediately—one interview, one job. I'm not sure that he even had to prepare a resume. Tom soon left Oats to go back to State College.

After several years in the tank business, Chris took a sabbatical. He went out to State College and helped Tom set up a factory to produce his line of low-cost furniture. Creating the necessary fixtures and machinery was fun for Chris after years of designing and building ever-larger tanks. In fact, many of the tanks were so large they had to be loaded onto barges for water transport to the customer. He had even taken business trips to Japan and Germany in pursuit of new business. Anyway, I thought Chris would enjoy living in State College, which is in great country for hiking, fishing, and hunting. However, he didn't like its remoteness from family and friends and came back to Delran once they'd gotten the factory up and running. Kitty and I went out to visit Chris, and Tom gave us a plant tour. His business seemed to be booming, and he was occupying two substantial factory-warehouse buildings. Tom was a classic workaholic and had little time for socializing, much to the dismay of Gina, his girlfriend.

Carol, Tom's mother, ran an outlet shop in Cherry Hill for his products and some others furniture lines. A new baseball park had been built in Camden for a minor league team called the Riversharks. Appropriately, the stadium was called Campbell's Field, as it was built on the site of torn down buildings of the Campbell Soup Company. Carol and Tom took what's called a skybox as a promotion for their business. The second opening night, we were invited to the game, and it was a wonderful occasion. The ballpark was beautiful and its setting on the riverfront was spectacular, and the game was exciting. Among those in Tom's skybox was Bill O'Donnell, engineering manager of another Camden tank fabricator related to the Joseph Oat Company. Bill was looking for a senior engineer and there was Chris, ready for full-time employment. Chris was soon back to work, seemingly without doing any of the hard work of finding a job.

I Become a Software Engineer

As my involvement with the FSVS program was ending, I met with Bill Lawrence, my manager, and his boss, Bob Chan, to discuss my next assignment. I asked for a software assignment. Bill was going to head up a study project for a naval shipboard communications system, and I went along with him. However, calling the technical shots was my old nemesis, Günter Ludwig, lead systems engineer. My task was to write the SRS (Software Requirements Specification), one of the documents usually prepared after the contract has been awarded. The SRS is normally prepared from the top-level requirements document, which did not exist because neither Günter nor the navy had done one. I didn't enjoy working with Günter, and the funding was iffy, so I found it difficult to commit myself to the project.

Early in 1989, I lost my unit manager position and became a senior engineer. I had hoped they would return me to the engineering staff as a principal engineer, but no. I had given up smoking in 1988, when I lost my enclosed office on the FSVS program. Furthermore, no effort had been expended to provide the tools I would need to carry out my new assignment. So I was not at all happy and asked for a new assignment after completing a rather perfunctory SRS. By the middle of 1989, I was assigned to the space station software group headed by Al Starr, a software guy and my first real software manager. I was finally a software engineer.

The software engineer plays a crucial role in the development of sophisticated systems. Fundamentally, he creates or implements a set of algorithms—rules—to meet the requirements of the project. Complex algorithms are often separately developed by mathematicians and scientists. Together with other members of the project team, the software engineer helps allocates implementation of the algorithm(s) to hardware and software. In the hardware realm, he may help choose the chip set—family of microprocessors and allied integrated circuits—to be used to store and execute—run—the software. In the software realm, he allocates the implementation to a structure of programmable modules. He usually chooses the computer language to be used by the programmers to implement his design. The programmers then decompose each module and write the executable code in the prescribed language. For this reason, programmers are sometimes called coders. The prevailing programming languages in my experience were C and Ada, the latter named for Ada Byron, the poet's daughter, who worked with Charles Babbage and is considered to be the world's first programmer. Because the C language is really down and dirty, some really obscure bits of code can be written. For this reason, the U.S. government preferred Ada, even though it was more cumbersome and expensive to implement. On a large C project, programmers sometimes tried to outdo one another in the generation of obscure code. This gave rise to the derisive appellation of C "weenie," as in software management saying, "Stop writing code like a C weenie."

In spite of the best research efforts of government and universities to develop estimation and management tools, large software projects usually substantially overrun the initial estimates. Sometimes this is the fault of software engineering and management, but more often is caused by the faulty specifications set forth for the project. Typically this was not anyone's fault; it came about simply because the full scope of the project was not understood by anyone at the outset. When I estimated software projects, I usually told management that the ultimate cost could easily be twice my estimate. But we always had to go with the lower figure in order to get the job and then hope that we get well by scope changes and add-ons. So to the management, software became the most risky element of any large-scale project. But oddly enough, they paid the least attention to it, saying something like "We don't understand it, so let's move on." I always enjoyed being responsible for software bids since no one dared to question my designs or my manpower estimates.

To bid on and execute most government contracts, software staff had to be certified for the complex Ada language. Accordingly, a bunch of us took the Ada certification course, which was given, as I recall, near the GE facilities in Pennsylvania. Once we started on the actual project, it usually became apparent that the job could not be completed within time and cost limits by using the Ada language. The cost of the Ada language processors themselves was prohibitive. There existed a large commercial library of software packages that we could buy and incorporate into our system, saving much time and effort. These modules were, of course, written in the C language. Under time and cost pressures, the government allowed us to use as many of these C modules as we could. To meet the Ada requirement, we simply wrote a top-level shell program that invoked the C software modules we needed. As I write this memoir, I don't know if the government is still pushing Ada. I suspect not, even though they had spent a lot of money to develop the world's strictest and most comprehensive programming language, not to mention that expensive tools needed to create, test, and maintain programs written in Ada.

On the Home Front

After some slow years, the mid-1980s saw a new influx of Irish lads and lasses seeking fame and fortune in the USA. One of the early arrivals was Deirdre Donnelly, daughter of Tommy and Nora Donnelly who were O'Donnell family friends from way back. We were told that she had to leave Northern Ireland because of her involvement with civil disobedience, and perhaps the IRA (Irish Republican Army), during the "troubles" that grew out of the Catholic civil rights movement. Deirdre eventually moved on to California where she married and has a family. After Deirdre came her cousin Dominic Ward. He was a carpenter—home remodeler who settled in the Philadelphia area. Next came Deirdre's brother Owen, who moved in with Dominic and worked with him for a time. I can remember driving Owen over to Dominic's place and could see that he was not impressed; apparently, Dominic had oversold his situation and as Owen looked around the squalid place, he said, "So this is where it all happens." Dominic and Owen soon split up, and Owen remains very active in home and business remodeling in the Philadelphia area. He is also involved in promoting Irish causes, locally and in Washington.

Cars don't last forever, and Kitty and I both went Japanese for the first time. I let Patrick use Pop Carlin's 1980 Chevrolet Citation that I'd taken from the Carlin estate and bought a 1986 Toyota Camry. It was a four-cylinder sedan that was really rugged. I was involved in several scrapes while I owned it, and the other car was always punished. In 1993, we took the Camry on a trip to Savannah and Asheville with Agnes and Jim Fletcher. Even with four overweight adults and lots of baggage on board, the car performed very well in very hilly country. Kitty had been driving her Ford LTD wagon since 1979, and all of the power features had gone bad—you couldn't even get the windows open. It was bi—three rows of seats—and was handy when we had lots of children to transport to and from soccer games. By 1987, Kitty was thinking of a new car—she had owned three successive station wagons and was thinking in that direction. I said that she didn't need a wagon any more and should get something that fit her new image as a businesswoman. Accordingly, she got a white 1987 Toyota Celica. While it was a good choice for her, we mistakenly bought the low-end four-cylinder model. It was quite peppy, but the more upscale version would have been a better investment.

In April of 1987, I was summoned to be a juror for the United States District Court in Trenton. Until then, I had always served jury duty in the state court in Mount Holly. Furthermore, I'd never actually sat on a jury. As an engineer whose father had been involved with the casualty insurance business, I was usually shown the door rather quickly. On another occasion, I'd known one of the attorneys and that served to excuse me. However, in Trenton I used the same ploys to be excused and they loved me. I was assigned to an interesting case and heard testimony for several days until the judge called a mistrial. She invited the jury into her office, showed us around, and explained why she had to reluctantly end the trial. I found federal jury service to be far more pleasant and better organized than state service. It also paid a significantly more handsome per diem. The trip to Trenton wasn't a problem because we didn't have to be there until 10:00 a.m. most days. I drove over to Pennsylvania and up I-95 to the Trenton downtown exit. It was an easy ride. Now that I'm retired, I'd love to do U.S. jury service, but I've never been summoned again. Perhaps the judge had arranged for me to get a bye. However, I continue to be summoned for state jury duty in Mount Holly, but have usually wriggled out on various grounds. Now that I'm over seventy-five, I have an automatic out.

Space Station

By 1989, our Camden GE operation had a large team on the sixth floor assigned to Space Station. As I write now, I cannot recall what they were all doing. It was no wonder that a year later NASA (National Aeronautics and Space Administration) cut back our funding, and there were layoffs. Al Starr was responsible for the onboard communications system and decomposed it to two subsystems, each with a software team. I led one team and someone I'll call Mark Wilson the other. Our initial task was to help systems engineering allocate the overall communications requirements to hardware and software, and then to each of the software subsystems. Then Mark and I would prepare the SRS documents as were mentioned previously. At that point, we would be able to make initial estimates of memory requirements and manpower for programming and test.

In those days, we received from NASA the requirements' documentation in the form of hard copies—they were many and huge. We then had to enter much of that same information into our word processors that ran on a large VAX multiuser system supplied by DEC (Digital Equipment Corporation). It was a steep learning experience to adjust to the idiosyncrasies and slow pace of the VAX. While I mastered the VAX word processor, I found that I was typing a lot of material that already existed on NASA printouts. Nosing around, I learned that the systems engineers had acquired a scanner that read documents and created text files on its host IBM PC. Although it was primitive, it did the job and had an automatic sheet feed. You could dump an entire document into the scanner and come back later for the computer file. You then had to upload the file to the VAX, and I don't recall how we did that. Anyway, I became quite expert in using the device and people started coming to me for help with it. It and I became quite popular when others realized how useful it was. Later, I had to give up this toy to a high-priority job on another floor.

The Space Station project was very complex, and I enjoyed working with Mark; and Al rarely bothered us. He had some other work that he had carried over from an earlier assignment and had some people assigned to that; the rest worked for Mark and me. I soon found that many of Al's software people were subpar and included some new hires that were virtually useless. They were amusing, however. Assigned to me was a young Russian Jewish gal who had just graduated from Pitt (University of Pittsburgh) with a master of computer science degree. All of her work had been under the direct supervision of a professor and on large central computer systems. She had no idea how to take an assignment and run with it. She was a bright young woman; and we had many wonderful chats about movies, plays, and books. She even got me involved with her love life and marriage plans. I heard all about her wedding trip to Hawaii and that she really enjoyed Maui "wowee," that island's potent marijuana. However, she agreed that she was totally unsuited to our unstructured environment and asked for a transfer. Because she was on GE's "gifted" program, she was quickly relocated. I hated to see her go, but it was for the best.

Another amusing young person I had was a recent electrical engineering graduate of Rutgers, who also happened to be Jewish. I've forgotten how he ended up on the Space Station software. He too soon realized he was out of his depth but hung around for a while before we moved him out. On a famous occasion, I made a presentation to NASA for the Space Station's UHF communications link with astronauts working outside. The hardware was not an issue, as an onboard microprocessor had been allocated to me for this application. My proposal was simplicity itself, and the NASA engineers were relieved as they weren't even sure that the job could be done. After the meeting, my young Rutgers engineer was so impressed that he went around telling everyone that I had dazzled NASA. To cap it off, he said that I'd worn a Hawaiian shirt for my presentation. I was sure I hadn't, but I was in dressing down mode in those days before casual dressing came to pass. It was a lot of fun being in software at GE because the old-line and hardware-oriented management were usually at our mercy. I spent a lot of time counseling both of those youngsters and enjoyed having them around.

Off to California and Bust

Late in 1989, NASA called a meeting of all Space Station software contractors. The meeting was held in Huntington Beach, near the facilities of our prime contractor. Al sent Mark and me to represent the Camden GE operation. He traveled alone, but Kitty went along with me. We stayed at the Marriott in Irvine, surrounded by a host of new black-faced Darth Vader looking buildings, many of which were empty during the California economic downturn. The Marriott Hotel itself was in bankruptcy. I don't remember if Mark and I ever got to give our pitches. However, it became apparent to the many attendees that software and its memory requirements would probably sink the Space Station. I recall leaving the meetings with a bad feeling.

Our first evening saw Mark, Kitty, and I having a nice dinner near the Marriott. Coincidentally, in that very restaurant, the Budweiser people were introducing a new nonalcoholic brew, O'Doul's, and had a big kickoff with free samples. It quickly became my brew of choice. After dinner, Mark left to visit some friends in the area while Kitty and I went to a first-run movie in Costa Mesa. The film was Michael Moore's controversial *Roger and Me*. I loved it, but Kitty was less enthusiastic; however, we both became Michael Moore fans and have seen all of his films. After the movie, we walked through the interesting park or commons in the center of town. It was a bit spooky at night but very much California. Rising early the next morning, I got to see the sun pop up from behind Camelback Mountain. It was truly awesome. I also saw that the San Diego freeway traffic was bumper-to-bumper at 5:00 a.m.

After the NASA meeting, Mark presumably flew back to Philadelphia while Kitty and I took some time off to drive up to Salinas to visit Tom. At the time, he was working at a really beautiful single-family housing project between Salinas and Soledad where he lived. His home was a cabin in an abandoned resort, popular in the 1920s, in the mountains west of the Salinas Valley. From his deck, he could look across the valley to the infamous Soledad prison complex at the foot of the eastern mountains. While visiting with Tom, we did some touring and enjoyed the beautiful Soledad Mission and the Pinnacles National Monument. The latter was the eroded caldera of a volcano, which had erupted many centuries before. We particularly enjoyed looking at the houses at Tom's project. We stayed the night at the Valley Key West motel in Soledad and next morning set out for San Francisco to get our flight back to Philadelphia. Along the way, we made the obligatory tourist stops in Carmel and Monterey. It was a quick but good trip to the West Coast.

Once back at work, I found that Mark was among the missing. I asked Al about him and learned that Mark's wife had called to say that he couldn't come to work, and she didn't know when or if he'd ever be back. Apparently, he had gone on an alcoholic binge while in California and ended up in a rehab facility. Mark never did come back to work, and his place was taken by a woman who was a protégé of Al Starr's. When she left to have a baby, I covered for her; but when she returned, she complained to Al that I had undermined her position. I truly had no interest in her position, but she couldn't accept that because had our positions been reversed, she would have done her best to undermine me. I had never encountered such a vicious and spiteful person in all of my work experience. She loved to whisper in Al's ear, but I'm not sure that Al really bought into her schemes. I really missed Mark.

It was all for naught anyway, because we were notified early in 1990 that our software contract was to be scaled back. After the California meeting, NASA determined that almost all software would be done by IBM at Houston. The reason given was that IBM would be able to reuse a lot of software they had developed for the space shuttle. This is a typical management decision that is based on false hope. However, I continued to work for Al Starr on Space Station and other tasks through 1991. I'm not sure what I accomplished in those years, but it looks very impressive on my annual performance reviews, so I got some decent raises.

While on Space Station, I was asked to run for ASPEP councilman or shop steward. I had become quite friendly with one of the Camden union higher-ups when he learned that I had been a union official back in the 1950s. A formal election was held, and I lost to a total idiot, 16 votes to 5. The victor was a very strange software noncontributor, but was better known to the rank and file. Anyway, he was soon laid off during the Space Station cutbacks, and I served in his place until I moved on. Because of the engineer's union, the company always waited until it had a legitimate excuse to get rid of deadwood as part of a layoff for a business downturn, rather than for poor individual performance. I lost a significant number from my small council group, but I shed no tears. Maybe I was a bad union man, but I never understood why they were on the payroll. One of them hardly ever showed up for work.

RCA had after-hours courses in various subjects, including the computer programming languages, C and Pascal. I threw my hat into the ring as a backup languages instructor and ended up with the Pascal course. I developed quite a course and gave it for several years. In it, I went beyond the language itself and taught some software history and issues. I enjoyed doing it and received a small stipend. While working with Al Starr, I asked him to approve tuition reimbursement for a BCC (Burlington County College) two-part course in the COBOL (Common Business Oriented Language) programming language. He was surprised that I was interested in a language that was used for high-level business applications, rather than for scientific and aerospace applications. I explained that I thought it would be good background for my course as people asked about it from time to time. He signed the approval, and I took the course at night and drew an excellent instructor. It turned out to be a really good experience; however, I never had an occasion to actually use COBOL.

I also got Al to approve my going for a computer science masters degree. New Jersey Institute of Technology (NJIT), in Newark, arranged for classes to be held locally at night. I completed a number of courses and then ran into a succession of really unqualified instructors. In the course on computer programming languages—my specialty—I had to take over from the stumbling instructor on many occasions. When I drew that instructor for my next course, I decided to withdraw. Again, I didn't really need the degree; I was just doing it for fun or personal satisfaction.

Up through the 1980s, almost all programming and word processing were done on a VAX multiuser central computer system. Because of the high cost of VAX software modules, only two computer languages were supported, FORTRAN (Formula Translator) and BASIC. The government had paid for FORTRAN because it was used on a large supersecret project in another building. FORTRAN was unique in that it had a large built-in library of mathematical functions. In other languages, we usually had to write our own math functions. BASIC was usually provided free of charge on most systems in those days and was widely used. Until I was assigned to Space Station, I had no occasion to use the VAX; I had always used a personal computer (PC) at home or at work. As the Space Station software contract was drying up, I was asked to help fix up an older program that had been developed using DEC FORTRAN on the VAX. By then, I had used a number of the newer programming languages, but naturally, not FORTRAN. I was well on my way to becoming a FORTRAN programmer when I was pulled off that job and given a new long-term assignment. Sayonara, Space Station! Happily, it turned out that I never again had to use the VAX system.

Off to the Southwest

Kitty and I entered the 1990s with the notion that it was time for us to do some serious traveling. In April of 1990, we flew out to Denver, rented a car, and drove to Santa Fe, and then on to Tucson. I'd always wanted to visit Santa Fe, and I wanted Kitty to experience Tucson, a place that I had discovered on my business trips to Fort Huachuca. I had ideas about moving to Tucson after I retired, so we looked at houses while we were there. We found the perfect house on a hill overlooking the city. However, it was out of the city proper and had no deli anywhere nearby. We stayed at the Ramada Foothills on the eastern side of Tucson, where the staff were mostly college students from New Jersey, and played tennis on courts located at the foot of a mountain that seemed to rise out of nowhere. A park ranger came by and asked if we had paid the court fee. We said we didn't know about court charges, and he asked where we were from. It turned out that he, too, was from New Jersey and let up play on without paying. We made some of the obligatory tourist stops in Tucson, including the beautiful campus of the University of Arizona, the Arizona-Sonora Desert Museum, and the town of Tombstone, home of the legendary shootout at the OK Corral. On our last evening, we dined at the Gold Room of the famous and romantic Westward Look Resort. The ambience and music were wonderful, but the view out over the city at sunset is incomparable. Early the next morning, we left for Phoenix and our flight back to Philadelphia.

From Denver, the trip to Santa Fe took us to Pueblo, a very nice Colorado town where we had lunch and drove around admiring the buildings. I believe that it was on the next leg of our journey that we got our first glimpse of antelope racing across the plains. Unfortunately for us, Raton Pass into New Mexico through the Sangre de Christo—Blood of Christ—range was socked in so that we couldn't see past the side of the road. Once over the pass, we made a coffee stop in Raton, a tiny railroad town that boasts an Amtrak stop on the main line of the Santa Fe. Raton has very little in the way of tourist accommodations other than an interesting junk shop. It was very cold, so we joined the few other Raton visitors warming themselves around a potbellied stove in the shop. Naturally, we had a fine bull session with the owners and our frozen fellows. I don't recall that we bought anything. I also don't recall being able to get even a cup of coffee in Raton. Oddly enough, an Amtrak train stopped in town and people got off while others got on. I have no idea where the passengers went or where they came from, seemingly out of nowhere.

In Santa Fe, we stayed in a motel near the center of town so that we could walk everywhere. I enjoyed getting up early and hanging out at a favorite coffee shop on the plaza while Kitty slept in. We visited the beautiful adobe New Mexico Museum of Art and enjoyed visiting the many art and furniture galleries. We bought two prints, which we've had framed—one is in Florida and the other in our Cinnaminson master bedroom. Our day trip to Taos was by way of the scenic mountain road along which we visited the quaint mountain village of Chimayo, which is twenty-five miles north of Santa Fe. It is the site of New Mexico's most famous Spanish colonial church, the 1816 Santuario de Chimayo. It is known as the Lourdes of America for the devotion of its many pilgrims. It's a round-shouldered, twin-towered adobe classic, which squats behind an enclosed courtyard. A pit in the floor of a small room to one side of the church holds the "holy dirt" for which the site is venerated. The nearby Ortega's Weaving Shop and Galleria is perhaps the finest establishment we've seen in the Southwest to date. Owners Andrew Ortega, whose family brought weaving to Chimayo in the early 1700s, and Evita Medina Ortega, whose family built the Santuario de Chimayo church, opened their shop in 1983 to showcase the arts and crafts of New Mexico. In addition to their custom weaving, the Ortegas offer a wide selection of New Mexican books, music, foods, and original artwork.

The art colony of Taos was also one of the places on my must-see list. We enjoyed visiting the shops on Taos Plaza and admired the magnificent Kachinas, dolls carved primarily by the Hopi and Zuni Pueblo Indians and used in religious ceremonies. Unfortunately, the Kachinas I liked were too expensive so I never got one. We also visited the Taos homestead of Kit Carson, the famous scout and scoundrel. Other than visiting the shops on the plaza, there's not much to see and do in Taos—there's probably more action during ski season. Leaving Taos, we decided to take the main highway back to Santa Fe and ran into a freak snowstorm north of the city. As we drove through the blinding snow, our car was continuously whipped by tumbleweeds that practically ran us off the road. Back in town, people wondered what we were talking about.

Another day we drove up to Los Alamos, Mesa, to tour the atom lab and remains of the boys' school that originally occupied the site. Of course, the lab is not open to the public; but we did visit their museum and saw the replicas of the first atom bombs, Fat Man and Little Boy.

We stopped at the Los Alamos County Museum, which is in the last of the old school buildings. There really wasn't much to see at Los Alamos, so I was very disappointed. On the way back to Santa Fe, we visited Bandelier National Monument. This park preserves a series of caves that were occupied by Indians centuries ago. The caves were formed as air bubbles in the volcanic soil. I took a great picture of Kitty climbing up a rustic ladder to one of the caves.

From Santa Fe, we set out for the Grand Canyon, following old Route 66 and the tracks of the Santa Fe railroad. Our lunch stop was at Gallup, New Mexico, a major town on the vast Navajo reservation. Sadly, we were pestered by drunken Navajos begging for handouts. Quickly departing Gallup, we made our next stop at the old railroad town of Holbrook, Arizona. There we were checked into a motel by a woman from New Jersey. It turned out that the motel and restaurant were owned and operated by a family from Ohio. I imagine they had gone to Arizona for health reasons. Anyway, she was a retired schoolteacher, and we had a great time talking with her. In fact, it was hard to get away from her. The next morning as we left after breakfast, she was obviously very sad to see us leave, as it was a rather lonely place. We did get a picture of the beautiful Navajo County courthouse, about the only building of significance in the town.

From Holbrook, our route to the Grand Canyon took us to Cameron and a pit stop at its trading post on the Little Colorado River. It was a sunny day for our arrival at the canyon, so there were lots of tourists and parking was hard to find. We enjoyed the view, which we shared mostly with Japanese visitors—they were everywhere in the American West. Lunch became a problem as the only restaurant on the south rim was full up, and there was no parking in any case. To find food, we left the national park and headed for Williams, Arizona. Just outside the park, we found a McDonald's and were served and seated just as a busload of Japanese pulled in. I hadn't seen the bus, so I was surprised when Kitty jumped up and ran to the ladies' room to beat the crowd. We definitely need more comfort stations in and around our national parks and monuments.

From Williams, we headed to Flagstaff where we planned to stay for the night. The downtown area seemed blighted, with many shuttered shops—just like Camden, New Jersey, a city of similar size. We drove around to try to find the real nice Flagstaff, to no avail. We couldn't even find the university, which was probably in a comfortable suburb. Flagstaff sits at seven thousand feet but close to the rim of the Arizona plateau. When you go south from Flagstaff, you quickly drop into the hot, dusty bowl of Phoenix.

Not knowing where to go, we settled for an overnight stay at a downtown motel. It turned out to be a bad choice—it lay alongside the main line of the Santa Fe railroad. So all night long, we heard the *ding-ding-ding* of the railroad crossing gates as great freight trains rumbled by.

The next day, we headed south to Phoenix and Tucson. Once off the aforementioned rim, we began to see the strange saguaro cactus plants—the ones with the up-reaching arms—dominate the landscape. We decided to make our lunch stop in downtown Phoenix, to see what the city was like. It was very disappointing, just brand-new office buildings with no obvious amenities—the kind of downtown that's deserted when the workers go home so that you feel unsafe walking around after hours. With difficulty, we found a place for lunch and quickly departed for Tucson. The latter is a much more livable city than Phoenix, partly due to elevation—Phoenix at one thousand feet and Tucson at over two thousand feet. As mentioned previously, we had a very enjoyable stay in Tucson. In fact, the entire trip was outstanding and fulfilled my longtime desire to spend some time in the Southwest. However, we decided not to plan on retiring to Tucson, mostly for family reasons.

Family Matters

While I was shuffling through a series of assignment at RCA and then GE, the family was not standing still. Patrick graduated from Holy Cross High School in June of 1986 and started that fall in the business school of Drexel University. His early co-op assignments were with a grocery brokerage firm that handled, among others, Coppertone products. Patrick got to go to some beach resorts to hand out samples as a Coppertone Kid. Our garage became a repository of Coppertone and other products of the firm. Patrick would graduate from Drexel in June of 1991 and go to work full-time with the Coppertone firm.

Son Jim graduated from Delran High School in 1989 and decided to go west. He rendezvoused with his older brother Tom in California's Santa Clara Valley—a.k.a. Silicon Valley. Tom was living and working on a ranch near San Martin. Jim was able to share an apartment in nearby Gilroy, a farm community at the far southern end of the valley. Jim pumped gas in the town of Morgan Hill and eventually went back to school—part-time—at Gavilan, the local community college. After a few years at Gavilan, Jim came back east and spent his final college years at Rutgers Camden.

Meanwhile, Kitty's secretarial career at Campbell Soup was advancing as she took a series of computer training programs: DisplayWrite 4 (1987), Harvard Graphics (1989), and Word Perfect (1992). Naturally, by the time she completed some of the courses, the company had moved on to other equipment with new applications. Through most of those years, Kitty took summers off so that she could enjoy the Tenby Chase Swim Club and vacation at Ocean City. In June of 1989, she was given a best secretary award with the added notation "See you in the fall." As mentioned previously, Kitty was highly sought after and finally succumbed to pressure to work full-time but with extended vacation privileges.

The 1980s decade ended with more deaths in the family. Etta Carlin, the last of Kitty's aunts, died in Moorestown in December of 1989 at age ninety-six. She was buried in the Carlin family plot in Moorestown's Mount Carmel Cemetery. During 1990, Mother began to fail from the ravages of her colon cancer operations. While I visited her on weekends, my sisters carried the day-to-day burden of her care. In March, we got unexpected help when Kathleen Casey arrived from Northern Ireland for an indefinite stay. Kathleen was the daughter of Pat's friend Josephine Casey (nee Sherry) and a trained nurse. The Caseys lived near the O'Donnell homestead on Gort Road, the Brantry. Kathleen was very jolly and visited Mother in her apartment every day. She was a real help until her boyfriend arrived on the scene. I imagine he convinced her to return to Ireland with him, and they left in September.

After Kathleen left, Mother asked to go to a full-care nursing home. After a month in Holmdel's Arnold Walter Nursing Home, she died on Friday, October 9, at the age of eighty-seven. That very weekend, Tom Gaffney, our wedding's best man, and his wife, Mary, flew in from Florida to join us for our fortieth high school reunion, a dinner-dance at the famous Glen Island Casino. Sunday morning, we all drove in my car to our hotel in New Rochelle, New York. From the hotel, we drove to the casino through the surrounding park, which was full of families enjoying a beautiful but windy day. The casino is located right on Long Island Sound, a beautiful spot, so we took lots of pictures. The casino itself was very popular during the big band era and had been substantially renovated in recent years. It was a great place for a large dinner-dance.

Our high school, Cardinal Hayes of the Bronx, commemorated classes on the "fives." In 1990, the honored classes were 1945 (the first), 1950 (mine), 1955, and so on. Of my class of 752 graduates, about sixty-five attended this particular reunion. Oddly enough, John Patrick O'Neill, my nemesis who sat next to me in class, was one of the sixty-five. As he was in our homeroom, Tom and I had to have our picture taken with JP as we called him. In 1990, JP lived in New Jersey and had never married. The Hayes orchestra played for us throughout the evening, and it was easy to see how the school had changed since 1950—all of the players were black or Hispanic. We also learned that the school's enrollment was under 1,000 down from 3,400 in my time. The decline reflected the whites' flight from the school's core Bronx neighborhoods, and the opening by the Archdiocese of New York of a new high school in nearby Westchester County. I was disappointed in the low turnout by my class, but it was an enjoyable evening.

The following day, Monday, we drove down to Lincroft for Mother's wake and funeral. The Gaffney's had flight reservations out of Philadelphia either that evening or early the next morning. They took our car back to Delran, left it, and took the local shuttle to the airport. Tom knew Mother quite well and regretted that he couldn't stay for the wake and funeral. At the time of Mother's death, our sons Tom and Jim were living in California, and we decided that they need not come back east for the funeral. Sons Chris and Patrick came up for the wake and funeral and stayed overnight with Sheila and her husband Pat McCormick. However, Patrick had to leave directly from the funeral after serving as a pallbearer—he had to get back to Drexel for a midterm exam. Monday night, Kitty and I stayed with my sister Pat and Jim Maguire. The funeral was held at St. Leo's in Lincroft, followed by the burial at Gate of Heaven, Valhalla, New York. Pat and Jim, my sister Maureen, Kitty, and I had a comfortable limousine for the long trip to the cemetery, never concerned that we had no car for the trip back to Delran from Valhalla.

The Gate of Heaven Cemetery is a huge place and a repository for many deceased notables in good standing, more or less, with the Catholic Archdiocese of New York. Right near the main gate is the prominent gravesite of slugger Babe Ruth. Mother was buried in the plot already occupied by my Dad, way in the back of the cemetery near the Harrison station of the old New York Central Putnam Division, now Metro North. As people started leaving after the burial service to either go home or to our luncheon at a nearby restaurant, we realized that we needed a ride. When we realized that our Chris was still around, we quickly attached ourselves to him. He was driving an older small Toyota pickup truck, with no springs and no air-conditioning. For the trip from Valhalla, I sat in the front while Kitty and our weekend luggage were jammed in the space behind the seat. It was a long, windy, and hot ride back to Delran, a distance of about 150 miles. Six years later, I went back to Gate of Heaven for a burial service for my second cousin Monica McShane (nee Browne). I haven't been back since.

In December of 1990, we decided to have a family Christmas in California. Chris and Patrick joined Kitty and I as we flew out to San Francisco where we spent a few days as tourists. We stayed in a very nice small family-style hotel. The weather was sunny and bright but chilly. The boys spent many hours riding the cable cars around the city. I squeezed in a courtesy call on the Seniornet folks at the University of San Francisco. I had been a discussion leader—moderator—on some of their Internet forums. After a time, I dropped out as the level of dialog was too low to sustain my interest. When we left town to drive south to Gilroy, we found horrible traffic caused by the closing of several roads due to the recent earthquake. A very fine motel in Gilroy became our base for a family reunion with Tom and Jim who were living nearby. Gilroy itself is a very nice farming community at the southern end of the Santa Clara Valley—Silicon Valley to some.

Surprise! Surprise! A cold wave hit the area, and the temperature in Gilroy dropped below freezing to 19 °F. It was so cold that water pipes froze all over the area. A Christmas Eve wedding reception in our motel's ballroom was interrupted by a ceiling collapse due to a burst water pipe. As an early riser, I liked to go to the lobby for coffee and a newspaper. Christmas morning, I went down and found the staff huddled in blankets. I got my paper and coffee and sat by the roaring fire they had started in their beautiful stone fireplace. Misreading the schedule, we missed Christmas mass by a half hour or so. Anyway, we had to be off to Santa Clara for dinner at my Cousin Patricia Browne Rupel's home. A divorcee, Pat had three boys and with our four, we had quite a male majority. Pat's mother, Anne Browne, usually lives with her, but she was back east that Christmas, so I didn't get to see her, and I can't recall when I last saw her.

Though it was cold—one of the coldest winters on record—we did get to visit some of the area's tourist spots. We went to the beach resort and college town of Santa Cruz, but nothing was happening as college was recessed, and it was too cold to walk on the beach. We did walk around the splendid little nearby beach town of Capitola, where I believe we were able to get some hot coffee and perhaps some doughnuts. But my favorite place in the valley is Felice's Western Store in San Martin, where Tom was living. San Martin is a ranching center located midway between Gilroy to the south and Morgan Hill to the north. Felice's Western clothing and saddlery is absolutely first-rate, but budget busting. I just love to walk around the shop and admire the stuff, particularly the beautiful custom saddles as well as the dusters and hats such as Clint Eastwood has worn in his many Westerns. Evenings, the boys went off to taste of the area's slim night life. One evening, we all went to dinner at a very unusual restaurant near the San Martin airfield. The place was called the Flying Lady, or something like it, and featured model planes that flew around the perimeter of the room on a track arrangement. Despite the extremely cold weather, we had a very nice family reunion at Christmas in California.

Back to Work

Back at GE, I was between jobs and helping out on proposals when a wonderful opportunity came my way. Somehow—I have no idea how—I began moonlighting with a firm in Cherry Hill. The business of the firm was computer scoring of the tests administered by school systems around the East Coast. My assignment was to determine why scores were being processed incorrectly and, if possible, make the necessary changes. But first I had to understand how the optical scanner worked and how it produced its output, which was fed to a large Hewlett-Packard system for final processing. To help with the mechanical tasks, I had a young woman whose job it was to feed the arriving raw answer sheets into the scanner.

I had never worked with anything quite like the programmable scanner, and it took about a month of evening and weekends for me to figure out how it worked and how to make it do what I wanted. Thereafter, I was able to fix several bugs and improve the reliability of the output data written to magnetic tape for the HP system. However, a major problem persisted—scores were incorrectly allocated to students with similar names. I found that insufficient storage had been allocated for names in the scanners memory. Thus, for example, the names Frances, Francis, Francesca, and Francine, might all treated as Franc. So in the Lopez family, the scores of Frances might be assigned to Francis, and vice versa, in an unpredictable way. I explained to the owner that the problem could be easily fixed by increasing the name storage allocation in the scanner. He decided that he didn't want to pay for reprogramming the HP system to handle the longer name fields and wanted no such changes to the scanner program. After all my work, he continued to produce faulty scores until he went out of business a few years later. This was a most enjoyable project, but like so many others I had worked on and would work on, the outcome was unsatisfactory to me on the personal level.

By 1991, most of the Campbell Soup buildings in downtown Camden had been torn down. Some of them I had watched as they were being built in the late 1950s. I particularly remember the can manufacturing facility as it was under construction. Campbell's used a lot of cans and decided to make them in Camden. It was a big deal back then. After the can plant was razed, GE and Campbell's did a land swap. We got the site of the old can plant to the north for our parking lot, and Campbell's got the old "Ferry" lot to the south, which they planned to build on but never did. Our new parking lot was much better—it was higher and with superior drainage. During rainstorms, we no longer had to wade through water to get to our cars.

In 1991, GE announced that we would get new engineering and manufacturing buildings with substantial state and local funding. The new buildings went up very quickly on a nearby site vacated by Campbell's. It was a big day when we moved into the new facilities from buildings that dated back to WWI. Shortly before we moved, I was assigned to a new, hot program with the initials LMD/KP. I've forgotten what LMD stood for, but the KP was "key processor." I was back into the cryptographic world with its cipher-locked doors and super security. My new boss was Al Anderson, a young chap from GE's Valley Forge unit, on his first managerial tour. Our group of six or eight was assigned to create the software for a PC-based test bed for the new product, a remote cryptographic key processor terminal, which was to be the backbone of an advanced NSA system. On our job, we were to use the C programming language in a rare concession by the government.

It was an interesting software project and my first where actual programming would be done by a team. As a group, we met to allocate tasks to each member, and I offered to take the input-output modules, the most difficult. Everybody, including Al, was happy with that and off we went to our individual assignments. In my May 1992 performance review, Al wrote, "Mr. O'Donnell is looked at as the computer language expert by his peers." I recalled that back in 1984, I was told not to get involved with software as the young hotshots would make me look bad. I now knew that it was Bob Chan's personal fear of software speaking.

When our software was ready, I started integration testing with the actual KP terminal and worked with Jim Carroll to solve the compatibility problems we found. Jim was a young fellow from Drexel who had worked in our group on Space Station. He and I were quite friendly and were in a foursome that played tennis after work, once a week. As part of the integration testing, I had to use a Sun workstation connected to our internal secure multiuser system and with a big Hewlett-Packard computer that the factory wanted to use for test purposes. Because the factory programmers were behind schedule, I had to write test programs for the HP computer in a BASIC language variant called RMB—Rocky Mountain Basic. I really enjoyed working with RMB; it was a nice change from the C programming language. To help, I had a young fellow I'll call Nuyen who was in the GE "honors" program.

In order to properly allocate manpower costs to contracts, the technical and project management staffs were required to log onto our secure system each day and post the day's hours to the jobs they'd worked on. This meant that every afternoon, the system bogged down, and we were unable to access it for our test purposes. That meant that Nuyen and I could accomplish nothing for much of the afternoon. Being very job oriented, I didn't like to sit idle and found a way to circumvent the security and directly access the system. Nuyen wasn't happy as we had been using his password while I waited to get my own. However, our work proceeded apace and all seemed well until a security audit found that I had basically hacked into the secure system, using Nuyen's password. We were summoned to meet with Phyllis Stevens, the new top manager of the software skill center. She was a real GE-type politician who gave us hell, saying that we'd made a serious security violation, and a notice would be placed in our permanent records, possibly causing loss of our security clearances. I'd never been very respectful of the security aspects of the government jobs I'd worked on, so I told Phyllis that it was my fault, not Nuyen's, but since he had allowed me to use his password, he was considered equally guilty. After we left her office, Nuyen asked me what it all meant. I said they put that kind of stuff in your file so they can take it out and use it against you whenever they want. He turned pale and shortly thereafter asked to be assigned elsewhere in the GE world. Oddly enough, my last manager when I left RCA in 1970 was a Don Stevens. I don't think Phyllis and Don were related except in my dim view of both.

Patrick Graduates and Off We Go to Iberia

In the fall of 1991, we left my chaotic work situation behind and took a Globus/Brennan package tour of the Iberian Peninsula—Spain and Portugal—with side trips to Morocco and Gibraltar. We flew in and out of Madrid, a splendid city notable for its beautifully dressed young women. After checking clothing prices in the shops, we wondered how they could afford such finery, probably by living with parents for years and years. In Madrid, we did an extended city tour with a major stop at the Prado Art Museum. I wasn't crazy about the museum's collection—too much of Madonna and Child. However, I did enjoy their fine collections of Velasquez and Goya works. Unfortunately, we didn't have time to see Picasso's famous *Guernica*, which is located in a separate building. I asked our Prado guide where I could see some impressionist and modern art, and he said in New York.

The Spanish countryside was a marvelous blend of rich browns and greens, the latter provided by the olive trees. We made a side trip to the Valle de los Caídos (Valley of the Fallen) with its huge basilica that was constructed by dictator Franco to honor the fascist who died in Spain's civil war of the late 1930s. The basilica and the surrounding tombs were built into a mountain north of Madrid and the overall effect is truly stunning. I understand that at some point, all of those who died in the civil war were allowed to be buried there. As we left Spain for Portugal, we made an overnight stop in the university town of Salamanca. As we walked around town, we saw many students wearing American college sweatshirts and talking, in English, about upcoming parties. So much for higher learning in Spain. Unfortunately, our hotel was located adjacent to the student quarter, so we got little sleep that night and had a 5:00 a.m. wake-up call for an early departure to Portugal.

In Portugal, we made the obligatory stop at Fatima, the purported site of an appearance by Mary, mother of Jesus. The site is an ugly space occupied by a huge arena for outdoor masses. Nothing was going on while we were there, but all along the roads, we had seen pilgrims walking to Fatima for services. However, there is a nice enclosure, which allegedly contains the very tree from which Mary "appeared" to the shepherds. Lisbon was a far more interesting place to visit. The highlight for me was a visit to the Plaza of the Navigators and the immense sculpture commemorating the major role the Portuguese school of navigation played during the age of discovery. Since its promotion by Ernest Hemingway, visitors to Portugal are expected to undergo fado singing. One evening, our group went to a restaurant for dinner followed by a fado performance, where fado is a kind of lament for those who have died at sea. It was the group's consensus that it was absolutely the worst kind of traditional singing that we had ever experienced. Some people felt that we had probably just experienced bad fado. Bad or good, if you go to Portugal, skip the fado.

From Lisbon, we were off to Seville, the capital of Andalusia, one of Spain's autonomous communities. Andalusia is the most populous and the second largest in land area, of the seventeen autonomous communities of the Kingdom of Spain. We stopped for lunch at a restaurant just over the Spanish frontier. While we lunched at the bar, we heard a party going in an adjoining room. It was, in fact, a wedding party, and our group was invited to join in and share the wine and dancing. Hanni, our marvelous tour director, had a hard time getting everyone back on the bus. We made an impromptu stop along the road to check out a marble quarry. It was interesting to see how slabs of marble were cut from a below-ground facing and lifted to the surface. As an engineer, I wanted to spend more time there, but the consensus was to mush on. Upon arrival in Seville, some of the group elected to go to a bull fight while the rest of us did a tour of the city. That evening, we all went to a restaurant and then on to a theater for a performance of Flamenco music and dancing. This is really good stuff, much more upbeat than fado. Flamenco si, fado no.

As a new member of the European Union, Spain was undergoing subsidized upgrades of its roads and facilities for the Seville world's fair, Expo '92. Consequently, some parts of the city were hard to access. However, we were able to visit the Plaza de España, a magnificent semicircular square that has become a symbol of the city. It was designed for Seville's 1929 Iberian American Expo in a style that has become known as Andalusia regionalist because it employed a mixture of architectural influences from the past including Renaissance and Gothic. The inner part of the plaza contains a series of glazed-tile depictions of historical scenes from every Spanish province. A canal runs around the perimeter of the square, and you can hire a boat for a short pleasure trip. Vendors' carts offer a variety of goods, including candy, ice cream, and souvenirs. The plaza reminds me of the southeast corner of New York's Central Park but was too isolated to draw the foot traffic that makes New York's park so popular.

We made the obligatory stop at the Seville cathedral to view the tomb of Christopher Columbus. Of course, no one knows for sure where he is actually buried. The local story was that he had been buried in Havana and dug up for a triumphant return to Seville, his jumping-off place for the Americas. His tomb is very beautiful, but in a dark corner of the cathedral. Our stay in Seville was very brief, and we were soon off to the southern seaport of Algeciras for our trip to Morocco.

For our overnight stay in Morocco, we were allowed to take only a small overnight bag. The rest of our luggage remained safely, we hoped, in Spain. The ferries that cross the Strait of Gibraltar to Tangier are operated by Arabs, so we were ready for anything. It was really great to make the fabled three-hour passage, and the weather was perfect. On the far shore, we could see the Atlas Mountains rising up behind the beaches; perhaps the very beaches that our GIs landed on during WWII. A bus met us on arrival in Tangier, and we were given a quick tour of the city. We then took a walk through the Kasbah, the old Arab quarter. There we were constantly pestered by peddlers and beggars; it was just awful. In a carpet shop, I mistakenly expressed interest in a beautiful rug and barely escaped with my life. They were determined that I should buy it once I touched it. Kitty and I decided to skip the belly dancing and had a private dinner in our hotel. When walking the hallways, we could hear mysterious tapping coming from some of the rooms. It was eerie.

I arose early the next morning and from the lobby could see no one outside the hotel, so I thought I'd take a short walk. But as I approached to the front door, figures suddenly appeared out of the gloom—a full complement of beggars and peddlers was ready to descend on me once I stepped outside. After an early breakfast, we were bused to the pier for the return trip to Spain. As we stood in line waiting to board the ferry, its engine was started and flaming debris from its smokestack fell on us. Standing with us were some men who'd arrived the day before with a ferry—not ours—that had been reconditioned in Germany and England. They told us that the Arabs ran the ferries with minimal care for the engines and mechanical equipment, hence the flaming debris falling all around us. Once under way, our group settled in one of the lounges where we met another group coming back from an extended stay in Morocco. The feeling in that group was that Morocco was the pits, and they were quite put out that we had to stay only one night in Morocco. As we approached the port of Algeciras, we could see the trash being dumped into the sea from Gibraltar due to the long-running territorial dispute between Britain and Spain.

Fortunately for us, the frontier was open, and we were able to visit the Rock of Gibraltar, a British colony of some sort. It was very crowded with tourists and the many Brits who have retired there. The road into the "rock" is part of an airfield used by the Royal Navy and can be closed at any time for air operations. Our tour took us into the deep caves the Brits had dug as part of the rock's defenses. No part of the rock is very far from a steep drop into the sea, and our local tour bus driver loved to pull in and park with the front bumper hanging over the edge—he was not popular. We enjoyed having a visit from one of the rock's famous Barbary apes. The legend is that Gibraltar will remain British so long as the apes they imported thrive on the rock. Gibraltar seemed to be a very nice place to live if the dispute with Spain were resolved along with their trash and sewage problems. However, it seemed to be too crowded to be a comfortable tourist destination.

From Gibraltar, we went to the Costa del Sol and a stay at a beachfront hotel in Torremolinas. Although there were a few people on the "topless" beach, it wasn't really beach weather. The town itself was in the hills behind the beach, and we took the hotel shuttle up there. The shops were pretty much like those we had visited elsewhere in Spain—very nice stuff, but very expensive. However, there was a McDonald's right on the main street, so we enjoyed some American coffee while seated at an outdoor table where we could watch the world go by. Getting back to the hotel proved to be a problem. We wandered around, trying to find the funicular to take us down to the beach. No one we met spoke English, and we found it quite by accident. Next day, we took a side trip to Mijas, the so-called white city in the hills behind the Costa del Sol. It was a wonderful place to visit. The burro carts, the local taxis, were memorable; each burro displaying its license number on a plate hanging from its neck.

From the Costa del Sol, we set out for Granada, the last city taken back from the Moors in 1492 by Ferdinand and Isabella—los Reyes Catolicos—the Catholic monarchs. I had hoped to visit the church where they are buried and asked to be dropped off the bus as we went by it. It was close to noon, so I raced to the door only to have it slammed in my face—it was siesta time in Granada. Disappointed, I walked to our hotel, dodging the construction barricades and the noisy scooters that echoed through the narrow streets. I found the center of Granada to be far too noisy for my taste. We made the obligatory visit to the Alhambra, the beautiful fourteenth-century Moorish red palace that is probably the top tourist destination in Spain. It is truly magnificent and well worth seeing, as is the adjacent Generalife gardens. From Granada, we set out for Toledo with a lunch stop in Don Quixote country—La Mancha. We enjoyed an al fresco lunch in a garden with prominent statues of Miguel Cervantes, the Spanish literary giant, and his creation. Off in the hills, we could see the famous windmills that Don Quixote thought were his enemies; hence, the expression "tilting at windmills."

Our first view of the ancient Spanish capital of Toledo, a World Heritage Site, was from the surrounding hills. The city lies in a bend of the Tagus River, which surrounds it on three sides, making it quite a stronghold. Our bus descended to a bridge over the Tagus and up to the city gate. There we had to leave our bus as the streets were much too narrow for it. Our group walked across the city with stops at a number of tourist spots. The city is known for its huge collection of paintings by El Greco that adorn many of the city's churches and palaces. I'm not a fan of his work, but his famous painting, the Burial of the Count of Orgaz, exhibited in the Church of Santo Tomé, is indeed outstanding. Toledo is the capital of the province of Toledo and of the autonomous community of Castile-La Mancha. It was interesting to walk past the multiflagged governor's palace, which was heavily guarded due to the threat of bombings by ETA, the Basque separatist group. In fact, all government buildings in Spain were closely guarded by well-armed men.

From Toledo, we mushed on to Madrid for an overnight stay prior to our morning flight back to the USA. For our last night in Spain, we had a group dinner in a restaurant located in a former home of the artist Goya. The setting was fine, but the food was not to my taste. I think everyone felt that the Globus-Iberia experience was outstanding. We all loved Hanni, our multilingual Jewish Italian tour guide, who played Irish comedy CDs and sing-alongs on the long bus legs. She was not afraid to step out of the bus and direct traffic whenever we got into a tight spot in the narrow streets of a city or town. Hanni told us that she would have a few weeks off to recover from our grueling adventure and then start on another entirely different tour. She had spent considerable time in the United States and was quite knowledgeable about American places. She could probably direct tours anywhere in the Western world.

Journeys to Points South

After six months back on the job, Kitty and I took another trip to Florida in April of 1992, this one a "fly-drive." We flew into Orlando and drove up to Orange City to visit with the Gaffneys—Tom was my best man. We spent some of our time looking at housing developments in and around the lovely nearby town of Deland. On the last evening of our stay with the Gaffneys, we had dinner at the Chart House in Daytona Beach. It was a really great spot right on the Intracoastal Waterway, but expensive.

From Orange City, we drove to South Palm Beach where we had accommodations at the beachfront Hilton. One of the Hilton's attractions was its tennis courts, so we went out and started to play, only to learn that they were a separate concession and that we would have to pay to play. I pointed out that no one was using the courts and that we would pay only half of what they asked or go elsewhere. They agreed to our terms, and we played for several days, usually by ourselves. We also learned that the beach was another pay-to-play separate concession. This was our first experience of this new way of doing business, and we were not happy about it. It seemed to us that if a hotel advertises tennis and beach that it should be included in the daily room charge. I guess we couldn't really complain too much as our travel package—air fare, car rental, and Hilton (three nights)—cost only $858.

While staying at the Hilton, we visited Aunt Mary and Cousin Jackie in Lantana. They didn't seem to be faring too well, and the house seemed to be failing as well. Mary's prized Oriental rug was worn down to the backing. They knew we were coming and had some tea and biscuits for us. We also went to Palm Beach and took a walk along Worth Avenue, the Fifth Avenue of the South. On the way, we passed Ivana Trump's house where we saw her chauffeur polishing her Bentley. We later saw the car outside one of the shops on Worth Avenue. We had lunch at the Breakers, then the preeminent, classic, Flagler-built hotel—the best in the Palm Beaches.

That summer, the Democrats held their presidential convention in New York City. Arkansas governor Bill Clinton became their candidate for the presidency. Bill had proven himself to be an outstanding campaigner throughout the primaries. During his acceptance speech, he promised that his team would get right on a bus and go across the country to listen to the people. His first stop was to be in Camden, New Jersey. We looked forward to seeing him, but on the day of his visit, the city was jammed and he appeared up town and out of sight from our building. The bus trip was a very successful campaign strategy and put him well on the way to the White House. He was not nearly as good a president as he was as a campaigner.

After a few months back at work from our Florida trip, in October of 1992, we took off for New Orleans where we stayed in a truly shabby hotel out on Canal Street. I had spent a lot of time in New Orleans and wanted to show it to Kitty. Of course, things had changed for the worse. The interstate highway ran right through the center of the city, blighting the adjacent blocks. I had seen the same thing happen to Syracuse, New York. We enjoyed the ambience of the French Quarter, particularly the beignets at the old market and the jazz, which flowed out onto the street from the many clubs. I had a particularly fine lunch at a small place near Jackson Square. I've forgotten what my entree was called, but I remember that it included the local favorite red beans. We splurged one night for dinner at Antoine's, one of the city's best known restaurants. It turned out that we did not meet their dress requirements, so we were placed in the balcony, not the main dining room.

I had wanted Kitty to see the beach along Lake Pontchartrain where we'd hung out forty years previously—it was kind of like Coney Island. Unfortunately, the lake beaches were not accessible due to flooding from a stormy hurricane season. Furthermore, it seemed that the expanded University of New Orleans now cut off much of the beach access. The Charles Avenue streetcar ride to the end of the line in the Garden District was probably the highlight of the visit for me. Along the way, we passed several university campuses, including that of the Jesuit school, Loyola of the South. When I was at Keesler, we woke up every morning to the powerful Loyola radio station, which every morning advertised the Sock Center in downtown Biloxi, Mississippi.

From New Orleans, we set out for San Antonio, Texas, with a detour through Louisiana's Cajun Country. We drove through Houma, a small but interesting bayou town, where—by pure chance—we found a bakery with world-class donuts. Our next stop was Morgan City, where we hoped to visit a wildlife sanctuary. Morgan City is on the Atchafalaya River, a major outsource of the Mississippi River, and is an important gulf port for the shrimp fleet and for offshore oil rig services. The road to town was littered with debris from Hurricane Andrew that had made a direct hit a month or so before our arrival. We found the wildlife sanctuary essentially abandoned, and the animals scattered by the storm. We didn't linger in Morgan City and pushed on, accidentally ending up on a road along beautiful Bayou Teche. The homes along the bayou were gorgeous, and the road took us into the town of Franklin, which looked like it belonged in Vermont. Living along Bayou Teche seemed to be about as good as it gets. Our last stop along the bayou was at New Iberia, the home of Tabasco sauce. I'd love to have toured the factory, but time was running out.

From New Iberia, we set out to pick up I-10, the main east-west highway, at Lafayette, the major city in Cajun Country. We didn't tarry there and pushed on to Lake Charles, then Beaumont, Texas, and, finally, Houston. It was dark by the time we got to Houston, and we decided to go through the city and spend the night on the west side so as to have an easier start in the morning. We chose a La Quinta Inn for our resting place. When I objected to the high room price, Kitty was mortified, but I did my thing and persevered. Finally, the exasperated clerk, who I believe was from New Jersey, asked if I had a Sam's card. I quickly produced mine and was given another $10 off the room rate. Kitty couldn't believe what was happening, but I felt I was justified in bargaining the price down. Next day, I called ahead to the La Quinta in San Antonio and explained that I had a Sam's card and wanted that special rate. The person I spoke to had no idea what I was talking about.

I was reminded of my last visit to Texas, a business trip to Dallas. It too was in the fall, but what a difference. On that occasion, the temperature was over 100 degrees each day; and I recall the blazing sunsets as we made our way back to our motel. My companions and I drove over to Fort Worth for dinner and a quick look-see around town. I came away more impressed with Fort Worth than with Dallas. Next day, I took some people to lunch in Dallas and was mortified by their slovenly appearance while the people around us were so well dressed, particularly the women. I think I concluded that we couldn't do business with them if they didn't know how to dress and behave in public.

We found San Antonio to be a wonderful place to visit. We stayed downtown near the famous River Walk, which runs along the San Antonio River. The downtown buildings are mostly of modest size, and many have been substantially restored—prettied up for the tourists. The hop on-hop off boats along the elegant River Walk provide easy access to many bars, restaurants, hotels, and shops. Away from River Walk, we visited a number of tourist spots, including an old Spanish mission where Teddy Roosevelt formed his Rough Riders cavalry unit for the Spanish-American War. Other stops included a visit to the museum of the Lone Star Brewery and to a fantastic arboretum that was built on the site of an abandoned stone quarry. But in downtown San Antonio, one has to visit the great Texas shrine known as the Alamo. As you go in past the Texas Rangers that guard the entrance, you realize that you are entering a place that is sacred to Texans, and you feel stirred yourself. It's the same feeling I get when I visit Gettysburg.

For our last day in San Antonio, we decide to drive north to Johnson City and LBJ's nearby ranch. Johnson City is totally misnamed; it's just a small town. We visited LBJ's boyhood home and arranged for the ranch tour. But first we had some lunch in what appeared to be the only eating place in town. As I looked at the menu, I saw that chili was the food of choice. I couldn't recall that I'd had any chili since leaving the service in

1952. I decided to risk a bowl, and it was delicious. I've been a chili fan ever since. In fact, I wouldn't mind taking all my meals in Johnson City. We took the group ranch tour, which wandered around the grounds and along the shallow Pedernales River made famous when LBJ would drive through it on his jeep and splash his guests. The main house was off-limits as it was still the residence of LBJ's widow, Lady Bird. Next day, we flew from San Antonio back home to Philadelphia. It was a short but wonderful and action-packed trip to Louisiana and Texas.

We Become Martin Marietta

The GE years had been awash with glorious pronouncements of how wonderful the company and its management team were, and how fortunate we were to be parts of it. In November of 1992, the GEers were shocked to learn that the government systems divisions throughout the company were to be "merged" with those of Martin Marietta (MM), a Baltimore-based conglomerate. GE management took great pains to explain that it was a merger, not a sell-off. In April of 1993, each employee received a letter from Norm Augustine, MM's president, welcoming us to the *new* Martin Marietta. The GE loyalists felt that they had been sold down the river. We knew how they felt because RCA had sold us down the river to GE some years before. That same April, I received my very last raise to $69,000, or about $96,000 in today's money.

As we wrapped up our portion of the LMD/KP project in Camden, I was asked to help with the portion of the system under development by the GEers in King of Prussia. They were having a lot of trouble with the input-output interface to the rest of the system, my specialty. I then spent many days a week on the road to Pennsylvania. Fortunately I was able to set my own hours since nobody knew where I was most of the time. I usually left home about 9:00 a.m., well after the rush hour peak, and left King of Prussia about 3:00 p.m., well before rush-hour peak. Typical of how these things go, they needed my help but were slow to provide the equipment I needed to work with. Working late one day, we finally got all of the necessary pieces but found that a key connector had come in with protruding ears that would not allow it to go into our equipment. I said let's go to a machine shop and cut of the offending ears. I was told that there was a machine shop in a nearby GE building but that it was undoubtedly closed. I said let's go and check it out. We went there and found that it was indeed closed and locked up, but in an enclosure with only an eight-foot wall. I said let's find a way to get over the wall and do the job we came to do. We were able to get one of the younger guys over the wall, and he opened the place up. We did the job, locked up, and vanished into the night. If we'd been caught, I would undoubtedly have gotten another black mark in my personnel file, again for leading young GE engineers astray.

After helping to get GE's input-output system back on track, I was asked to pick up a LMD/KP software component that had been done by another senior engineer I'll call Hal, but needed a major upgrade. In those days, we had very nice two-man cubicles with separate computer workstations and a shared center table. So Hal was moved into my cubicle and started briefing me on his design, which I found to be very clever but very complicated. As I got into it, I wondered why the job was being taken from Hal, and why to me. I got to know Hal quite well and began to understand why. Hal was not well-regarded by the management, so raises hadn't been forthcoming. Through a chain of unfortunate events, he found himself in severe financial difficulties—he and his family were living in a rental apartment. He was raised in the Mormon faith and had served as a missionary for two years in Ireland, of all places. He was a wonderful cube mate, and we had many enjoyable bull sessions. Hal seemed to be jealous of my status, which he said was definitely in the top 10 percent, while he felt that he was in the lowest 10 percent. Looking at the quality of his work, I told him that I thought he too should be a 10 percenter.

Unhappy with his treatment, Hal had run-ins with management and with the engineer's union, ASPEP. I guess he expected the union to give him more backing than they were able to. I was quite friendly with one of the key ASPEP representatives who would often stop around to see me. I think he was very surprised to see that I was "bunking" with Hal. One day, when Hal was away, the union rep asked me about him. I said that Hal did a very clever work but seemed to be very unhappy at work and at home. Some days later, I was talking with Hal, and he seemed very despondent. He said he'd bring in a gun if he thought it would do any good. I was shocked because there had been a number of workplace shootings by disgruntled workers, and Hal was surely disgruntled. I was fearful of my own safety should he decide to bring in a gun. I quickly met with our union rep to discuss this development. We agreed that Hal had to be let go immediately, and it was done. I don't know about his departure terms; I hope he got some financial help. I felt awful about ratting Hal out as I liked him very much; but my personal safety and that of others was at stake. I often wonder what became of him.

After finishing up the job that Hal had started, I was finished with the LMD/KP project and started working on major proposals for new business. I enjoyed the challenge and working with the other members of the proposal team, usually all top people. The downside was that we didn't win any of those I worked on. I also worked as an internal consultant, wherein I helped engineers who were having hardware-software problems on their assignments. This was far more fun than doing proposals. In May of 1993, I received notice that I would receive an award of $1,500 for my work on the LMD/KP project, and in July, I received a letter of commendation from Mike Barbee, the overall project manager.

Less Work, More Play

By 1993, the Reagan defense spending spree had run its course, and the frugal Clinton years were upon us. It seemed to me that our business would fall off; and consequentially, good assignments would be hard to come by. I had reached the age of sixty and started thinking about retirement, but when in doubt, take another vacation. Fortunately, by 1993, I had five weeks of paid vacation time and had no pressing assignments, so why not take off. Kitty had a very flexible arrangement with Campbell Soup that allowed her to take extra vacation but mostly without pay.

I'd always wanted to visit Savannah and some other cities in the south. Kitty's cousin Mary Ruff Murphy and her new husband, Vince Schenk, had recently moved to the Savannah area. It seemed like a neat idea to drive down, stay in Savannah, and visit them, and then do a side trip over to Asheville, North Carolina. We also wanted to visit our Jim who was then at Elon College near Greensboro. Kitty's sister Agnes and her husband, Jim Fletcher, thought it was a great idea and joined us. We arranged for Mary and Tom Gaffney to come up from Florida to join us. We all stayed at the beautiful Hyatt Regency, which sits right on the Savannah River. Next day, we all did a city tour followed by dinner at the Pirates' House Restaurant. Savannah is a beautiful old Southern city with many splendid homes, the best of which were occupied by General Sherman and other high-ranking Union generals during the punitive campaign that followed the burning of Atlanta.

On the following day, we left the Gaffneys behind, and the four of us drove east to the Schenk's home at the Landings on Skidaway Island. They had a terrific small house in that wonderful gated community that had every imaginable amenity. Mary, a realtor in her own right, gave us a fine tour of the community; and I—for one—was ready to buy and move there, but cooler heads prevailed. We learned that the Landings folks had a sister community near Asheville, which we resolved to have a look at when we got there. After a nice visit with the Schenks, we set out for the long drive across North Carolina and up into the high country. As I mentioned previously, I was quite impressed with how our four-cylinder Camry handled the hills despite the load of four adults and their luggage. As we neared Asheville, we visited the Landings community at Hendersonville and found that it was much too vertical to suit us. I couldn't imagine how one could leave the house in winter.

In Asheville, we chose a motel that had a nice card table setup in the lobby for our evening pinochle games. Next day, we made the obligatory tourist stops, including the graves of O. Henry and Thomas Wolfe. Best of all for me, I got to visit the Wolfe homestead, a beautifully tended state historic site. A son of Asheville, Wolfe moved to New York City where he worked and died young. He was one of my favorite writers during my college years. While walking around downtown, we met a woman who said that she was the retired secretary of General Sarnoff, RCA's founder. We had a nice chat, and she directed us to a wonderful restaurant in a nearby village. It was called something like Ten-Cent Cotton. We didn't visit the famous Vanderbilt Biltmore estate because it required extensive walking, no trams for the weary of knee—me—and back—Agnes.

From Asheville, we drove east toward Elon with a stop in Greensboro, the site of a famous battle during the American Revolution. At that time, the place was known as Guilford Courthouse. The National Military Park has a splendid memorial to General Nathaniel Greene. The battle was fought at this then small North Carolina back county hamlet in March of 1781. It was the largest, most hotly contested action of the Revolutionary War's climactic Southern Campaign. While technically a British victory, their losses were severe. The park is located within the present-day city of Greensboro and is well used by walkers and joggers. It is really a special place. We went on to Elon where we met our Jim who joined us for lunch, or possibly dinner. I have a vague recollection that we stayed overnight near Elon and pushed on for home the next day. It was another quick but interesting trip.

In July of 1993, Kitty turned sixty. The boys planned a dinner celebration at the Café Gallery in Burlington City, one of our favorite restaurants. One of its plusses is its location on the Burlington waterfront with its small boat landing. Kitty and I had never been in Chris's power boat, and it was decided that we would take it to Burlington. As we boarded the boat in Delran and headed toward Delaware, I noticed that the river was very choppy. Chris said that the river was the roughest he had ever seen it. As we turned north for Burlington, we were heading into the wind and waves started coming over the bow. Kitty and I were dressed for this special occasion and were quite wet when we docked, mercifully, at Burlington. We had a fine dinner and eagerly accepted a ride home by car with a friend of Patrick's who had met us at the restaurant. That was our one and only outing in Chris's boat. However, in a few months, we would be on another rough boat trip.

Off to Europe Again

While looking through the Globus 1993 tour catalog, we noticed a new offering called "A Taste of Europe." It offered an eleven-day romp from London to France, Switzerland, Germany, and Belgium, and back to London. We signed up for a mid-October session and flew to London from Philadelphia. In London, we did the Tower and saw the famous ravens and the Crown Jewels. Dingy downtown London buildings had been cleaned up considerably since my last visit—Westminster Cathedral positively sparkled. I've forgotten what else we did but recall that we were given a boat ride on the Thames. From London, we motored to Dover and caught the hovercraft to Calais, across the English Channel. Riding in a hovercraft is much like riding in an airplane, you can't see anything through the small windows, and the boat's huge wake pretty much obscures everything.

From Calais, we motored to Paris where we were given a city tour and then had a day or two on our own. There were two highlights for me, with the Eiffel Tower the number one. It is an engineer's dream, and the French have taken great pains to provide explanatory panels, which detail its design and operation. Because of my fear of heights, we only went to the first viewing stage, which is at about the 300-foot level. From there, we had a panoramic view of the Place de la Concorde, the Louvre, and the Tuileries Garden—where, in 1792, the Swiss Guard was massacred as they tried to protect the royal family from the Paris mob. The tower's passenger lifts from ground level to the first level are operated by cables and pulleys driven by massive water-powered pistons. As they ascend the inclined arc of the legs, the elevator cabins tilt slightly—but with a slight jolt—every few seconds in order to keep the floor nearly level. The elevator works are on display and open to the public in a small museum located in one of the four tower bases. Maintenance of the tower includes applying fifty to sixty tons of paint every seven years to protect it from rust. In order to maintain a uniform appearance to an observer on the ground, three separate tones of paint are used on the tower, with the darkest tone on the bottom, and the lightest at the top.

My other Paris favorite is Monmartre, near the famous Pigalle nightclub district of Moulin Rouge fame. Monmartre is the highest point in the city, and the summit affords a wonderful panoramic view. The Monmartre summit boasts the famous basilica of Sacre Coeur as well as many charming eateries and ice cream stands. Weekend crowds come to watch the many artists at work and to simply enjoy the ambience. We had taken a tram ride to the summit from Place Pigalle but decided to return to the Metro by walking down the steps. Along the way, we met another couple from New Jersey who had a sad song to sing. Her purse had been snatched in the Metro. We had been warned about this by Globus and were very wary, particularly in the Metro, which seemed full of suspicious-looking characters right out of old-time melodramas.

From Monmartre, we took a taxi to Les Invalides, the monument at the old soldier's home where Napoleon lies buried. It was interesting to see his magnificent tomb and those of his marshals. One in particular, possibly that of Marshall Ney, had a touching sculpture of his coffin being carried on the shoulders of his men. We were a bit taken back by the shabbiness of the building's interior—it seemed dirty. Time did not permit a visit to the adjacent military museum, and we set off by taxi to the museum of modern art in the old Quai d'Orsay palace. When we got to the palace, we found a very long line, waiting for admission to a showing of the Barnes collection from Philadelphia. In lieu of standing in the cold and rain, we visited the smaller collection in the Orangery, which houses the huge and famous *Water Lilies* mural of Monet. We later learned that with a tourist card from our hotel, we would have been able to bypass those standing in line at the palace.

As we motored southeasterly from Paris toward Switzerland, we made a brunch stop in the town of Beaune, the capital of Burgundy's wine country. The town offers walled fortifications and innumerable shops, restaurants, and cafés for the hordes of tourists; but we didn't stay long enough to enjoy them. We ate our lunch, sitting on a wall so that we could watch the world go by. Some passersby were eating good-looking French pastries that Kitty identified as "Napoleons." We made a tour of the nearby bakeries, asking for Napoleons and mystifying the shopkeepers, one of whom told us sternly that Napoleon was in Paris. We realized that they were not called Napoleons; and worse, we didn't see any of whatever they were in their showcases. Leaving Beaune for Switzerland, we passed the flooded plain of the Saone River and drove along the north shore of Lake Geneva; we could see the city on the far shore. Our first stop was the Swiss capital of Bern—meaning "bear." We had a brief stroll through the center of town to the bear pits along the River Aare, a tributary of the Rhine. The river looked cold but beautiful, with flanking houses hanging over the water. Bern looked like an interesting place to visit, but we were soon off to Interlaken.

Unfortunately, the weather was bad during our stay in Switzerland so that the Alps were never visible to us, nor could we see the far shore of the many lakes we passed. Interlaken is named for the fact that it is on the land bridge between two lakes and is on the River Aar, which connects them. It is beautiful place—perhaps my favorite of all places. Because of its location, all roads and railroads pass through Interlaken. As a railroad buff, I enjoyed seeing the beautiful and colorful Swiss trains standing as passengers entrained and detrained. We enjoyed a fine lunch and toured the many splendid shops in town. They had absolutely the finest merchandise for hikers and skiers, but to me it all seemed very expensive. I wanted to buy something, a hat I think, but decided to wait and find it cheaper elsewhere; of course, I never saw anything quite as interesting anywhere else. Kitty didn't think I would, and she was right. Her idea is to buy it when you see it, but she has become more frugal of late.

On the journey from Interlaken to Lucerne, we made a stop at the Swiss Open-Air Museum near Brienz. It has some eighty typical houses and farm buildings from all over Switzerland scattered across its alpine pastures. Traditional crafts are demonstrated by costumed workers during the season—we were too late. The interesting thing about the stop was the presence overhead of Swiss aircraft obviously on practice bombing runs. Because of the overcast, we couldn't see them; we'd just hear them fly by followed by a boom that echoed through the surrounding hills. It was totally weird.

We stayed a few days in Lucerne (Luzerne), a German-speaking Swiss city, also on a river with an adjoining lake. One of the city's major tourist destinations, a shop-laden bridge over the river had been destroyed by fire a few months previously, so scratch that one—also scratch the glorious mountain views. Next to our hotel was a shopping plaza with exquisite shops—my favorite sold the most beautiful and expensive musical instruments I'd ever seen. I suggested to Kitty that we visit the Rolex shop; but once inside, we got cold feet when we were ushered to a private viewing room. The most impressive thing about the city was the main railroad station and the hundreds of bikes parked outside. Around 5:00 p.m., we watched as men and women got off the trains and mounted their bikes for the ride home in the rain and fog. We wondered why the station was heavily patrolled by police with drug-sniffing dogs—what were the Swiss into.

A highlight of our stay in Lucerne was a ride to the top of nearby Mount Pilatus. You make the trip on a wonderful Swiss-engineered cog railway that hugs the cliffs all the way to the 7,000-foot summit. It's thrilling to look out of the window and see another train coming down toward you, knowing that you will have to pass each other somewhere, but where. As we rose, we broke through the cloud deck into the sunshine and, voila, we could finally see the Alps off in the distance. It actually seemed warmer on the summit than in the city. We took some pictures at the same locations where Kitty had been photographed on her 1956 trip to Mount Pilatus. We had few encounters with Swiss people who are not known for their friendliness, so I really can't say anything about that. The country is blessed with natural beauty—lakes, rivers, mountains, farms, and forests. Their brown dairy cows are the best. However, its terrain presents a real challenge for highway and railroad builders, so I think they are also blessed with outstanding engineers.

Leaving Lucerne for Germany, we got to the Rhine (Rhein) River at the Rheinfall. The river is the Swiss-German border except where the border swings north so that the waterfall is entirely within Switzerland. The Rheinfall is Western Europe's highest waterfall despite its small main drop of some seventy feet. From the Rheinfall, we drove through Germany's Black Forest, also socked in, and lunched in the pretty town of Offenburg. Our next stop was Heidelberg where we visited the famous schloss that overlooks the city and the university. Unfortunately, it was just about dark by the time we got to Heidelberg, so we didn't see very much. It became apparent to us that this tour tried to cover too much ground for comfort. So we went off to early dinner and bed, to prepare for an early wakeup for the next leg of our odyssey.

From Heidelberg, we motored to the Rhineside town of Bingen where we boarded a boat for our sixty-plus mile cruise down the Rhine. In Bingen, they were setting up for market day, and I'd love to have been able to stay for that, but off we went. Fortunately, the weather cleared a bit so that we were able to enjoy the views along the Rhine—beautiful castles (schlosses) and villages. I also noticed many riverside railroad stations with large car parks. I wondered where the owners worked. It was cold on the river, but I was well prepared—for me—with cap and scarf. We debarked near Koln (Cologne) where we visited the magnificent cathedral, the seat of the Roman Catholic Archbishop of Cologne and one of the largest churches in Europe. The nearby University of Cologne is one of Europe's oldest. But we had no time to visit as we had to move on to Brussels.

The trip to Belgium took us through a corner of the Netherlands. It was nearly dark by the time we got to Brussels. We made a perfunctory tour of the main square and then took off for our overnight stop at Bruges. The city is the capital and largest city of the province of West Flanders in the Flemish region of Belgium. An inland port city, the historic city centre is a prominent World Heritage Site. Bruges, with its quaint old buildings along its many canals, looked mighty inviting and was advertised to be one of the most interesting places on our tour. This proved to be another mistaken notion in a long list of mistaken notions.

Next morning, Kitty and I set out to visit some of the shops but found that they didn't open until 10:00 a.m. As we waited around, we were intercepted by one of our group and told to get back to the hotel as we had to leave early to catch the ferry to England. We learned that the English Channel has become too rough for the hovercraft. We made a mad seventy-five-mile dash from Bruges to Calais in order to catch an early ferry, so as to keep on schedule. The channel crossing was indeed rough, and many were seasick, including Kitty. I had gone to the bar to get us some coffee when the ferry started to roll. With a cup and saucer in each hand, I had no way to grab anything, so I sort of pitched across the room and landed near Kitty without spilling a drop. It was great to see the White Cliffs of Dover looming out of the gloom, indicating that our perilous voyage would soon be over. From Dover, we motored to London, where we had a quick meal in our hotel room, and went off to the theater for a play whose name escapes me. Next day, we went to Heathrow for our flight home.

We had been the guinea pigs for Globus, the first to take eleven-day taste of Europe and also the last. The company realized that the idea was noble, but the reality was anything but. There was just too much travel time and too little fun time. However, for me it was an opportunity to visit places I'd probably never get to again. I loved the few days in Paris and our taste of Switzerland, even though the weather did not cooperate.

It seems to me that the Swiss have somehow ended up with the best part of Europe, along with much of the world's money. The fact that they avoided the world wars enabled them to rake in cash from both sides, and they won't share with anyone.

Moonlighting and Layoff—At Last

In the absence of a firm new assignment, I started thinking about leaving the company. They had a scheme wherein you could submit a letter stating that your job was no longer needed, and if accepted, you would be laid off. That was a far more lucrative way to go than just quitting or retiring. It really was a weird notion because engineers did not have formal positions in the company; they worked on projects as needed. Anyway, early in 1993, I wrote the first such letter, and my request was rejected on grounds that I was needed. After a few months, I tried again, same result. In December of 1993, I started moonlighting on a project that had been secured by an ex-RCA senior management type I'll call Tom. He had worked closely with NSA and was able to secure a small development contract from them. He needed someone to do the input-output hardware and software, and I had been recommended by Clancy Smith, a fellow unit manager on the big FSVS project who was already working with Tom and his team.

When I found myself working two jobs, I increased my efforts to get laid off from MM. Early in 1994, I succeeded on my third try. When asked when I wanted to go, I said tomorrow. The bureaucracy required more time, so my departure date was set as February 25. While waiting to go, I put in token appearances with the group to which I'd recently been assigned. They were nice enough to hold a modest farewell party for me, even though I'd only been with them for a few weeks. At retirement, among the many options I had as a result of working for three companies was a lump-sum payout of my accrued RCA pension. I grabbed it and ran. I also got half a year's layoff pay, plus medical coverage and small monthly payments for my shorter accrued GE and MM pensions. Suddenly I found myself cast adrift from the corporate world after twenty-five years with RCA-GE-MM. It felt good to be on my own and setting my own hours. However, Kitty continued to work for Campbell Soup Company. In March of 1995, I received a letter from MM's president, telling me about the *new* Lockheed Martin Corporation. If I'd waited another year, I would have worked for four companies without leaving the building.

With Tommy, circa 1984.

Tommy dressed to kill in Italy, circa1985.

Tom stopped to visit in Ireland when he left Italy for the USA in 1985. That's one of the Donnelly boys with him.

Two Toms—Donnelly on left, O'Donnell on right—at Altmore, near Pomeroy, County Tyrone. Tommy Donnelly befriended our family on our many visits to Ireland. He and Nora visited with us on a trip to the USA.

The FSVS engineering unit managers with our boss Phil Liefer who is seated. To my left are Tom Sassena, Clancey Smith, and Bob Gordon. Al Stromback was away when the picture was taken. Phil was fired shortly afterward. RCA was very tough on outsiders.

No sooner had our deck been installed than I had to tear up sections to allow access for termite treatment.

From right to left: Kitty's uncles Tom and Vincent, along with Vincent's recent bride Dorothy, circa 1985. Marietta Carlin Sullivan is at the far left of the picture.

Our family in the backyard, late 1980s.

Kitty's 1987 Celica.

Mother sits with Kitty and my sister Pat in our yard at Tenby Chase, late 1980s. I'm standing between two Jims—Jim Maguire at my left, and our Jim at my right.

The Manual Gateway System cabinet with senior engineers Tom at left and Ron at right.

A rare photo has all five of the Carlin sisters with their husbands. It was taken at the Sullivan home in Springfield, Pennsylvania, with my tripod-mounted camera on timed release. *Seated left to right:* Kitty, Joanne Hoops, Marietta Sullivan, and Agnes Fletcher. *Standing left to right:* Phil Hoops, me, Pat Davis, Richard Davis, Ed Sullivan, and Jim Fletcher.

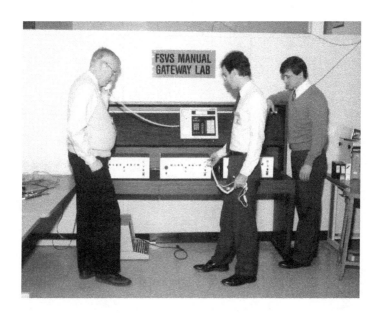

Craig Martin, center, demonstrates the gateway system while one of the mechanical engineers observes.

My group has grown as everyone who has helped in any way at all shows for the photo session. My key engineers were Ron DiUbaldi (seated far left), Barry Lada (standing behind my right shoulder), and Craig Martin (to Barry's right). I've forgotten the names of the others.

Beautiful downtown Raton, New Mexico. It's the first stop after you cross Raton Pass coming from Colorado.

The mysterious Amtrak station in Raton. We saw actual people getting on and off a train. The passengers seemed to appear from and disappear to nowhere.

This site commemorates those who were killed, primarily the Franciscan friars, in the Santa Fe Pueblo Indian uprising in 1680. The natives drove out the Spaniards, but unfortunately for the Indians, they returned some years later to stay.

The beautiful Santuario de Chimayo, the Southwest's most famous church, is on the high road from Santa Fe to Taos. On Good Friday, pilgrims known as Los Penitentes carry wooden crosses to the church and leave them. Some can be seen in the picture.

Taos Plaza, 1990. There were wonderful shops selling
Pueblo Indian artifacts. Sadly, by a return visit in 2000,
all the good shops were replaced by dross.

The last of the original—pre-Manhattan Project—boys' ranch
school buildings is now the Los Alamos County Museum.
Among the school's famous students was writer Gore Vidal
who hated it.

Kitty explores the Indian cave dwellings at Bandelier National Mounument near Los Alamos, New Mexico, 1990.

I'm looking at what's left of a kiva at Bandelier. Kivas were roofed and the site of tribal ceremonies.

The beautiful Navajo County courthouse at Holbrook, Arizona, where we made a fun overnight stop at a motel run by people from Ohio.

Kitty takes a break at Hopi Point on the Grand Canyon's south rim.

Kitty takes a break at the Ramada Foothills Inn, Tucson, Arizona, 1990.

I'm about to be swallowed by organ-pipe cactus at the Sonora-Arizona Museum near Tucson.

Tom Gaffney and I got together for our fortieth high school reunion at Glen Island Casino in New Rochelle, New York, in October of 1990. In the middle is JP O'Neill who sat next to me in home room and tried to moderate my behavior, much to my annoyance.

My sisters and I on the occasion of Mother's wake, which was held in Red Bank, New Jersey, in October of 1990. The famous Molly Pitcher Inn can be seen behind us.

Kitty and I enjoyed our cozy San Francisco motel, Christmas 1990.

Chris and Patrick flew out with us and enjoyed roaming around town—Golden Gate Bridge in the background.

One of my all-time favorite shopping experiences was in this store in San Martin where Tommy was living.

I'm on the beach at Capitola, California, December 1990.

We're all together in the very strange Flying Lady restaurant in San Martin. Note the plane flying by in the upper foreground.

Chris and Pat are dressed for success outside of our motel in the farm community of Gilroy, the garlic capital.

This photo of the boys was taken on Christmas morning, 1990, in our Gilroy motel lobby. Gilroy is located at the southern end of the Santa Clara Valley, a.k.a. Silicon Valley.

Kitty and I on Christmas morning, 1990.

My cousin Pat Browne Rupel had us over for Christmas 1990's dinner at her home in Santa Clara. Her mother, Anne, and my dad were second cousins.

Patrick graduated with a business degree from Drexel University in June of 1991.

Busy O'Donnell Street in the professional district of Madrid.
The O'Donnell and O'Neill chieftains were driven from Ireland
and settled in Catholic Spain around 1600.

We had to get off the Metro at the O'Donnell Street station.

The old city of Segovia still has its Roman aqueduct.

A "home depot" in Lisbon, Portugal.

This is a monument to the people who led Portugal's age of exploration. The leading figure is that of Prince Henry who conducted the famous school of navigation.

Kitty and I take a break on the Lisbon waterfront near the plaza, which commemorates Portuguese exploration and across from the "home depot."

Kitty checks out a souvenir seller at Seville's magnificent Plaza de España. It was built for a 1929 expo.

We are now on the ferry from Algeciras, Spain, to Tangier, Morocco. That's the Rock of Gibraltar in the background.

Kitty is getting up close and personal with one of Gibraltar's Barbary apes. They love to jump on the tourist buses.

It's getting chilly as we wait to board our bus at Gibraltar.

We are with our terrific Globus guide Hanni, a multilingual Italian Jewish woman. She loved to play tapes of an Irish comedian on long trips and also favored sing-alongs.

The beach in front of our hotel in Torremolinas on Spain's Costa del Sol.

This is a stop on the road to Granada. I just liked the picture.
The terrain looks much like Southern California.

The famous Court of the Lions in the Moorish palace of the
Alhambra in Granada. You just have to take this picture.

Here I am with Don Quixote in La Mancha. We had a fine al fresco lunch in this beautiful stopping point on the road to Toledo.

Kitty looks down on the old Spanish capital of Toledo on the Tagus, one of the Iberian Peninsula's major rivers.

The Toledo capitol building of Spain's autonomous community of Castille-La Mancha.

For our last night in Madrid, we had a Globus group dinner at a house once owned by the 1800s artist Goya. I didn't like much of the food in Spain, so this meal didn't disappoint. The décor wasn't that great either.

This house on Jackson Square in New Orleans is where William Faulkner lived when he wrote *Soldier's Pay*, his first novel.

We had a really fine lunch of local goodies at this restaurant in the French Quarter, a.k.a. Old City. We flew into New Orleans in October of 1992 and then went on by car to San Antonio.

I'm checking out one of the aboveground mausoleums in St. Louis Cemetery in the French Quarter.

I'm standing at the American position in the battle for New Orleans, which actually took place in Chalmette, on a big bend of the Mississippi, east of the city.

Shopping in the New Orleans's Banana Republic on the Mississippi isn't much different from your local store.

We made a nice detour along the beautiful Bayou Teche in Louisiana's Cajun Country.

This beautiful home called Shadows on the Teche was open for inspection. While this was probably the best, there were many other wonderful homes along Bayou Teche. That's Kitty standing near the front door.

The steps from this bridge lead down to San Antonio's River Walk, October 1992.

One of the barges that carry passengers along the San Antonio River with frequent stops along the River Walk.

Kitty is sitting on the rear porch of the Lyndon Johnson homesite in Johnson City, Texas.

Agnes and Jim Fletcher joined us on an April 1993 motor trip to Savannah, where we met the Gaffneys who drove up from Florida.

The group toured the Savannah art museum "frames" show.

Kitty's cousin Mary Ruff and her second husband, Vince Schenk, at their home in the Landings, a marvelous gated community east of Savannah.

Agnes and Jim on the road to Asheville, North Carolina.

We visit the home and memorial to Thomas Wolfe, a local who went to New York and became a famous writer. I was an avid reader of his voluminous books during my college years. He died young and is mostly forgotten now.

We meet a friend while strolling in downtown Asheville. In fact, we did meet a retired secretary of General Sarnoff, RCA's founder, and she directed us to a wonderful restaurant out in a country village.

Kitty and Jim Fletcher take a break on the veranda of Asheville's famous Grove Park Inn Resort. The place was built in 1913 and sits in the hills overlooking the city.

Agnes Fletcher and I have a laugh at a family party in the early 1990s.

We play partnership pinochle with the Fletchers almost every Saturday night. This shot was taken in our Tenby Chase living room in the mid-1990s.

I'm at the Tower of London trying to sneak into a picture with one of the "beefeater" guards, October 1993.

This is the hovercraft we took across the English Channel to France. It is fast, with seating like a jet airliner, but you can't see anything out of the windows once she is up to speed.

Here we are at the Eiffel Tower, an engineer's engineering marvel, October 1993.

Here we are inside I. M. Pei's new pyramid entrance to the Paris Louvre museum. It is an impressive piece of work and serves its purpose very well. A section of the museum can be seen through the glass.

We really enjoyed our weekend visit to Montmartre, the highest point in Paris. It's a sort of high-class Coney Island with a great view of the city.

On the stairway down from atop Montmartre, we met some people from New Jersey who took this picture for us. It was a bit chilly.

In the fog, we pass a descending train on the cog railway to the top of Mount Pilatus, Switzerland.

This picture was taken at the Swiss Open-Air Museum near Brienz on the road from Interlaken to Lucerne.

This photo was taken on a restaurant terrace atop Mount Pilatus, near Lucerne. Kitty had a picture taken at this very same spot on her 1956 trip to Europe.

The view from our Lucerne hotel window.

Here I am cruising on the Rhine in Germany. It got chilly, so
I went inside for the next picture.

Cruising the Rhine (*Rhein* in German) is fascinating. There are lots of castles (*schloss* in German) as well as commercial boat traffic, along with neat small villages and rail stations with full parking lots. That's a schloss framed in the window.

The beautiful old city of Bruges, Belgium, was our last overnight stop on the European continent, October 1993. From Bruges we drove to Calais, France, to get an early morning ferry across the channel to England.

PART 7

Retirement Years (1994-Present)

Work and Play

After leaving Martin Marietta, I was able to devote a lot more time to Tom's enterprise that I'd mentioned in the last chapter. I worked on a contract basis wherein I was paid only for the time I actually worked. This suited me very well as I could get in some early morning tennis before rolling into the office after 10:00 a.m. My specific task was to develop a SCSI—an industry standard—interface for the unit under development. This was a natural assignment for me as I had been more or less the input-output expert at RCA, then GE, and finally, MM. Tom's hardware team had chosen a low-end microprocessor with which I had no prior experience, and I would be writing code that would run on the "bare metal." This was software lingo, meaning my code would run directly on the computer chip without benefit of an operating system or any supporting software; in other words, I was on my own. I also had to choose a SCSI chip to work with the microprocessor, one with a development kit, which would allow me to run and debug my code prior to integrating my SCSI component with the rest of the unit. After a slow start, I made rapid progress on my tasks, but Tom was having trouble getting funding to continue all facets of the job. I offered to take some time off; it was spring and a good time to take a vacation, so off we went. Tom never did get the necessary funding for me, so there was no job to come back to.

In April of 1994, we took U.S. Airways to Jacksonville, Florida. From Jacksonville International Airport, we drove south to visit with the Gaffneys at their home in Orange City and then went on to the Beach Club at St. Augustine Beach. One of our purposes was to explore the beautiful city of St. Augustine and then do further research on the area's home offerings. We enjoyed the tram ride around the city and the excellent commentary of our guide, a former officer in the Marine Corps. I particularly enjoyed visiting Flagler College, which occupies the old Hotel Ponce de Leon, a famous luxury resort built by magnate Henry Flagler in 1887. The gorgeous building was wired for electricity by Edison himself, and the windows in the dining room were designed and installed by Louis Tiffany.

The Gaffneys came over to St. Augustine, and Tom and I had a long bracing walk on the beach. We all toured the silly Fountain of Youth attraction, which was inspired by Ponce de Leon's alleged search for same. We had some really fine dinners at popular restaurants right in the center of the city. Our search for housing led us to a new development called

Marsh Creek, which was located near the Beach Club. The houses were exactly to our taste, and the community offered all of the amenities, particularly golf and tennis. Nearby, off-site, were a Catholic church and a bowling alley. Who needed anything more? We chose the house model and lot, and I was ready to make a deposit, but Kitty objected. She wanted to think about it some. Actually, she wanted time to talk me out of the whole notion of pulling up stakes and moving to Florida. As far as I was concerned, we had seen just about everything north Florida had to offer, and this community on St. Augustine Beach was our first choice. We left a solid maybe with the Marsh Creek salespeople and flew back home to sort it all out.

Upon return from Florida, Kitty went back to work, and I began to seriously pursue downsizing as opposed to total relocation. To that end, I engaged some local realtors to help find us a smaller house. One of them worked in Moorestown for BT Edgar, and the other was our friend Liz Oberschmidt who covered Delran and Cinnaminson for Lamon. I told both agents that I wanted a one-story house—or rancher—that was not on a corner lot. Apparently every rancher built in Moorestown was on a corner lot, so that agent quickly tired of working my case. Liz persisted, and I saw many places with her, including the relatively new condos in Cinnaminson. Since I was free during the week, I had lots of time to look at houses, but there wasn't much to see. Liz too was tired of the unrewarding showings, and we left the house hunting in abeyance.

While not looking at or for houses, I was job hunting—looking for a part-time or contract work in my field. I solicited all of the firms who were running ads in the *Philadelphia Inquirer*. While I got some good feedback, nothing promising came up in 1994, so I had lots of time for tennis and reading. Meanwhile, Kitty was still on the job at Campbell Soup's headquarters in Camden. Her full-time position was working out well for her, and she seemed to enjoy her work.

Over the years, I had accumulated a substantial collection of LP (vinyl 33 rpm) records, mostly jazz and Broadway show albums. With the advent of music on CDs (compact disks), I started to replace my LPs with their equivalent CDs as they became available. Unfortunately, many of my favorite LPs never became available in the CD format. Thus, I was forced to do my own transcriptions from LP to CD. This is a time-consuming process, well-suited to a retiree. Because the original LPs were worn and scratched, the resulting CDs were often rather poor, but better than nothing as I planned to dump all of my LPs; and did so. In playing my

Broadway albums, I've become fascinated with the powerful soprano voice of Doretta Morrow. She appeared in many shows and a film, but I was particularly taken with her singing in the show *Kismet.* I looked her up on the Internet and found that there is very little about her life other than her birth in Brooklyn (as Doretta Marano) and death from cancer in London in 1968 at age forty. She remains a lovely will-o'-the-wisp to me.

Fateful Journeys

In January of 1995, we left for an extended motor trip to Florida. We were making a last swing through possible areas of full-time relocation in the area north and east of Orlando. Our first stop was at the Marsh Creek community, which was still my first choice. We moved on southwest toward Orlando and visited a number of promising communities along the way. Our goal was to get to Sarasota where we had beachfront hotel reservations on Lido Key. The area was so changed from my visits in the early 1950s that very little was immediately recognizable to me. We had an enjoyable stay in the Sarasota area and particularly enjoyed sitting on our balcony and watching the sun set over the Gulf of Mexico.

As we talked over what we'd seen of Florida, it became clear that Kitty was not willing to pull up stakes and move to Florida. She thought we should consider buying a winter place such as our friends the Dennens had bought in Boynton Beach. They had suggested that we come by and check out their community on one of our Florida trips. Well, it now seemed like a good idea, but we had plans to visit one of Kitty's recently retired coworkers who'd bought a place north of Tampa, and we couldn't do both. We had a nice visit with Ann and Bill Gillespie at their home in Spring Hill. I was very envious of them as I sat on their porch, which was right on the golf course. Leaving Spring Hill, we drove north to I-10, which we took east to Jacksonville where we picked up I-95 for the long ride home. This was one of our longest stays in Florida, almost two weeks.

A week or two after we got back from Florida, my sister Pat received a call from Florida saying that Cousin Jackie had died. The caller was Mr. Anderson, one of a couple who were members of Aunt Mary and Jackie's Evangelical Methodist church. The Andersons were looking for a family member to take charge of the funeral, the house, and Aunt Mary who was in a nursing home. I immediately called the Andersons and asked them to hold the funeral until I could fly down to be with Aunt

Mary. Thanks to the Andersons, it all went off quite well. It turned out
that their church had been sold to a Finnish congregation, but we were
able to have the funeral conducted by their old Pastor Lundberg in the
old church. I picked up Mary from the nursing home and took her to
the church and the nice reception afterward. We met a few of Mary and
John's old friends from New York, people that I hadn't seen or heard of
for more than fifty years.

The Andersons were a really nice and helpful couple. They took
me to the Sander house in Lantana, and it was in a terrible state. I told
the Andersons to take what they wanted, and they took the television
set and perhaps, some pictures. They told me that Jackie had worked
around the old church, mowing the lawn and tending the shrubbery,
and was popular with the parishioners in spite of his acute alcoholism.
One could see the evidence throughout the house—empty Southern
Comfort bottles piled up in his closet along with case after case of
empty Budweiser cans. He had called me several times to warn me that
he might not be around for my next visit, so I'd better get on down
to see him. I felt bad that I hadn't seen him before he died. I searched
through the house and took everything of value that I could fit into my
luggage for the flight home. You couldn't stay very long because of the
fleas that quickly found you to be a desirable host. Meanwhile, I found
a local realtor to take the house off my hands for $22,500, and some
local car buffs to take away the two cars that stood in the yard full of
junk, like the house itself. I told the car guys that they could have the
cars so long as they took them with the stuff in them. They were not
thrilled with the deal, but I explained my problem, and they were good
enough to help me out.

The Andersons had mentioned to me that they would be selling their
villa in order to move to an assisted living facility near Orlando. I had
to fly back to Florida to sign some papers, which transferred the house
proceeds to Aunt Mary's nursing home, Boulevard Manor in Boynton
Beach. Kitty decided to come along to check out the Anderson home as a
possible winter place for us. We liked the house very much, two nice-sized
bedrooms and two baths and in a nice community. We drove around to
check out the clubhouse and pool areas and found them strangely empty
on a beautiful March day. We did see a few people on foot but with canes
and walkers. In fact, both of the Andersons' neighbors were on oxygen.
Kitty felt that we wouldn't be happy in such a geriatric setting, so we
passed on the Anderson home but treated them to a nice dinner at their

favorite Thai restaurant. Though we would remain in touch by phone and mail, that was the last time we saw that wonderful couple.

Since we were in Boynton Beach, the time had finally come to visit the Dennens and check out their community. We drove to Hampshire Gardens and saw that it was nicely situated on the Intracoastal Waterway and had much younger and more active residents than the Andersons' community. We were presold when Marion Dennen had Pat Downey show us a selection of one- and two-bedroom units that were for sale. We chose the one-bedroom unit closest to the water that was closest to move-in condition. Because Hampshire Gardens is a co-op, not a condo, we had to be interviewed by an owner's committee. Since our plane was leaving in the morning, a committee was quickly assembled for the interview by Hank Zaranski, the building president. That evening we dined with the Dennens at Gentleman Jim's, which was right next door. If approved, we agreed to pay $28,500 for the unit, with a $1,000 deposit and a closing in October of 1995. We assumed that we would be approved and received a formal letter to that effect in early April. We were really excited about this unexpected turn of events.

Back to Work for Me

Sometime after we returned from Florida, I was offered a software engineering position through Texcel, a contract engineering firm located in Lafayette Hills, Pennsylvania. My assignment was with Telesciences, a Moorestown, New Jersey, firm founded by ex-RCA people and located on New Albany Road, which was very close to home. The company made customer billing systems for telephone utilities. It was no coincidence that my task was again to develop the software to allow a new piece of Telesciences equipment to be connected to existing telephone taping systems. Everything about the assignment played right into my strengths, telephony and input-output interfaces. It was great to be back working in a hardware-oriented company, where I could do the software the way I thought best, that is, without interference from management. It was hard going at first, as they didn't have the tools I needed to test the code as it was developed. In fact, I had to bring in some software from home to use while my purchase requisitions dragged through their system. But it all came out well and proved to be a most enjoyable job.

What was really interesting about the whole experience is how they happened to choose me for the job. Manny Fard, the young Iranian engineer at Grim with whom I played tennis, was now at Telesciences. He was trying to do the software but had other assignments as well and wasn't making much progress. When they decided to get someone from outside the company, they had Manny screen the resumes. When he saw mine, he told management to go after me. They did, and so Manny and I were now working together. He was not a trained software person in that he didn't know how to set up and manage a software project. He did have some familiarity with the C programming language, so he tried to follow along with what I was doing. Just for fun, I made life difficult for him by taking advantage of C's loose syntax to write segments of obscure code. Manny would say, "Jimmy, don't write C 'weenie' code." Eventually, Manny moved on to tackle field service problems and left me on my own.

By the fall of 1995, I had successfully concluded my assignment, and I started to collect unemployment payments while I looked for a new job. By then they had changed the unemployment system so that it was no longer necessary for me to report to the office every two weeks. That would later make it feasible for me to collect unemployment payments while I was job hunting in Florida. In October, we had a "closing" on our Florida apartment. Closings of Hampshire Gardens co-op units were really informal. We sent a check down to the building president, and he sent back an executed stock certificate and proprietary lease, which granted to us the right of occupancy.

Our First Winter(s) in Boynton Beach

On January 10, 1996, Kitty retired from Campbell Soup with an adjusted ten years' service, earning a small pension and medical benefits until she reached the age of sixty-five. Her retirement made it possible for us to spend our first winter at our Florida place. Off we went on the long trip down I-95; it turned out to be 1,170 miles. We spent the first night in South Carolina, finally choosing Santee as our preferred stopping point, having done 640 miles the first day. That usually got us to our Santee motel around dinner time, unless there was a major stoppage on the highway. In the first years, we stopped at Clark's motel, which had a decent restaurant. In later years, we've been staying at the Best Western in order to earn points for a free night.

We are usually among the first arrivals for the Continental breakfast, which starts at 6:00 a.m. We then leave the motel about six thirty and drive south through South Carolina in the dark—the sun's up by the time we reach the state line. From Georgia's south, it's usually a fast trip all the way down through Florida, except for the city of Jacksonville. I-95 through that city, Florida's largest, has been under construction since we began driving to Florida. God knows when it will be finished. We generally arrive in Boynton Beach about 3:00 p.m. We then stop and pick up some breakfast items at the Publix supermarket before we go to the apartment. After unloading the car and making coffee, we have to start cleaning all surfaces as they accumulate an ungodly amount of grimy dust. I came to believe that the dirt was a consequence of the exhaust of many boats plying the adjoining Intracoastal Waterway and the diesel locomotives on the nearby Florida East Coast Railroad (FECL).

It takes a while to get used to Florida and communal living, but we did pretty well. I was anxious to get started playing tennis, and the Delray Beach courts were recommended to me. The first years I was able to get games regularly and enjoyed Delray very much. However, a new management took over and raised the fees for out-of-towners. At the same time, the Canadian dollar went south, and many of those players went elsewhere. It became nearly impossible to get games, and after some six years at Delray, I left for the Boynton Beach courts where I'm now playing regularly for less than half the cost of Delray.

Although I no longer play tennis in Delray, it is a beautiful city with fine shops and restaurants and an excellent esplanade along the ocean. In early years, we enjoyed the art shows and other amenities on weekend evenings. However, the city has been transformed by the construction boom, with pricy townhomes rising all over the downtown area. Though we still visit some of the shops, Delray is losing its charm for us. With the rising price of movies, we've started watching DVDs at home and no longer go to the Regal 18 Cineplex in Delray Beach.

Agnes and Jim Fletcher came down for a visit that first winter, and we were able to keep up our pinochle competition. They were able to visit again the next few years, but the steps to our second floor unit eventually became too much for Agnes. One year we went to Everglades National Park via the Shark River entrance, which offered a tram ride out to a viewing station. The day was nice when we started in our summer clothes, but rain and cold moved in, and the tram offered no real shelter. When we got back to the base, we raced to the car and cranked up the heater and made an immediate stop for hot coffee at a nearby McDonalds. Another year, Agnes and Jim arrived in time for a nor'easter that blew for about three days. Jim and I enjoyed going out to the Boynton Inlet Park to watch the huge waves crash ashore on the breakwater. Over the years, we took the Fletchers to some other tourist venues, including the Flagler Museum located in the 1913 Flagler mansion in Palm Beach and the boat tour of the Palm Beach waterfront from the Lake Worth lagoon. Jim and I visited the West Palm Beach zoo. Although it is small, it has a wonderful river otter exhibit.

Off to Memphis

Back in New Jersey, I resumed my job search, and we resumed our search for smaller quarters. We saw quite a few houses around Cinnaminson but nothing of interest. My job search also proved fruitless. In September my sisters and I, along with our spouses, attended a reunion for graduates of St. Margaret's School. The occasion was either the seventy-fifth or the seventieth anniversary of the opening of the school. The dinner-dance was held in Yonkers, and there was quite a nice turnout. However, I was a bit disappointed that there were so few from my class, probably only six of us. It was nice to see those who did attend, particularly Chris (Finn) Fox and John Gallagher.

We had often spoken to my sisters about making a trip to Memphis to visit Graceland, home of the late Elvis Presley, with side trips to other places along the way. Pat and Jim Maguire expressed interest, and in October of 1996, we set out in their Toyota Camry for Chattanooga, Tennessee. On that leg of the trip, we stopped in Staunton, Virginia, to visit the birthplace of Woodrow Wilson, our twenty-eighth president.

I'd always wanted to visit Chattanooga because of its famous Civil War Battle Above the Clouds. In the Union attack on the Confederate positions on the heights of Lookout Mountain, the Union commander lost contact with his men because they went into the clouds, which covered the rebel positions. We drove to the top of Lookout Mountain from which the entire city lay at our feet in a bend of the Tennessee River. We could see all the way out to the Missionary Ridge battlefield, which lies well to the northeast of the city. We stayed at the Chattanooga Choo Choo, a hotel, which was attached to the beautiful old Union Station. Rooms were also available in railroad sleeping cars parked on the tracks behind the main waiting room, which now serves as a restaurant and lounge. It was all very nice, indeed.

One evening we went out to dinner at one of the recommended steakhouses, something like an outback. We were seated in a booth alongside the bar where we were exposed to constant traffic—waiters carrying trays of drinks and food. I was sitting on the outside and aware of the possibility of a disastrous collision, when it happened. A waiter carrying a large tray of drinks dropped the whole tray contents onto me. I was covered with red wine, Bloody Marys, you name it. I screamed that my brand-new shirt was ruined, and the waiter was most apologetic, saying that he was new and so on. I think Kitty and Pat thought that I overreacted, but I was wet, and they were dry. The management gave us our food and drinks for free and offered to pay for a new shirt. I don't remember if I took them up on it or not; however, my shirt proved to be fine after a washing. Next day we visited the excellent Chattanooga art museum, which sits on a bluff overlooking the river.

From Chattanooga, we set out for Nashville with a side trip to one of Tennessee's famous distilleries that of the George A. Dickell company in Cascade. The grounds and company store were very nice, but we didn't get to see the production process nor the facilities. Perhaps we were out of season. In Nashville, we did many of the principal tourist stops, including the Hermitage, the beautiful home of General, then President, Jackson. I was not too impressed with the Country Music Hall of Fame, but a visit to the famous RCA Studio B where many classic country recordings were made was interesting. The studio has changed physically to accommodate the changing sounds of contemporary.

I'd always wanted to experience the *Grand Ole Opry* and visit the famous Ryman Auditorium from which the show was broadcast every week for years. Now the *Opry* originates from its very own Grand Ole Opry auditorium within the Grand Ole Opry theme park outside of town. We went to lunch at the Opry hotel and purchased tickets for that night's performance. The place is huge, and we took the boat ride around the interior of the hotel and that is pretty incredible. The Opry auditorium is also huge and our show featured some top performers including Vince Gill and Sawyer Brown. However, for me, the highlight of our visit to Nashville was a show at the Ryman Auditorium. Although the Ryman is too small and no longer home to the Opry, it has been beautifully maintained. At the Ryman, we saw a show called *The Lonesome Road*, the Hank Williams story with lots of his words and music. It was beautifully done; to my mind he was the all-time best country artist.

From Nashville, we pushed on west to Memphis, which is a port on the Mississippi River. We stayed in a hotel on a downtown main street, not far from Beale Street with its blues joints and monument to WC

Handy. There were no formal sightseeing tours, so we visited the major attractions in Memphis on our own. I was most interested in the scale model of the Mississippi River system built by the Corps of Engineers on Mud Island. There's an aerial tramway out to the island where we walked the quarter-mile length of the model, which shows all of the system's input and output waters, with exact flow measurements. It is an engineer's dream. Different, but just as impressive, was the Danny Thomas memorial at St. Jude's Children's Research Hospital. Using his show business popularity, Danny had founded and raised a great deal of money for the hospital, which has remembered him with a beautiful memorial chapel on-site.

Of course, we had to see the ducks at the Peabody Hotel. The elegant hotel's famed ambassadors are five Mallard ducks—one drake with his white collar and green head and four hens with less colorful plumage. The ducks are raised by a local farmer and a friend of the hotel. Each team lives in the hotel for only three months before being retired from their Peabody duties and returned to the farm to live out the remainder of their days as wild ducks. At noon the ducks ride down on the elevator and step off for a short walk on a red carpet to a fountain in the center of the hotel's courtyard. There they enjoy a romp in the water before being taken back upstairs. Their foray draws a large crowd every day. The crowd is so big, and the ducks so little, it isn't easy to find a vantage point from which you can see them. Our spot was on the mezzanine overlooking the courtyard.

When I said we had to leave early the next morning for Graceland, my sister objected. She felt that since we were on vacation, we should be taking it easy with a leisurely breakfast and so on. I explained that we were not on vacation; we were on tour with a tight schedule. It was a good thing that we got to the Graceland visitor's center early as we were able to get right on a bus to the actual homesite. I found my visit to Elvis Presley's home and its many outbuildings to be very evocative of the man and his times. I was moved, particularly so at the family gravesite. By then it was raining, increasing the sense of something lost. By the time we got back to the visitor's center, we saw crowds waiting for the buses and agreed that it was good that we had gotten there early.

Although, overall, I was not happy with our treatment by service personnel at our hotel and elsewhere in the city, we did find some interesting eating places on our own. The Arcade Restaurant was a favorite of Elvis while he was in Memphis. We went there for lunch and had his booth, which now has an Elvis bust overlooking the table. We also found, with difficulty, a place right on the Mississippi that was different and gave us views of passing trains and workboats on the river. On our last night in Memphis, Kitty and I went to dinner by way of the streetcar while Pat and Jim took the car to a gambling casino in Mississippi. Unknown to us, the streetcar stopped running at an early hour and was not available when we left the restaurant. However, we were able to get a taxicab back to the hotel, so no harm done.

On to Arkansas, St. Louis, and Falling Water

From Memphis, our tour took us across the Mississippi to rural Arkansas. It was cotton country, and large covered mounds of cotton bales awaited pickup all along the roads. We stopped for lunch at a new but old-style drive-in restaurant in the little town of Blytheville. A pretty young carhop took our orders, but we got out of the car to eat at the tables. On the way back to the main highway, we spotted a John Deere showroom. I stopped because I wanted to buy a John Deere green hat. Of course, everyone came with me—the staff brewed up some coffee for us, and we had us a real down-home visit. My sister Pat bought some John Deere toys for her grandchildren; I don't know how we found room for them as the car was already fully loaded.

Once we got into Missouri, I wanted to go to the Trail of Tears State Park just north of Cape Girardeau and view the Mississippi from a bluff in the park. But when we got to the park, it was raining and my passengers felt that the road to the bluff was too hazardous; hence, I was outvoted and went back to the main highway, U.S. 55. Rather than mush on to St. Louis and arrive during rush hour, we decided to stop for the night in the riverfront town of St. Genevieve, Missouri. We asked at our motel for a good restaurant right on the river. They sent us to one that seemed quite satisfactory until we saw that a twenty-foot floodwall lay between us and the river. After dinner we walked on down to the flood gates that were open and gave access to the levee and the Mississippi.

We arrived at St. Louis on a cold, rainy, and windy day. After checking into a hotel, we went down to the waterfront and toured the very interesting Museum of Westward Expansion, which essentially commemorates Jefferson's Louisiana Purchase and explorations of the west. All except my sister went up into the arch in the very claustrophobia-inducing transporters. Between the height and the sway of the arch, I felt very uncomfortable on top and was glad to get back to the ground. We took a drive out to one of the large city parks and visited the old original Catholic cathedral. There I was quite taken with a statue whose base was engraved with the words "St. Louis/King." We spent a couple of hours on a gambling boat tied up along the levee. The hill section of the city was a well-known Italian enclave. Jim Maguire had made many business trips to St. Louis and knew an excellent restaurant and, it proved to be first class. No trip to St. Louis would be complete without a visit to the Anheuser-Busch brewery, home to the famous Clydesdale draft horses. The plant tour was probably the best I've ever taken, and the beautiful round stable with horses, dogs, and cats is totally unique. The plant grounds are quite beautiful, and the tourist facilities excellent.

Pat felt that we should visit Lincoln's home in Springfield, Illinois, on the way home, so we had to leave St. Louis right away to keep on schedule. In Springfield, we toured the Lincoln Memorial/Tomb and his final home, had lunch, and moved on. The city of Springfield seemed like a nice clean and orderly place with an imposing state capitol building and a beautiful new Lincoln Library. Leaving Springfield, we had to drive east through much of Illinois, then all of Indiana and Ohio, before our next stop in Pennsylvania. My sister Pat recalls that we stayed overnight near Dayton, Ohio, and that she had an excellent steak at a nearby steakhouse.

Our final tour stop was at Falling Water, the famous Frank Lloyd Wright house on Mill Run, near Bear Run, Pennsylvania. The house is cantilevered out over a waterfall in a most picturesque setting, especially so when the woods are in their fall colors. The house was built in 1935 as a vacation home for a Pittsburg department store owner who had many tussles with Wright over the design and furnishings. While the house has leaks and aging problems, it is undoubtedly the most iconic home in the USA. Touring Falling Water with us was the University of Notre Dame glee club. While we and they were on an upper balcony, I asked them to do a song for us and they did. I've forgotten what they sang, but it was totally appropriate for the setting and the occasion. Although we had many miles to travel home from Falling Water, the glee club provided a fitting climax to our long tour.

Florida, Key West, and Back to Work

The remaining fall months passed uneventfully as we prepared for Christmas, New Year, and our second winter in Florida. The winter of 1997 brought a visit from my sister Pat and Jim. They drove down from

New Jersey because Jim wanted to have his golf clubs at the ready. He did get to play with the Hampshire Gardens men's group. We all wanted to visit Key West, so Kitty, with difficulty, got us hotel rooms through the tourist office. Upon arrival, we found that our rooms were in a "gay" hotel called the La-Di-Dah. We were politely told that a mistake had been made, and we were shunted off to another hotel, which was very nice but not in the center of town. However, we were allowed to park at the La-Di-Dah during our stay. We enjoyed our brief visit to Key West but found the two-hundred-mile drive back to be most tedious, so much so that I won't be making the drive again, I hope.

During the Maguire's visit, we played some Canasta, but it didn't seem to me that it was as much fun as I'd remembered it. That the ladies were winning may have been another reason for my discontent with the way the game was going. One day Jim and I went to the Boynton Beach library and looked up the rules for Canasta. The missing ingredient was icing the deck, which was not covered in the simple rules we were using. Icing is the main strategy of the game, and Jim and I wanted to incorporate it in our play. The ladies went along rather reluctantly, and of course, they never won another game. They didn't like the cut throat nature of Canasta when played with "icing."

When I got back from Florida, I resumed my search for a temporary software assignment. I directly contacted the people at Telesciences and signed a contract to expand the work I had done in 1995. Between then and 1997, Telesciences had been taken over by a British telecom firm called Securicor. In the summer of 1997, we had a big outdoor celebration when Securicor Telesciences became Axiom, and I got a T-shirt. I don't recall much about that summer at Axiom, but I know that I enjoyed working with the hardware engineers to arrive at a successful product.

Another family milestone was reached in June when Jimmy graduated from Rutgers with his degree in computer sciences. The graduation ceremony was held in the Tweeter Center in Camden, a very nice partially open venue. Afterward Jimmy, Patrick, Kitty, and I had a late dinner at the locally popular Pub restaurant. Jim had interned at several local firms and had a choice of jobs. I tried to steer him toward Computer Sciences Corporation, a big government systems software firm in Moorestown. However, he chose to go with a Moorestown firm that provided security systems for retail establishments. After graduation, he moved out to live with some friends in a rental house in Riverton, or perhaps it was Palmyra.

When Jim moved out, he left us with an "empty nest." For some time, Patrick had been working for Merck-Medco in North Jersey and living in the upstairs apartment of the Michael Quinn homestead in Clifton. The house was owned by Eileen Quinn Dodd and her husband, Joe, and had been in the Quinn family for almost sixty years. I had stayed in the house back in 1956, when on a five-week assignment with RCA's plant in Harrison. Merck subsequently divested Medco, which became Medco Healthcare Solutions, a move that proved to be a bad one for Merck's cash flow. By 1997, Chris had been out of Drexel for nine years and living in a house he had purchased in nearby Riverside. Chris, ever the entrepreneur, remodeled the basement into a nice small apartment, which he rented. He then went on to redo his kitchen with new cabinets and appliances. But the best thing about the house was that it had a large two-car garage, which easily accommodated his automotive hobbies. We were glad to see all of his cars and car parts eventually move from our garage to his. He even took the sailboat with him.

Off to the Pacific Northwest and Canadian Rockies

In the fall of 1997, we took another Globus tour, this time to the Pacific Northwest and Canadian Rockies. The tour originated and terminated in Seattle. This gave us a chance to see some of the city, including a rare glimpse of far-off Mount Rainier. From Seattle, we were bused east through the magnificent Cascade Mountains and then followed the Columbia River north into Canada. We caught our first view of the ice field-covered Rockies appropriately enough at Glacier National Park. From there we went on to Banff and Lake Louise in the province of Alberta. The latter was pretty much fog shrouded as were most of the other lakes we passed. Our highly-touted boat ride on Maligne Lake proved to be a bust. It snowed a beautiful snow overnight, and we could see very little from the boat, so we had a snowball fight with another tour boat. The highlight of the trip thus far was the ride up Athabasca Glacier so that we could get out and actually walk on an ice field. You make the trip up the glacier in a special vehicle called a Brewster Coach. It was very cold and snowy, so I did not linger outside the coach.

From Jasper National Park in the Canadian Rockies, we set out for Vancouver on the coast of British Columbia. We made an overnight stop in the town of Kamloops, which is located on a tributary of the Fraser River. The town is elk—actually wapiti—heaven, with the beasts roaming the streets in herds. While we were eating dinner in the hotel dining room, elk were looking at us through the window. After dinner we went for a short walk through the center of town, dodging elk droppings all the way. As darkness fell, the herds started moving through town toward wherever they spend the night.

Our next stop was our hotel in Vancouver, a major Canadian port on the Pacific coast. We did lots of sightseeing around the city and had an excellent salmon dinner at a waterfront restaurant. The cold and foggy weather obscured some of the best views, all of which were across Burrard Inlet to the north. The sky ride to the top of Grouse Mountain yielded no view of the city as promised. We made a very informative visit to a salmon weir and hatchery and to the gorge at Capilano Park. The main feature of the park is a long suspension bridge some 230 feet over the gorge of the Capilano River. We were promised a certificate if we made it across the bridge and back. Kitty and I got about halfway across the swaying bridge when fear set in, so we decided to forego the pleasure and the certificate and scrambled back to safety. For giving it such a good try, we were given our certificates.

From Vancouver we took a long ferry trip to Vancouver Island and the provincial capital of Victoria. It seemed to be a beautiful and interesting city, but heavy rains limited our activity. Between showers, we did manage to have tea in the elegant Empress Hotel, the city's finest. The whale-watching boats were not permitted out of harbor due to rough seas. So while Kitty shopped, I took the guided tour of the capitol building. My tour group included a five-time Jeopardy champion who asked me not to mention it to the others in the group. I have seen him on many Jeopardy programs that feature former champions. After the tour, I set out for our hotel and had difficulty finding it. When I arrived, I was soaked to the skin and cold. I took a hot shower and went to bed, which is where Kitty found me when she finally got back to the hotel.

Next day we set out for the ferry back to the mainland but stopped on the way to visit Butchart Gardens. This magnificent arboretum occupies an abandoned quarry. We were the first arrivals of the day, and fog still lay in the lower elevations. By the time we left, the sun had broken through, and Japanese tourists were pouring into the gardens. It was a beautiful place, and we had a commemorative framed print shipped to us; it now hangs in our dining room. From the ferry, we went directly to Seattle for the flight home. We had decided to stay overnight and go back east in the morning. When we were dropped off at the Globus hotel, rain was coming down like a monsoon. We needed a taxi to get to the hotel where we were staying, and they were hard to come by on a rainy Friday evening in downtown Seattle. I was thoroughly soaked again by the time we got to our hotel. We inquired at the desk for a good nearby restaurant, and one was suggested, allegedly a few blocks up the avenue. We asked for a taxi and were told that it wasn't necessary, rain had stopped, and it was nearby. Well, it turned out to be many blocks up a steep hill in the University Heights part of the city. We were much annoyed by the time we finally found it. It turned out to be a very nice restaurant, and the hosts took our picture and mailed it to us. They also drove us back to our hotel and said they would have picked us up if we had called. Sob, choke! Thus, our trip ended on a relatively high and, finally, dry note.

Two Out of the Blue

Back in Delran—to an empty house and garage—we became even more determined to find a smaller house. We had been looking seriously at Holiday Village East, a fifty-five-plus Hovnanian community in Mount Laurel. Though not thrilled with the idea of moving out to Mount Laurel, we seemed to have no viable alternatives. A two-bedroom model with optional loft and two-car garage was what we had settled on. What finally turned us off were so-called lot premiums that surfaced at the last minute. None of the salespeople had ever mentioned that all of the remaining lots were "premium" lots, priced at $25,000 or so. The lot premiums put the new home price out of our comfort range. We took another look at the two condo communities in Cinnaminson but could not imagine that we could be happy in either. So as we prepared for our third winter in Florida, we had no plan for downsizing.

Out of the blue, we got a call from our son Tom who was a patient at the Veteran's Administration (VA) hospital at Coatesville, Pennsylvania. On the phone, he told us he had a small job in or around the hospital and wanted to see us. He got transportation into the city where we picked him up and took him to the apartment he had been sharing in northeast Philadelphia. He wanted to clear out his stuff and have us hold his things for him. He had told us that the VA had found him a better job but that he could only take it if he had a car. Our Jim had an old car that he was willing to sell to us, and we bought it for Tom. That was in the winter of 1998, and we haven't heard from him since.

By the winter of 1998, I was president of 2460 Corporation, the stock company that owned our building of twenty units—fifteen one-bedroom and five two-bedroom. Since we were a co-op, one of my duties was to interview prospective new stockholders when units were being sold. One of the first couples to be interviewed and approved was the Minnekers—Katie and Gordon—who were buying the unit next door to us. Originally from New Jersey, they had been living most recently near her daughter in the Williamsburg, Virginia, area. Originally, Katie had been in the real estate business in New Jersey, and Gordon had worked for Ford. After extensive renovation of their unit, they finally moved in, and we became very friendly with them. Katie was very competitive and loved to play games, as did Kitty and I. Gordon usually sat out the games, preferring to read or watch television. Katie became very involved with community activities, especially the annual show.

Kitty's sister Agnes and husband, Jim Fletcher, arrived on a Monday for a short stay with us. In addition to their baggage, they brought news of a house they had seen over the weekend at an open house. As described by them, it seemed perfectly suited to our needs and was in a great location close to their house. We had Agnes call Sandy Fletcher—married to her son Michael—and ask her to inspect the house, make a video tape, and send it overnight to us in Florida. As we sat around and viewed the tape, it was readily apparent that the Cherry Lane house and grounds were great, but that the thirty- to forty-year-old house would need a lot of work to be acceptable to Kitty and I. In particular, the windows were horrible and would have to be replaced with larger and more modern types. A new kitchen, master bath, and roof would be required. The oil heating system, ancient air-conditioning system, and electric hot water heater would all have to go.

Apparently lots of people had seen the place but were turned off by the work it needed. Jim Fletcher was confident that we could do much of the work ourselves. With that in mind, we made a $165,000 offer over the Internet, and it was quickly accepted by the attorney representing the incapacitated seller. We flew up to New Jersey for our first inspection and to sign the sales contract. Kitty had lots of misgivings but went along with the program. We quickly put our house on the market with our friend Liz Oberschmidt who was with Lamon Realty, a Cinnaminson firm. While back in Florida, it became apparent that, contrary to what we thought, our house was not going to be a quick sale. That meant that I might need a swing loan in order to make the mid-May closing on the Cherry Lane property. While in Florida, I started making calls to possible lenders and learned that since we had no income, we were not likely to get either a swing loan or a conventional mortgage.

Once back in New Jersey, we did everything we could think of to get our house presentable to potential buyers while at the same time planning for our move to Cherry Lane. The people at the firm representing the seller were anxious that I get a mortgage commitment as the May 15 closing date drew near. One of the Fletcher boys, probably Dennis, put us onto a broker whose specialty was getting "problem" mortgages. We were quite relieved when Carlos reviewed our assets and saw no problem getting us a "no-income" loan. We had pretty much completed the kitchen-dining area redesign while in Florida. On closing day, Jim Fletcher came by to get the house keys and immediately started tearing out the old steel GE kitchen cabinets. They were quite beautiful and extremely well made. I kept a few for a workbench in the garage, and we dumped truckloads of them at a nearby metals dealer.

In my new and unintended role as general contractor, I had many concerns. As a first order of business, I had to arrange for the replacement of the old air-conditioning, heating, and hot water systems with up-to-date gas appliances. A real concern to me was the huge 1,000-gallon inground oil tank. Before closing on the property, it had been tested, and no leaks were found, but I wanted it out of the ground and gone. Since there was no gas service to the house, I had to choose a contractor who could get the necessary permits, remove the tank, and schedule a gas hookup with Public Service Electric & Gas (PSEG). I contracted with All-Ways of Maple Shade—recommended by Mike Fletcher—to handle all of this, and it took months to accomplish. Meantime, we worked through the summer on other problems without the benefit of air-conditioning.

Another major consideration was windows. We did not like any of the existing windows—neither type nor placement. We wanted much larger windows in entirely different locations. I got bids from several places for Anderson double-hung, tilt-wash windows, with the colonial-style mullions. We were buying sixteen large windows and wanted them delivered as needed. Peter Lumber, of Medford, offered good pricing and free delivery and helped with some design details. So while we were set with windows, we would be making substantial changes to the exterior of the house; thus, we would need a supply of the cedar shakes that had been used for the original siding. Andy Fletcher picked some up for us at Diamond M, outside of Mount Holly. We decided not to apply for a building permit for this work on the grounds that we were simply replacing existing windows. Fortunately, the building inspector never raised any questions about the entirely new look of our exterior.

We decided that the screened in the porch had to go as it darkened the kitchen and dining room. It was a tough job as it had been built into the house, and its removal meant structural changes to the roof. It was a big job and done by the Fletcher Brothers under the direction of Jim Fletcher. The roof affected roof area was shingled temporarily as the entire roof was to be done while we were in Florida. We had added a sliding glass door off the kitchen dining area, but it did not open onto the existing patio, so we couldn't use it. One day Dennis Fletcher came by and asked if I'd like a deck to go over the window well outside the sliding door. He had taken one away from a job they were doing, and it was on his dump truck. With that temporary deck in place, we were able to use the new door. We eventually added a new deck large enough to provide access for both kitchen and dining room sliding doors.

After soliciting bids from several kitchen vendors and installers, we decided to go with KraftMaid cabinets. They were available from Home Depot, and they had made a design and given a price, but then lost everything in their computer system. That turned us off Home Depot. We finally went with Contempo, a small Cherry Hill firm. They were very competitive price-wise, but miles ahead on customer service. The person we worked with came out to the house several times before, during, and after installation, which was done by Chris and Jim Fletcher. Chris made a jig to enable him to accurately install the thirty-plus cabinet knobs that cost over $8 each. Contempo also arranged for fabrication and installation of new kitchen countertops. Once the vinyl floor was put down, we had ourselves a great kitchen-dining area. Everybody loved the pendant lamps we hung over the counter separating the two parts of the room.

Bathrooms also presented a challenge. We completely gutted the master bath, which meant that we had to buy everything needed for the new bathroom. Consequently, Kitty and I had to spend an entire morning at a bathroom supply house choosing shower, vanity mirror, plumbing fixtures, and more. Our neighbor tiled the floor for us, and Fletch and Chris did the rest of the job; it turned out very well. After much thought, we decided to keep the old, but rather handsome and high quality, existing American Standard coral fixtures in the guest bathroom. We were so happy with Contempo that we had them do a new countertop and that too went off without a hitch. Lastly, while in Florida, we had the tub area redone and a beautiful floor installed using large eighteen-inch ceramic tiles in a diagonal pattern. That room turned out very well, and we were glad we kept the old sink, toilet, and tub. Both bathrooms had large widows, which made them very bright. We eventually replaced the original canopy windows with colonial-style sliders.

We Finally Sell, and I'm Called Back to Work

Jim Fletcher and I worked every day on the house; he did the skilled jobs and I the more labor intense. We always took midmorning and midafternoon coffee and donut breaks. Meanwhile our house in Tenby Chase was languishing—we'd had no offers. Kitty felt that we had to upgrade the kitchen, but I dismissed this idea as foolish. Now that we had a mortgage and were keeping up two houses, something had to be done. At Kitty's insistence, I convinced Fletch that we needed to take time out to replace the kitchen countertops, range, and range hood at the Tenby house. We ordered an L-shaped countertop at Home Depot in Philadelphia, which we picked up in Fletch's truck and installed without much trouble. Along with the countertop, we put in a new sink and faucet set and the aforementioned range and range hood.

Not long after we upgraded the kitchen, we got our first offer, and it was a low one, much lower than we had hoped for. But with no other buyer in the wings, we settle for $195,000 with a mid-July closing. Naturally the young couple brought along her father who found all kinds of problems for us to fix before closing. Fortunately, their engineering inspection found only eight problems, of which we agreed to fix four. Thankfully, the rotted out living room window and the destroyed parquet floor under the family room carpeting were never mentioned. We came away from the closing with enough money to pay off the mortgage on Cherry Lane but not much to contribute to our costly projects. Based upon recent sales and our very desirable Tenby cul-de-sac location, we had expected to clear about $25,000 more than we did. But the house had been well used by our family of six for twenty-nine years, and I guess it showed it. In the final analysis, it was good to be rid of it.

Back on Cherry Lane, Fletch and I we were working away when I got a call from Axiom, the company I'd been working for. They wanted me back to do the design for a SCSI interface for the product I'd worked on the previous year. This situation knocked my summer plans for a loop, but I had to keep in Axiom's good graces, so I went back to work with a ninety-day contract. They had moved to a new building in an office park at NJ Route 38 and Marter Avenue. I was dazzled when I reported for work—their new quarters were the nicest I'd ever experienced. Right from the start, it seemed to me that the company was spending far too much for its plush quarters and employee perks, but I quickly adjusted to them.

They brought me in early enough on the job so that I could participate in the top-level design wherein tasks are either assigned to hardware or to software. I felt that much of the job could be done with a microprocessor and software but that would not leave sufficient work for the hardware staff. So we parceled out the work—I got to choose the SCSI chip and do the software to work with their hardware. We got through the successful design and lab test phases, and I prepared to go out for the first actual installation and system test. The chief engineer called me in to tell me that the company was in financial difficulty, and they could not extend my contract to cover field testing. In fact, all contract personnel were to be terminated. Most of Axiom's customers were telephone utilities located in Asia and South America. The mid-1990s economic crisis that started in Asia was causing a serious business contraction throughout the telephone industry, and Axiom was hard hit. I later learned that the company was taken over by a Norwegian firm with all engineering to be done in Norway. So ended my Axiom career, and though I didn't know it at the time, I would not work again. I did apply for unemployment compensation, which I continued to collect even after I'd gone to Florida for the winter.

When I got back to working full-time on the Cherry Lane house, we decided to tackle the front hall. It had been finished in rough flagstones, which were very hard on the feet. Since the master bedroom opened onto this hall, we found the flagstones totally unacceptable. The boys were available, and I put them to work, tearing out the stone. Once the stone was removed, we found no subfloor; it was now open to the basement. Furthermore, the joists had been cut so that the stone surface was the same as the rest of the house. Before we could lay a subfloor, we had to scavenge boards to the joists to make it come out level. Chris—with help from Pat and Jim—was the lead mechanic on the front hall job, one which turned out to be a lot more difficult than it appeared at first. Once we got the subfloor down, we were ready for wall-to-wall carpeting in the front hall, living room, and dining room.

As we got into the fall season, we made a last-minute decision to have the floors refinished in the master bedroom and closets. There was a lot of nice exposed oak floor area, and we didn't plan on carpeting it. We engaged John's Hardwood and moved into the spare bedroom while they worked. It took them three or four days to sand and apply multiple finish coats. They did a fine job, but the sanding dust got into everything, including places that we'd already cleaned and painted. It was a nightmare, and we decided not to do the other rooms. In fact, we decided that we'd had it—we'd been working day and night on the house for months. There was lots of work still to be done on the house, but it could be done while we were away for a while.

A Trip to Chicagoland

I had made several business trips to Chicago and found it to be a most congenial place to visit, but Kitty had never been there. We decided to fly to Chicago and stay in a downtown hotel and then travel through the Upper Midwest with a rental car. We enjoyed several days in the city and took the boat trip along the Chicago River. The boat goes through the lock into Lake Michigan and then back down the river between the downtown buildings. From the river, the city looks beautiful, much nicer than any comparable water view of New York City. We had dinner in a restaurant whose windows overlooked Rush Street, the center of Chicago's nightlife. I particularly enjoyed visiting the Art Institute, which has my favorite painting, Seurat's huge work known colloquially as *Sunday in the Park*. We were able to visit the spectacular Egypt exhibit at the Field Museum (of natural history). Most people were looking at the dinosaurs, so we had the reconstructed ancient Egyptian village pretty much to ourselves. It was very reminiscent of the American Indian pueblos of the Southwest.

From Chicago we drove north to Michigan's Upper Peninsula where we stopped for lunch in the quaint old city of Marquette, which is right on Lake Superior. It was my first view of this mammoth lake, and I was very impressed with its seemingly unlimited expanse. Marquette was an important shipping point for iron ore during the years of active mining in the Upper Peninsula. Now it seems abandoned. From Marquette we drove eastward along the lake for a while and then cut cross-country to Sault Ste. Marie where we spent the night. Next day we took the boat ride along the St. Mary's River which is the boundary between Michigan and Ontario. It is also the water connector between Lake Superior and Lake Huron. Across the river lies Sault Ste. Marie, Canada, a much larger and more beautiful city than its ugly Michigan counterpart. It is home to Canada's largest steel mill, which we were told is heavily subsidized by the government to create employment in this remote corner of the country. Our boat took us into the mill's huge docking areas for raw materials carriers.

The boat trip took us in both directions through the famous "Soo" locks. Because there is so much traffic, primarily iron ore boats to and from the Mesabi Range in Minnesota, there are actually five parallel locks. It is all quite spectacular. We were a bit apprehensive when a rusty Ukrainian freighter was put into the lock alongside our tiny tour boat. Actually the ore boats do not carry raw iron ore, as it would be too expensive to ship. They carry a processed iron product called taconite. It looks and feels like small coffee beans and can be used directly by the steel plants along the Great Lakes. It is also shipped overseas by way of the Great Lakes and the St. Lawrence Seaway. Of course, I was far more interested in all of this than was Kitty, so off we went.

Heading south from Sault Ste. Marie, we made our first stop at the little town of St. Ignace, which lies in the shadow of the famous bridge over the Straits of Mackinaw. Facing the waters is a memorial chapel dedicated to the French Jesuit explorers Frs. Joliet and Marquette, who passed through the straits from Lake Huron to Lake Michigan. Fr. Marquette died in what is now Illinois, and friendly Indians carried his body back to St. Ignace for burial. Once across the Mackinaw Bridge, we stopped for lunch in Mackinaw City. The city seemed to be in the midst of a festival of some sort, and we couldn't wait to get back on the road and down the "thumb" of Michigan.

Somewhere along the way, we picked up a brochure for an interesting looking place called the Cross in the Woods Shrine. We detoured off I-75 to the Indian River, site of St. Augustine Parish and its famous cross. The extensive site has indoor and outdoor churches and is staffed by Franciscan Friars (OFM) of the Sacred Heart, St. Louis Province. The woodsy site is dominated by a 55-foot tall wooden cross, which sits atop a man-made Calvary Hill. It was a Sunday and lots of worshippers were leaving the indoor mass as we arrived. The shrine has a museum, which houses what is claimed to be the world's largest collection of "nun" dolls. I was fascinated by the dolls, as my aunt Alice Quinn used to make them to be sold to benefit her Dominican motherhouse in Columbus, Ohio. Aunt Alice made beautiful large Dominican-dressed dolls of museum quality, and I think my sisters each got one. I would imagine that they would be quite valuable. The grounds also house a separate shrine to Blessed

Kateri Tekakwitha, the seventeenth-century Mohawk Indian maiden who had become a Christian. The surrounding woods reminded the shrine's founders of the area around the Mohawk villages in New York State. Since Kateri had not yet been canonized, her name could not be given to the parish, although that was the original intent. Our little side trip to the Cross in the Woods proved to be something special.

After leaving the shrine, we headed south on I-75 to the faux Bavarian village of Frankenmuth. It seemed to be a popular spot with lots of tourists walking about on a nice Sunday afternoon. We checked into the Bavarian Inn Lodge right in town and joined the crowd, although I didn't see anything very interesting to look at. Next day we paid an obligatory visit to the world's largest Christmas store and, again, didn't see anything of interest to see or buy—essentially just more junk. Our trip ended in the Detroit area at Henry Ford's recreated nineteenth-century Greenfield Village at Dearborn. Perhaps fatigue had set in, or I'd seen it all before, but I was very disappointed with it. So instead of staying overnight, as planned, to visit the Henry Ford Museum next day, we caught the first plane out of Detroit for Philadelphia. Although our trip ended rather abruptly, we had accomplished pretty much all of what was intended. Mainly, we had seen the magnificence of the largest of the Great Lakes—Superior, Michigan, and Huron.

Emergency Hospital Visits

Once back in Cinnaminson, Kitty and I resumed our work on the house. There was always something to arrange for—carpets, wallpaper, paint, etc. I guess the stress and strain finally got to me in mid-November. One night I woke up with severe chest pains, and Kitty called our family doctors—the Basara-Epstein practice at the time. Dr. Basara returned her call and told her to get me to the hospital, and he would call ahead. I was admitted to what is now Virtua Memorial in Mount Holly and put in the care of Dr. O'Neil, a heart specialist. After a battery of tests, I was released after a one night stay—false alarm, probably stress.

By December we were totally fed up with the house project and looking forward to a respite in Florida. But before we left, we had some things to do. First of all, I had to quickly schedule the colonoscopy that I'd been putting off all summer. Second, we had to squeeze in an open house party for friends and family. My internist at the time, Dr. DeLacy, set it up for the procedure to be done in about a week at Rancocas

Hospital in Willingboro. I was instructed to stop using the ibuprofen I was taking for my bad knees. This was my second colonoscopy with Dr. DeLacy, and it was done in the hospital's same-day surgery unit and only with sedation—Demerol, I believe. DeLacy was rather amusing and the procedure, while very uncomfortable, went well. The doctor told me that he had removed fourteen polyps and felt that they were all benign, which proved to be the case.

A day or two later, on the day of our party, I started losing a lot of blood from the colon. I was really scared, and Dr. DeLacy told Kitty to take me back to the hospital, immediately. I was admitted and after signing away my life, given several blood transfusions. Happily, Donna, one of my caregivers, was a former girlfriend of Tom's—it's nice to have a friend on staff in the hospital. Once I was stabilized, Dr. DeLacy performed another colonoscopy to seal off a bleeding polyp site. It was likely that I had not been off the ibuprofen, which has the side effect of thinning the blood, long enough—only one week as opposed to the two usually recommended. The bottom line was that I missed a great party at our "new" house—Kitty and the boys did very well without me. Some people wondered at the lengths I'd go to avoid a party.

Fortunately, I recovered quickly, and we were soon off to Boynton Beach, Florida. We left behind several projects to be managed by Jim Fletcher while we were away. Chief among them was a new roof by the Fletcher Brothers. Another major project completed in our absence was renovation of the guest bathroom, including new floor and wall tile, and a new countertop. We chose the tile after we got down to Florida and had it installed by Classic Two Tile Setters, then in Moorestown but now in Medford. In April of 1999, we came back to find that they had done a great job. Our guest bathroom turned out to be very nice. Kitty made it even better by turning the linen closet into a display of her photos and knickknacks.

In the fall, we had solicited bids for a new large rear deck to cover the existing concrete patio and extend considerably beyond it. A Medford architect-builder gave us a detailed plan for an absolutely perfect design. Now that we were ready to go, I could no longer get in touch with him. We decided to go with a local builder, Cole Construction, who gave us the best price for a pressure-treated wooden deck. Mr. Cole made sure that we understood that the wood would warp and splinter and require lots of maintenance. We liked his price of $3,600 and went ahead in May of 1999, using the detailed plan we had in hand. It turned out to be quite a project, requiring several in-process checks by the building inspector. We were very pleased with the result; and the boys bought us a nice deck set—table, umbrella, and four chairs—to go with it. While we've enjoyed the deck now for nine years, the wood has certainly warped and split, and I do have to make repairs every year. We've also been reapplying a deck finishing stain every few years. Nowadays there are more choices of deck materials than we had in 1999, so if I had to do it again, I would use one of the new synthetic products, definitely not wood.

My Tennis Career

One of my main reasons for going to Florida was to play tennis all winter. I've already mentioned that in South Florida, I first played at Delray Beach and then moved to Boynton Beach where I play now. I'd like to tell how my interest in tennis evolved. I first picked up a tennis racket—or racquet originally—in 1942 or 1943. That's when Tommy O'Brien and I played in Fort Tryon Park, where the courts overlooked the Hudson River. After moving to Riverdale, I only played once or twice at the Riverdale Neighborhood House with classmate Bob Foran, whose family were members. After moving to New Jersey and marrying Kitty, she and I played occasionally, mostly when on vacation. During our stay in Santa Monica, I played most Saturdays with Jim Cairns, our Coral Gables neighbor. Back in New Jersey, I resumed very occasional play with Kitty until we moved to Tenby Chase.

The Tenby Chase Swim and Tennis Club had an active tennis program and not long after we joined—probably in 1972—an instructor was engaged to give a course of tennis lessons. Kitty and I were eager participants and learned how to really play the game. The instructional program generated a lot more players and greater interest in tennis at the club. Tennis activities included round robin play on weekends for men and weekdays for women. We also had singles, doubles, and mixed doubles tournaments. Evening play usually drew a party crowd on the tennis patio, which adjoined the courts. It was a great fun through the 1970s and into the 1980s. Thereafter participation started to dwindle as general interest in tennis waned, and many of our players turned to golf.

I'm not sure how it came about, but I joined a group in a seasonal contract for play at our local indoor facility, Millside Racket Club. Our indoor season ran from October through March or April. The Tenby Chase group had many tennis parties at Millside, and they were great fun. Our Patrick took tennis lessons at the club and toured with the team for matches around the area. He became a fairly competent player, but I could beat him until about ten years ago. Over the years, the composition of my contract group changed and finally dwindled until only Mitch Sheairs and I were left. At the same time, the group on an adjacent court was undergoing the same decomposition, and we decided to merge. The new foursome consisted of Joe Pearson, Tom Adams, Mitch, and I. Joe was the owner of Cherry Hill Toyota and was the group's captain. Shortly after we merged groups, the Millside Racket Club folded, and the building was sold for use as a warehouse—what a shame. We were forced to relocate our contract and found a new home at the Crossroads Racket Club in Burlington Township.

Our foursome was rather imbalanced, Joe and Tom were better players than Mitch and I, and they insisted upon playing together. Mitch and I met other players at Crossroads and were invited to play against other pairs at Crossroads and at Arrowhead in Medford. We did quite well because Mitch had a wicked lefty serve that resulted in weak returns that I could put away at the net. However, we did not fare well against Joe and Tom, and I could see that they were unhappy with us. My drinking was a problem, the two double martinis I had with dinner definitely adversely affected my play. Finally, when it became time for a new contract, Joe called to say that he and Tom wanted me to drop out, or that they would drop out. I really had no choice, and drop out I did. I believe that Mitch continued with them for one more season. I saw Joe one last time while Kitty was buying her Celica at Cherry Hill Toyota, and he was not very friendly. I bought my Toyota Avalon elsewhere.

After our contract group broke up around 1984 or 1985, several things came to pass. In 1986, I gave up drinking altogether and collected my MBA from Rutgers. I also started playing tennis after work with an RCA/GE/MM group. We usually played for hours on a lighted court on Maple Avenue in Moorestown. That group finally broke up when job changes and job transfers intervened. My partner Mitch had also given up tennis for golf, so once the RCA group broke up, I didn't get much play until I retired.

When I drove to work in Camden, I drove down Branch Pike past the tennis courts at Cinnaminson's Wood Park. Every Wednesday morning, weather permitting, I saw a tennis foursome at play. I was very envious and resolved that I would try to break into the group when I retired. Shortly after retirement, I dropped by, introduced myself, and gave them my phone number. In a few weeks, Jack Murray called to ask me to fill in for a player whom was out with a hand injury. From that point on, I played at 7:00 a.m. every Wednesday morning that I was in town, even when I was working. The original foursome was Jack, George Rogers, Bob Seymour, and I. When Ron Selm recovered, he bumped Bob, and I remained in the foursome until the group disbanded due to death and physical impairment.

The Wednesday morning tennis foursome was nearly wiped out in a freak accident. We played outdoors all winter except when ice, snow, or rain shut us out—in which case, we played indoors if we could get a court. We usually played at the Mount Laurel Racket Club where we got a cheap rate if we played at 7:00 a.m., which was fine with us. One winter, we started playing at an indoor facility in Deptford Township. There was a substantial snowfall the night before the fateful morning, and I noticed quite a bit of snow on the roof of the facility when we arrived. As we were playing on the end court, I started to hear strange cracking noises, almost like gun shots. I looked around and didn't see anything amiss. All of a sudden, the roof trusses gave way and started to fall on top of us, taking the roof with them. What I'd been hearing was the popping of the rivets connecting the roof trusses. We fled to the end wall of the building, which kept the roof from falling directly onto us. I knew that the courts were heated with gas and was afraid that a leak might cause an explosion and fire. I said lets get out of here and well away before we are trapped by fire and rescue vehicles. We just made it out of the driveway as they arrived en masse, blocking all exits and roads. From a nearby gas station, we watched for a time, and thankfully, there was no explosion. We heard that people on the court next to us had some minor injuries, and that was it. The accident doomed the facility, and it was sold; the site now is home to a Wal-Mart.

Jack Murray had lived in Pennsauken where he had been a regular during the heyday of Pennsauken tennis. Although he and Barbara had moved to Mount Laurel, Jack still played with the Pennsauken group and invited me to join in their Monday and Friday round robins at the facility on River Road. When I became a Pennsauken regular, we had two hard courts and three all-weather courts. Over time, the latter were in such poor shape that we could no longer use them, and they were converted to a terrific roller hockey rink complete with a grandstand and electric scoreboard. At the outset of my participation, the Pennsauken group consisted of twelve players who shared time on the available courts. I was the youngest member and arguably the best player. Over the years that group steadily dwindled so that only two of us are still alive and active—Roland Glasson and I. It was a lot of fun while it lasted, and the group had a lot of fine, but aging, players.

Although by this time I was playing tennis regularly in Florida, I no longer had regular play back in New Jersey. Finally, Roland Glasson called to say that they needed me to fill in with their Saturday-Sunday group that played at the Pennsauken members-only facility. I played with Roland's group for many years, playing a full three sets each day. Here too injuries and illness took a toll, and the group's composition changed over the years. Even Roland, next youngest after me, was unable to play for some time, and it became more and more difficult to find players willing to commit to every Saturday and Sunday morning. In 2001, I myself had to sit out several months prior to and after my knee replacements by Dr. Schoifet at Virtua Memorial Hospital in Mount Holly—more on this later.

By 2005, although my knees were fine, my back really started bothering me. I found it difficult to play a full three sets and had to take longer breaks as we changed court positions. That made my partners unhappy, but what could I do. My back became worse, and in 2006, I went to Dr. Atlas, an orthopedic back specialist. After MRIs of my back and both hips, I was diagnosed with spinal stenosis. Dr. Atlas then sent me to Dr. Gupta, a pain specialist in Voorhees Township. In July I had my first epidural injection at the Burlington Surgical Center in Willingboro. It helped a whole lot, and I had a second in August to treat the pain in my right hip. After the epidurals, I was able to manage the pain with NSAIDs and Vicodin when I played tennis. But I could no longer play three sets and had to gradually drop out of my weekend group.

After moving back to Cinnaminson in 1998, I started to look for games at Wood Park where I knew a few players, but I've forgotten how I knew them. Joe Morgan and Leo Genovese helped me to get into a few games, but I was by no means a regular. One day in 2006, while I was sitting around waiting for a game, Stu Hada, who ran private clinics, asked me to help him with a group of women who were taking lessons. He wanted me to play doubles with them because while they had been taught how to play tennis, they didn't know how the game was actually played and scored. I worked with this group of women for several months and had fun while becoming better known to the people at Wood Park. Stu even put me on his Wood Park Web site phone list, making me a "regular."

I gradually learned how to go about getting games—show up at nine thirty (summer hours) and one (winter hours) and hope that at least three others show up. It isn't the best system, but it is the only system. George Fresco, the Wood Park tennis "captain," was very instrumental in getting me into foursomes, often with him. Sadly George died in 2007, when he was in his midsixties. He went into the Lourdes-Rancocas Hospital for ulcer surgery and got an infection that killed him. That hospital has always had a bad reputation. There was a huge area-wide outpouring of tennis people for George's wake and funeral. We all felt his loss. The year-ending doubles tournament was held in his honor and, oddly enough, was won by George's son David, who came over from the West Chester area to play. So as I write this, I'm hoping that when I get back to New Jersey, I'll be able to get some springtime tennis play at Wood Park.

Off to England and Scotland

While looking through the 1999 Elderhostel overseas catalog, I noticed a program called Industrial Archaeology of the North. It was being given at the University of Durham in Northumbria (Northern England) and appealed to the engineer in me. I was somewhat surprised when Kitty said OK, let's do it. On July 30, we flew to London's Heathrow airport and there got a flight to Newcastle. Doing a UK-internal plane transfer at Heathrow is a true horror story for non-citizens. At Newcastle, we took a shuttle train from the airport into town and to the main north-south rail line. While the airport and shuttle train were very accommodating to travelers with luggage, the railroad station was the opposite. We had to lug our baggage up and down long flights of stairs to get onto the platform for the train south to Durham. Apparently the British rail system was designed for a time when travelers had servants or porters to lug their bags. At Durham, we again had a problem getting our bags off the train and onto the street where we got a cab to the university. The Durham station sits high above the surrounding low area so there is no way to avoid a long flight of stairs down to the street.

Upon arrival at the university, we were warmly greeted at Trevelyan College, and students offered to help with our bags. The school was pretty well shut down for the summer, so we were put up in the college's newish student residence. The building had received an architectural award, and I couldn't understand why the hallways were dark and narrow, with unnecessary steps, and the plumbing was terrible. Our beds were little more than army cots, but we did have a nice view onto a garden where we could watch a tribe of rabbits eating the vegetation. We had breakfast and dinner in the dining room, and the food was very good, particularly the huge English breakfasts. At dinner we had lots of fun passing the dishes up and down the long tables. Since we were on the road every day, we picked up box lunches after breakfast. There were several types of sandwiches in the boxes—some really awful—so it paid to be an early riser.

The weather was absolutely glorious—actually very warm—the whole time. Stafford Linsley, BEng, PhD, was our excellent "course tutor" as each day we set out to visit some aspect of the birth of the Industrial Revolution in Northern England. This meant we visited abandoned mines and railways, coal ports, pumping stations, and places where workers and bosses lived. Almost none of the places we visited had any facilities for visitors—no parking, no paved paths, no toilets. It's a very beautiful and hilly part of England, so there was lots of hiking over hill and dale. My knees were in bad shape, so I was often unable to keep up with the group. One day I noticed that a man in our group had an odd-looking walking stick. I asked him about it and learned that he had a German hiking staff that could be adjusted to serve as a staff or a cane. I resolved to get one as soon as possible.

It turned out that the program was actually conducted by the UK Summer Academy; consequently, many of the participants were from England. They were a jolly bunch, and we had a lot of laughs, particularly during the after-dinner games and lectures. It would fill a book to describe all that we did during the program, so I'll just mention one outing. We drove west to the beautiful Pennines to learn all about lead mining and visit an abandoned mine. We got on location and had to wait for another group to clear the mine. It was lunchtime anyway, so we broke out our boxes and lunched while seated on rocks in a dry stream bed—it was quite a picturesque setting.

After lunch we were outfitted with slickers, miner's hard hats, and Wellington boots for our foray into the mine. For some reason, Kitty and I were at the tail end of the group. Anyway, upon entering the mine, I found that the headroom was only about 5½ feet, and I was almost six feet tall. My head kept hitting the ceiling, knocking my hat askew, and I worried about how long I'd have to walk bent over on my bad knees. Furthermore, we were walking between tram rails in very cold water several inches deep. All of a sudden, my lamp went out, and I found myself in total darkness. I yelled to Kitty to hold up until I reached her so that we could share her working lamp. It seemed forever, but we finally reached a working chamber where we could stand. At the time the mine was worked, miners used small white candles for illumination, and they had to pay for them out of their earnings. It seemed ridiculous to think that mining could be done by candle light, so our instructor gave a demonstration. We all doused our lamps, and he lighted a single candle, which did, in fact, light up the chamber. I was very apprehensive about the trip back through the low tunnel, but it seemed to be easier than the entry. I found out later that some of the wise birds in our group did not go into the mine.

After a wonderful—but tiring—week, romping through the hills of Northumbria, we decided to make a short trip to Scotland. Back we went to the impossible Durham station for a train to Edinburgh. We had reserved seats in car C, but when the train arrived, there was no car C; and the train was packed with standees, mainly young people. We pushed our way onto one of the cars, and the young folks helped with our bulky luggage. Once aboard, we heard an announcement that car E was now car C, and we seemed to be in F or G. So we started to move forward, trailed by our luggage, which was being passed from hand to hand. Once in car E, we found our seats occupied, but they were quickly vacated when we presented our tickets. In an unlikely ending to our boarding adventure, we found that our seatmates were from New Jersey. They were a father and son heading to Scotland for the big Frisbee championships. That, plus the annual Edinburgh festival, accounted for the morning's chaos on the railroad. Another cause for celebration in Scotland's capital was the recent return of Scotland to self rule.

Once in Edinburgh, we had the usual problem of getting our luggage upstairs and out to the street to a taxicab. Upon arrival, the weather was rainy and miserable and remained that way during our short stay in the city. At our hotel, the inappropriately named Apex International, we found that our request for two beds translated to a double bed and an uncomfortable one at that. Nothing else was available as the city was jam-packed with tourists for the festival, so we were stuck with that bed. We did the usual sightseeing including a visit to the castle, which overlooks the city and is the site of the annual Tattoo. In an outdoors shop, I was able to buy a hiking staff like the one I'd seen in England. Food, however, was a problem as all of the restaurants and pubs were jammed. Out of desperation, we decided to return to our hotel's dining room only to find that it was closed for a wedding. It was a beautiful formal wedding, with the men in their kilts and sporrans. As guests in the hotel, we explained our predicament and asked to be served either in the lobby or in our room. They reluctantly came up with something, so our hunger was assuaged, at least for a time.

On to Ireland

Part of our overall trip planning was to attend the O'Donnell clan reunion to be held in Donegal. But first we were going to visit relatives in County Tyrone. We were rather glad to leave chaotic Edinburgh on a plane for Belfast, Northern Ireland. There we rented a car and drove to the Dungannon area where we stayed in a rather nice motor inn called the Oaklin House. Either from fatigue or general ineptitude, I had trouble keeping the car on the proper (left) side of the road. I'd keep hitting any curbing on my left or far side and tended to go around traffic circles the wrong way. Fortunately, traffic was light once we left the Belfast area, or we might have had a serious smashup. Furthermore, the Irish are very much aware that summertime visitors have trouble on their roads. Also helpful was the good weather that prevailed during our stay in Ireland.

I had alerted the relatives to our arrival, and they started showing up even before we unpacked—perhaps they had tipped the staff to call them. Our first visitors were my first cousin Eddie O'Donnell and his wife Margaret, nee Murphy, of County Fermanagh. They lived right in the center of Dungannon. Unfortunately, Eddie suffers from sudden epileptic seizures, and Margaret gave up her teaching position to look after him. They filled us in on family happenings and arranged to pick us up next morning to visit O'Donnell country in and around Eglish. Our first stop was to visit Aunt Mary, Eddie's widowed mother. She was living in a caravan (trailer) nearby while her new government-supplied house was under construction. Her homesite is just down the lane from my grandfather's place, now Eddie's, which is now vacant and going to rack and ruin. However, Eddie leases the farm and outbuildings to an adjacent farmer. At today's prices, the property is undoubtedly very valuable. Aunt Mary was very helpful and had dinner for us in her caravan after our day of visits.

Mary had arranged for us to visit Sean O'Donnell, a second cousin, who occupied a newish home on the site of the old homestead of my great-grandfather Edward. Sean and his wife, Sadie, were a lot of fun, and I got to see what was happening in his many mushroom sheds. He mentioned that mushroom prices were falling, and sometime after our visit, he got out of the business; apparently raising livestock is much more profitable. Sean's sister Peggy and her husband, John Campbell, of Armagh stopped by to see us. By the time we left, we'd accumulated a large contingent of cousins and neighbors, all of whom had to be in pictures. We got all of the cousin's names and birthdays, but I cannot now put the names to the pictures. That evening we visited Eddie's sister Claire and her husband, Anthony Daly, at their beautiful new home. My, how times have changed. Claire now works for the local Mercedes dealership and drives one. Anthony also has a job, but runs a farm as well. Their four-year-old son Michael wanted to be a farmer and had a complete set of farm toys. His favorite was his John Deere tractor.

I don't remember how we worked it out, but we did spend a day with the Quinn relatives. We drove over to first cousin Sean Quinn's place on our own, getting well and truly lost on the back roads around the nearby town of Pomeroy. Sean has the house his father, Joe, built on the homestead of my grandfather Patrick Quinn. The old stone farmhouse-barn of my mother's time wasn't used any longer but was abused by the Brits during the more recent strife. They thought the IRA were storing arms in it. Sean has added onto the original house and, though dated now, is quite comfortable. Sean and his wife Maureen "Rena" McKenna took us to see our mutual first cousin, Gerry Donnelly. He and his wife, Jean Mallon, have a new house and were looking after his ailing mother, my Aunt Celie, who was chair bound at age eighty-eight. Celie also had, next door, a beautiful brand-new house provided by the government, but I doubt that she had ever stayed in it.

One evening Patrick Donnelly, a second cousin, came to see us at the hotel. Patrick is the last of the Donnelly line that came down from my great-grandfather Murtha. He and his wife, Christina Nugent, had a drink with us; and when I asked what he did for a living, he responded with "We raise mad cows." Mad cow disease was a hot topic in the UK that summer. Patrick took us out to their place near Galbally via a circuitous route that took us to Altmore Chapel and the graveyard of my Quinn and Clarke forbears. Unfortunately, it was too dark to read the headstones, so I didn't learn a whole lot more about the Clarkes. My Florida Aunt Mary had told me where the Clarkes were from, but at that point, I knew very little about them. We had a nice visit with Pat and Christina and met some of their many children. I also went out to the barn to watch as the boys brought the cattle in for the night. Patrick took me to the site of our great-grandfather's homestead, now abandoned. As I looked across the road, I saw another sign of the times. On the highest point in Clonavaddy Townland stood a cell phone tower.

I had read about the new records center of the Catholic Archdiocese of Armagh. They had been given a government grant to put all of their parish records in a central database. I had hoped to be able to access their records to fill in gaps of my family history. Kitty and I went to Armagh and quickly found the center. We walked in and, at first, didn't find anyone about. Finally, a little elf of a man showed up and asked what we wanted. I explained that my Donnelly great uncle had been a priest of the archdiocese, and I wanted to research his family. The man led us to a computer and tried to access the database, without much success. I mentioned that I was a computer expert, and he left me alone to work with the system. I accessed many Donnelly records but failed to get any more information than I already had. Their records did not extend back past the inauguration of civil birth registry in the 1860s, and I needed information back to 1800.

When I spoke to the custodian about this situation, he suggested that the priests were reluctant to part with their own records and hadn't been too diligent about keeping them, anyway. He did go up to the bishop's residence and get us a free copy of the 1999 archdiocesan directory, a book that I've found incredibly useful. It was through this book that I was finally able to get in touch with my elusive Clarke relatives. I later learned that the records center is not open to the public; that access is via a paid commercial service center in Armagh. Fortunately, my priestly connection got me in and directly online.

On the day the O'Donnell clan was to assemble in Donegal, we set out from Dungannon. It was a beautiful drive through counties Tyrone and Fermanagh and on to County Donegal in the Republic of Ireland. Our destination was a new and spiffy-looking bed and breakfast (B&B) on Donegal Bay. The Water's Edge was my first B&B, and it was new and spiffy all right. But our room was tiny, too small to be comfortable, and nothing else was available. Basically, one person had to stay in bed to allow the other to access the closet or the bathroom or the door to the hallway. You had to sit on the toilet to use the sink. It was a disappointing and uncomfortable welcome to Donegal.

The clan gathering was to celebrate the four hundredth anniversary of the crowning of the last of the O'Donnell chieftains. The chief ceremony was a reception in Donegal (O'Donnell) Castle followed by the raising of the O'Donnell flag over the town center. The gathering included a number of bus trips to places of interest throughout the beautiful and hilly county of Donegal. The highlight was a visit to Doon Rock where the chieftains had been crowned during the O'Donnell ascendancy, roughly from 1200 to 1600 AD. Unfortunately for me, because of my bad knees, I was unable to make the final rocky climb to the summit—yet another disappointment. I had picked up a cold during our stay in Ireland. That, and accumulated wear and tear on the body, caused us to leave the clan festivities a day early and head back toward Belfast.

On our eastbound trip, we detoured through the beach town of Bundoran. I wanted to see it because my mother had often spoken about her visits there. The setting is spectacular—striking cliffs rise up just east of the town. Though it was the middle of summer, we saw no one on the beach itself. There seemed to be a lot of activity off in the center of town, but it looked to be too challenging for us to venture there. I later read that Lonely Planet has described Bundoran as "one of Ireland's tackiest resorts." It certainly looked tacky to us, though a magnificent resort hotel sat on a seaside cliff just outside of town. From Bundoran, we drove to Enniskillen, a truly beautiful town located at the junction of the two large Lake Ernes. We had lunch and strolled about town for a bit before pushing on to Dungannon.

We decided to stay in Dungannon to rest up for our flight the next day. Consequently we did not want anyone to know we were there. So instead of staying in the Oaklin House, we search for another hotel where we would be unlikely to meet anyone. We had seen an interesting-looking hotel in the Moy, a small village south of Dungannon on the road to Armagh. While we were on the street trying unsuccessfully to get into the hotel, we were suddenly face to face with Charles McKillian who was leaving a nearby pub. I couldn't remember when I had last seen him, but I recognized him—he is married to my first cousin Geraldine. He recognized me as someone he knew, but couldn't place—thankfully. We decided to quickly get out of town and to a nearby lodge that we knew was rather expensive.

Upon arrival at the stately manor house, I asked for a room with a minimum of two comfortable beds and bath. They said no problem, and led us to what was probably the former master's suite. It was a huge room with many windows overlooking nearby farms and, gloriously, with three luxurious beds and a strange but workable bath. Furthermore, there was a lot of really nice old comfortable furniture in the room. It was a really big house, and I got the feeling that we might have been the only guests. That evening we took a chance and had dinner right in the center of Dungannon, a heavily Protestant town. Naturally, we were seated at a table near a chatty couple who were clearly of that stripe, and I felt that we could easily have come to blows. Safely back at the hotel, we enjoyed a really restful night and a magnificent Irish-English breakfast. We felt the lodge was well worth the extra cost. We had an uneventful drive to Belfast where we checked in our car and caught our flight to London and on to Philadelphia. It was a memorable, but exhausting, eighteen days of travel to England, Scotland, and Ireland. We've since found our limit is ten days, and that's a stretch.

The New Year and We Go Off to Indian Country

The year 2000 became quite eventful for us. During the winter of 2000, we had a first visit by any of our children. Chris and his friend Danielle De Bow spent a couple of days with us and then went north to visit her grandmother. While in Florida, we looked at small four-door cars to replace Kitty's aging two-door Toyota Celica. Kitty decided she liked the sporty version of the Mazda Protégé. Back in New Jersey, I was looking through the car ads and spotted exactly the car she wanted. I called Faulkner Mazda and asked for their price and, without the usual gamesmanship, got a price that was very reasonable. I told the salesman to hold the car, and we'd be right over to check it out. By the time we got to their place on Roosevelt Boulevard, the salesman had the car cleaned up and sitting by the front door. Our contact was a good salesman who never got in the way, so we gladly made the deal with him, and Kitty drove home her new car. We put a sale sign on the Celica and parked it in front of the house. It was quickly grabbed up by a neighbor.

While in Florida during the winter, I was looking through an Elderhostel catalog and saw a program that immediately commanded my attention. Sponsored by the College of Santa Fe was a course which would take us to many of the important Pueblo Indian sites in the Four Corners region of the Southwest. As a fan of the Tony Hillerman novels, which are set on the big Navajo reservation, I'd always wanted a reason to experience that part of the country. Consequently, we signed up for the Elderhostel and in June left for Santa Fe by way of Denver. There we rented a car for the drive south to New Mexico. This was a motor trip we had taken ten years before, but this time, we planned to make some stops along the way.

The last time we got to Colorado Springs, Pike's Peak was not visible from I-25, so we didn't stop. This time we could see it and decided to drive to the top, a terrible mistake. The perilous ascent is definitely not for an almost seventy-year-old driver with a fear of heights. When we got to the 14,000-foot summit, we were exhausted and quickly succumbed to altitude sickness. We struggled from the car to a bench and collapsed, and all I could think of was that we had to do it again to get back to town. When I recovered enough to look around, I saw that there was a cog railway that we could have taken up to the summit. What idiots we were, doubly so because clouds had moved in, and we couldn't see much of anything anyway. Fortunately, on the trip down, we were on the inside of the road; and it was much easier. They even have a brake check partway down to be sure that you can proceed safely.

While in Colorado Springs, we drove out to take a peek at the upscale Broadmoor resort, one of the best in the country. It looked really nice but not for those on an Elderhostel budget. Heading south on I-25, we made it as far as Walsenburg where we decided to stop for the night. There wasn't much around except a sign pointing west to a motel. We had no choice but to go off in that direction or continue south on I-25 with even worse prospects. We got to the motel, checked in, and found that they had a restaurant next door. It all looked pretty cheesy, but the restaurant turned out to be a gem—their Wiener schnitzel was excellent. We learned that a family from Poland ran the place; at least three generations of them worked in the motel and in the restaurant, and it was fun talking with them. We learned that the road they were on led to the Spanish Peak's ski area and was very busy in winter but dead in summer. We were very happy that they were there in the off-season to rescue tired travelers such as Kitty and I.

Next day we went over Raton Pass into New Mexico, with a brief stop in the railroad town of Raton. As was the case last time through, clouds obscured the view from the summit of the 7,800-foot pass. In Raton, we were quite amused to pick up a glossy brochure extolling the town's many attractions. However, they did have a grand new visitor's center that had been built since our last visit. After Raton, our next stop was the really pretty town of Las Vegas, where we enjoyed a nice lunch and walked around the plaza taking many pictures. It was an important stop on the old Santa Fe Trail that started in Missouri. I felt that I could be very happy living in this Las Vegas; however, back into the car and on to Santa Fe. Our group was staying at the Loretto, a very fine hotel right in the center of the city. We arranged with the hotel to leave our car for the week that we would be gone. At our group's dinner meeting, we learned that we would be traveling in three fifteen-passenger vans, each piloted by one of our leaders, two of whom were professors and the third a Tewa Indian guide named White Eagle Tail, a.k.a. Richard Montoya. It looked like a fun group. and we were not disappointed.

After Sunday's breakfast, we headed north to Taos via the "high road" with several stops along the way. In Taos, we had time to leisurely wander around town until dinner. We dined and spent the night at the Fetchin Inn, Taos' finest. It was fun to sit on the porch and chat with the passersby. It became apparent that this was not going to be a budget trip when we were allowed to order food and drinks from the menu. From Taos, we went north along the Rio Grande into Colorado where we turned west for the San Juan Mountains, the river's headwaters. We had an outdoor box lunch at South Fork and then crossed the Continental Divide at 10,500-foot Wolf Creek Pass. The road then goes through Pagosa Springs, Durango, and finally Cortez where we stayed for two nights. After dinner each night, one of the professors traveling with us gave a talk on a topic related to the life and customs of pueblo peoples. I hadn't realized how many pueblos were still occupied. Our Indian guide was from one of the active pueblos we passed on the road from Santa Fe to Taos.

In the morning, we visited the magnificent pueblo ruins at Mesa Verde National Park, actually a series of heavily wooded mesas southeast of Cortez. Even though I had my hiking staff, my bad knees limited my access to much of the site. However, I did work my way down to and up from one of the major cliff dwellings, one that was occupied from about 600 AD to 1300 AD, some seven hundred years. Visible to the west from the park and from the Cortez area is Sleeping Ute Mountain. It is the dominant feature on the Ute Mountain Indian Reservation and Tribal Park that adjoins Mesa Verde. The shape of the long mountain ridge is exactly that of a sleeping Indian woman. There was much to see at Mesa Verde, and we spent most of the day there, returning to Cortes for dinner, lectures, and beddy-bye. Next day we drove up to Lowry Pueblo, about twenty-five miles northwest of Cortes. This was an interesting and accessible site, and we were able to take lots of good pictures. From Lowry, we made a long southwesterly trip across the desert on unpaved roads to Hovenweep on the Utah border. This pueblo site was very barren and really not very interesting. After seeing beautiful Mesa Verde, one wonders why Indians ever chose to live at Hovenweep. They must have been an outlaw band.

On to Canyon de Chelly

From Hovenweep, we set out for Canyon de Chelly, Arizona. The first leg of this long trip was through extreme southeastern Utah and into the huge Navajo reservation. The states of the Four Corners are all big, so there is no way to avoid long drives. Our vans kicked up a lot of dust on the unpaved roads, so we could undoubtedly be seen from afar. As we turned south at Bluff, Utah, we had our first look into Monument Valley, which was off to the west. The fantastic eroded rock formations in the valley are truly spectacular. After crossing into Arizona, we made a rest stop in the weird town of Mexican Water. Weird in part because the surrounding desert was dotted with oil wells pumping dollars into the Navajo economy. We had also begun to see scattered Navajo homes and hogans. Every family—even if they live most of the time in a newer home or trailer—must have the traditional hogan for ceremonies and to keep themselves in balance. It is extremely dry work walking, and even riding, in this part of the country, so our vans necessarily carried a large supply of bottled water. At the infrequent rest stops, we could pick up candy bars and cakes to augment our meager rations, and smokers could light up.

From Mexican Water south, we had a good road all the way to Chinle, the gateway to the canyon. We pitched our tents at the Thunderbird Lodge, a Navajo-run operation with a trading post where I was able to buy an inexpensive Navajo campaign hat. We dined at the lodge's restaurant, and it was fairly decent—surprise, surprise. Next day we rode to the entrance to the Canyon de Chelly National Monument where we transferred to the Navajo trucks that would take us on into the huge canyon, which is on Navajo Tribal Trust Land and sacred to the Indians. Riding in what seemed to be old army trucks on the rutted sand, which passed for a road, was a painful experience; and it went on for hours. Even so, we never got to the end of the southern arm that we had taken. No one was more relieved than I when we finally stopped for a break at the White House pueblo, which was built into the canyon's northern wall. I was even happier when I learned that this was as far as we'd go.

Painful as it was, the excursion into the canyon was the high point of the entire trip for me. Canyon de Chelly definitely has the mystical aura that I'd read about. Some Navajo families live and farm within the canyon, but it is mostly arid and sandy except where water seeps down from the canyon walls. After an outdoor box lunch at the lodge, we drove back to the canyon to view it from the southern rim road. From the overlooks, we had a wonderful view into the canyon and could see where we had gone in the morning. Really weird rock formations stick up and can really only be appreciated from above. The 800-foot pinnacle called Spider Rock is particularly well-known. It was certainly a lot easier to experience the canyon while standing on the rim as opposed to seeing it from the seat of ones pants on a Navajo truck.

After a second night at the Thunderbird, we set out for Aztec, New Mexico. This was a very long drive that took us through the principal Navajo towns of Shiprock and Farmington. Entering the former, we could see the famous 7,000-foot Ship Rock off in the distance, and it does indeed look like a vessel running with all sails set. The rock is quite spectacular as it rises straight up from the desert floor. The town of Shiprock is home to a Navajo Community College, a Navajo Chapter House, and a brand-new Wal-Mart, among other newish-looking businesses. We did take time out for a coffee break and then pushed on to Farmington and Aztec, much larger towns but east of the Navajo lands. We visited the ruins at the Aztec National Monument, but by then, the ruins were all starting to look pretty much alike to me. However, the site's great kiva—ceremonial gathering place—was far and away, the best we'd seen. I later learned that it is the only reconstructed great kiva in the Southwest. We did have an excellent dinner that evening in a private restaurant, which was situated on a pretty river that runs through the town of Aztec. We stayed the night at the Step Back Inn, Aztec's finest.

Saturday morning, we set out for the Chaco Culture National Historic Park, which is well south of Aztec. Upon arrival at these vast ruins, it became apparent that a lot of hiking would be required. I decided to remain with the vans as did White Eagle Tail, our Indian guide, who happened to be a Tewa tribal leader. He spoke one night about the problems his pueblo had with young people and drugs. I had noticed that he had not always joined us as we hiked through and discussed various aspects of the ruins we visited. We were told that he felt that the professors did not treat some of the ruins with the reverence they were due, particularly the ceremonial Kivas. While guarding the vans with Richard, I learned that this was indeed the case. It seems that the Indians often disagreed with their stories as told by the white professors. A common Indian joke was to give a professor the ceremonial name of Walking Eagle, meaning that he was "too full of s—to fly."

From Chaco Canyon, we had a long drive back to Santa Fe and stopped for dinner along the way. During this trip, we had a lot of laughs about things that happened along the way. As they say, it was a hoot. We had our last group dinner in Cuba, the first town with a restaurant worthy of the name. Bruno's turned out to be a very good eatery, and the predinner drinks flowed, giving rise to even more laughter. Dinner was so good that I couldn't help wondering how the college could afford the expenses of our lavish off-the-menu dinners. I suspect that this was the first and last of these Indian Country Elderhostels. Well after dark, we got back to our hotel in Santa Fe. After breakfast the next morning, we said our farewells and thanked the professors, particularly our leader Jay Peck. He had taken good care of us—provided plenty of water and steered us clear of the potential disasters that awaited along many of the trails. This was one great Elderhostel!

Next day, we left Santa Fe and returned to Denver via the western route through the San Juan Mountains. The first part of the trip took us along the same route we had traveled with the group. However, at Durango, Colorado, we headed north through the tourist towns of Silverton and Ouray. It's an exciting ride, which involves crossing several 10,000-foot-plus mountain passes. At Silverton, we stopped for ice cream and happened to be present when the town band put on a show in front of their beautiful new town hall. The drive was nerve-racking, and I was tired by the time we got to Ouray, so we spent the night at a very homey Best Western motel at the foot of a mountain. There were lots of tourists staying in town, so we were quite fortunate to get a dinner reservation at a very nice restaurant. Ouray is much bigger, with more to do, than Silverton; and we would like to have stayed longer, but onward, ever onward.

From Ouray, we took the main road in a northwesterly direction as far as Delta. At that point, we took the scenic secondary roads that join I-70 at Glenwood Springs. This too was an exciting drive. We stopped for lunch in the charming town of Redstone, which is perfectly situated on fast-running Crystal Creek. The town has few amenities other than its perfect location. But it has a fancy resort hotel and a fine art gallery, whose rear sculpture garden overlooked the creek. As we pulled into Glenwood Springs, we saw an unusual sight, a crowd of people standing in what appeared to be a large swimming pool with vapor rising around them. I soon realized that we were seeing people enjoying the natural hot springs for which the town is named. As we sped down I-70 toward Denver, we saw signs for Vail and decided to get off and have a look. We found it difficult to get around town as it seemed jammed into the mountains with very narrow and sometimes gated streets. It was clearly upscale, but we couldn't find a place to park, so we did not linger.

We had planned to stay overnight in Denver and have a look around. But as Denver came into view, it was rush hour, and the city seemed wreathed in smog. Furthermore, we had done some three to four hundred miles of hard driving that day and needed rest more than we needed to see Denver; thus, we decided to hole up in a hotel near the airport. Unfortunately for us, the new Denver International Airport is located on the northeast side of the city, meaning we had to drive all the way through the congested Denver area to get there. The airport is basically set down in the middle of a vast and empty plain with a few motels scattered here and there. We were at wit's end by the time we found a motel, the Ramada, that had a room for us. At that point, we checked in and crashed before tackling the next problem, where to eat. We did find a nouveau diner nearby and settled for that. Next day, we were off to Philadelphia. Our twelve-day trip to the southwest was wonderful but too long by a couple of days. We relearned the lesson that ten days touring is about our limit.

Summer, Winter, and New Knees

Not long after returning from Denver, we drove down to Delaware to stay with Pat and Dick Davis at their beachfront rental on Fenwick Island. They always took the same place for the week that included Independence Day, and we'd been joining them there for a few days for many years. We were always amused by the many young friends of Trish and Susan Davis who camped out on the living room floor. One summer, Phil Hoops and I stayed up all night playing Trivial Pursuit as Trish's friends tried, in vain, to beat us. After dinner we'd usually take the ladies shopping and then to a nearby bar for drinks. Because I didn't drink, I'd usually insist that we stop at Dairy Queen for ice cream. Joanne always backed me up on this move. Now, however, Pat and Dick's grandchildren were dominating the scene. On one occasion, we decided to go to dinner at a restaurant at the southern tip of Ocean City, Maryland. After dinner we wandered over to the nearby amusement park. I suggested to Kitty that we try an interesting-looking ride, and we did. When we got locked into the car, I became panicked from claustrophobia—it was built for kids, not adults. The ride was fine, but was I glad to get out of that cage. Pat and Dick thought it was very brave of us but funny.

A few days after we got back from Delaware, we got a surprise call from a Clarke cousin, Brigid Boyle. She and her daughter Marie were in New York, visiting her brother Patrick, and they'd like to come down for a visit. Patrick got on the phone, and I gave him directions by way of the New Jersey Turnpike to NJ Route 73 to U.S. Route 130. If they left promptly. they should have arrived no later than 2:00 p.m. Well, at 2:00 p.m. there was no sign of them. Eventually Patrick called on his mobile phone to admit that he was lost and seemed to be in the African jungle. When I finally realized that he was on his way to Trenton on I-295, I was able to redirect him toward Florence and then Burlington. We put out our Irish flag, and they turned up in Patrick's Mercedes SUV around three or four o'clock. We had a nice visit, took some pictures, and enjoyed some grilled steaks. I also gave them a brief tour of the Riverton area, which really looked great as the Fourth of July flags and buntings were still out. Patrick Clarke, a second cousin and very successful New York contractor, brought some fine Irish spirits.

Despite all of the summer traveling we had done, in October of 2000, we decided to spend the early fall in Florida. We drove down, making our usual overnight stop at Clark's in Santee, South Carolina. That fall I made my first visit to Dr. Glenn Berkin, a Florida podiatrist who has since become a valued member of my medical team. We drove back to New Jersey for Thanksgiving, and upon return, I took my bad knees to Dr. Brill, an orthopedist, who gave me a shot in each. He didn't suggest surgery, and I wonder why. I began to notice that in pictures I was looking increasingly bowlegged because I had been losing the cartilage in my knee. I looked like a cowboy even though I'd never been on a horse.

After the holidays, we left in January of 2001 to drive back to Florida. I soon found that I was in real trouble with my knees as I needed to use both hands to climb the stairs to our unit. Thus, when carrying home groceries, I had to place them on the steps above me and use my hands to pull myself up. Then I'd repeat the process until I reached our level when I'd be able to pick up and carry the bags to our door. Even worse, I could not play tennis at all. I'd been seeing an orthopedic doctor, Eric Shapiro, off and on for various ailments, including torn Achilles tendons and a torn rotator cuff. Dr. Shapiro looked at my x-rays and pronounced my knees beyond repair. He recommended replacing both knees and suggested that I use the knee surgeon on their staff. If I had the surgery done in Florida, I'd have to contend with our stairs, so I explained why I'd wait to have the job done in New Jersey.

Once we arrived back in New Jersey, I started searching for a knee surgeon. The obvious choice was the Rothman Institute at Jefferson Hospital in Philadelphia. They were well-known and advertised on television. However, some of my tennis associates were not pleased with their Rothman experience, perhaps because Dr. Rothman, himself, didn't do the actual surgery. I didn't want to go to Philadelphia anyway, so I looked for someone in New Jersey. I had recommended to me a knee surgeon who'd left the Rothman group and operated in New Jersey as well as in Philadelphia. I made an appointment with his office. Meanwhile, on a visit to our family doctor, I asked for his recommendation. Dr. Epstein spoke very highly of a Dr. Schoifet who operated at Virtua Hospital in Mount Holly. While I was trying to decide what to do, I ran into a former neighbor, Ron Pollock, who had been our chiropractor until his retirement. Ron had a knee replaced by Dr. Schoifeit and felt he was the best in South Jersey.

That clinched it for me, and I made an appointment to see Dr. Schoifet in his Lumberton office. I was very pleased to hear him say that he personally did all of his knee surgeries. It took several months to go through the drill required to get to the operating table, which I did on July 30. Coincidentally, his surgical assistant was Dr. Mutch, who had been a partner of Kitty's obstetrician Dr. Van Meter who had delivered Patrick and Jimmy, as well as Kitty's sister Joanne. Some years after, Dr. Van Meter died; Dr. Mutch gave up his obstetrical practice, probably because of the costly malpractice insurance needed for the obstetrical specialty. The knee surgery itself went fine, but afterward, I became very ill and required intensive medical care in the hospital.

The lesson is that while you carefully research knee surgeons, you'd best be sure that a good medical staff is available in case of emergencies such as mine. After eight days, I was finally released from the hospital to residential rehab, the final step before going home and outpatient rehab. A few months later, Kitty and I were invited back to the hospital for lunch and a focus group session. I had spent over twice as much time in the hospital as anyone else in my focus group—people don't get to linger in the hospital any more. Naturally, I had a lot to say about my experience, which was, on the whole, quite favorable. I was asked to name the most outstanding person that cared for me, and I quickly cited my night nurse, a mature woman whose name I believe was O'Connor. Based upon my experience, I'd recommend Virtua Mount Holly to anyone facing a hospital event.

By the end of September, I was back on the tennis court for some light doubles. We didn't take any trips that fall but had wonderful news when Chris and Danielle De Bow became engaged at Christmas. Danielle was a coworker of Patrick at Merck-Medco in North Jersey. I believe that she and Chris met on a ski trip. Danielle lived with her family at their lakeside home in Wayne, not far from the Quinns' where Patrick was staying, so they became quite friendly. Merck eventually divested Medco, which became Medco Health Solutions. Coincidentally, both Patrick and Danielle would move to Pennsylvania and work out of Medco's Blue Bell office.

Florida, Winter 2002

Our drive down to Florida in January of 2002 turned into a real nightmare. The weather forecast called for a snowstorm along the coast, but we discounted that report because we'd found the forecasts to be alarmist in the past. Well, this time they got it right. By the time we got to South Carolina, the snow was too heavy to fight through. So instead of going on to Santee, our usual overnight spot, we decided to get off I-95 in Florence and seek shelter. The secondary roads in the Florence area were an icy mess, with accidents in all directions. We made our way safely to an Econo Lodge but had a difficult time getting across the road on foot to the nearest restaurant, which was across the way. Right in front of the motel, the Florence police were struggling to untangle fender benders and keep traffic moving. Apparently South Carolinians had little experience with snow because I'd never seen such a traffic mess. But then I remembered I'd seen the same kind of panic as Kitty and I scrambled to get out of Manhattan one December when snow started to fall.

Weather reports the night of January 4 told of another storm approaching from the west. Kitty and I had to decide whether or not to stay in Florence or to push on, hoping to beat the new storm. We decided on the latter and determined to get an early start. I woke up about 4:00 a.m. and took a peek out the door. What I saw was an ice-coated scene such as I hadn't seen since I'd left New York. I quickly dressed and went out to my car—my trusty Toyota Avalon turnpike cruiser. The ice was so thick that I couldn't get the door open, so I couldn't get in to start the engine and get the heater going. I also realized that I had no ice scraper.

Fortunately a fellow traveler, also from New Jersey, was out working on his car with a scraper, which I borrowed when he finished. After lots of scraping, I finally forced my way into the car and got the heater going to start melting the thick ice. I then went in to get Kitty ready for departure. I think all we had was coffee when we left the hotel and made our way over to I-95. Once on the interstate, we proceeded south very carefully. In the darkness, we saw that many vehicles, including tractor-trailers, had run off the icy road; and we were all alone. After crossing the state line into Georgia, the sun was out and conditions improved. We were then able to get off I-95 to a McDonald's for coffee and breakfast. The rest of the trip was uneventful, but we began to think that driving to Florida in January was a bit risky, and I resolved to always carry an ice scraper.

I hope I'm not repeating myself when I mention that the highlight of the winter season at Hampshire Gardens was the annual show performed by the residents. Though the show mostly consisted of short musical acts, which were mimed, the format allowed for a wide variety of performances. Acts included ensemble dancing, individual singing or dancing, and ensemble skits—some of which were quite clever. For a number of years, I served as sound technician, which meant that I had to attend rehearsals and secure tapes of the music needed for each act. The performers tended to choose excessively long pieces, so I had to help edit them down to a length commensurate with their ability. On show day, I had to set up the equipment and place the loudspeakers around the pool area, which served as our playhouse. Management of the many microphones used in the show was a perpetual problem for the director and me.

When I took over the show's job, I found that much of the equipment on hand was inadequate and got the men's club to allow me to purchase and install upgraded electronics and loudspeakers for the clubhouse and pool area. The clubhouse sound system was also used for parties, bingo, and meetings. The new equipment was more complicated than the old stuff, and as a consequence, I was often called in whenever untrained operators had trouble with the system. Eventually I found involvement with the show to be tiresome, especially as the quality of the show began to slip. I believe that by the 2002 season, I had relinquished my position and expected to be a mildly interested spectator for the show. However, the ladies had developed a "nun" act in 2001 and planned to reprise it for 2002. This time they needed a priest and prevailed upon me to appear with them as Father Jim. I went to Goodwill and picked up a dark shirt and trousers to go with a Roman collar made by Fran Kuta, one of the sisters. In my role, I had to interact with Mother Superior as played by Alice Tuttle. I did a poor job of memorizing my lines and kind of faked my way through the role. It seemed to come off reasonably well, but I was not invited to appear in the ladies' skit the following year.

That winter I signed up for the Boynton Beach Citizen's Policy Academy. It was a twelve-week course designed to acquaint citizens, in depth, with the workings of the police department. It was extremely well done, and I came away with a much better understanding of policing a good-sized waterfront city such as Boynton Beach. The weapons, police boat, and K9 segments were particularly interesting. True to the police code, we had coffee and donuts every night. The real highlight was riding along with a patrol officer on the night shift. That's a story for a different book. I received my citation from Chief Gage on April 12, 2000. I later wrote to the chief to express my gratitude for the fine job done by his staff and to commend the officer with whom I patrolled. It was a really special once-in-a-lifetime occurrence.

I'm not sure exactly when Liz and Leo Oberschmidt, our Tenby Chase neighbors and good friends, bought a place in Hampshire Gardens. However, they appear in pictures taken during the 2002 season. They had sold their house in New Jersey and planned to spend most of the year in Florida. They bought a two-bedroom unit near the highway and had the kitchen totally revamped. It turned out to be a very nice unit, and we often played pinochle there after enjoying one of Liz's special taco dinners. Leo had suffered one or more strokes, and his medication was closely monitored by the veteran's medical service at the hospital in West Palm Beach or at their clinic in Delray Beach. Leo and John Dennen, another former Tenby neighbor, usually attended mass every morning. It was very convenient having John and Leo living nearby as, among other things, we were able to share airport pick up and drop off duties.

Once back from Florida, both Kitty and I had a heavy schedule of visits to doctors and dentists. Some were routine visits to my urologist and podiatrist. But my visit to our family doctor led to, thankfully inconclusive, chest x-rays. However, I was due for a colonoscopy and decided to have it done by Dr. Tom Kayal who had treated me in the hospital the prior year. The new way of doing them in office with anesthesia makes the colonoscopy a totally painless procedure. Fortunately, there were no polyps. In August, I had my year-after knee surgery checkup with Dr. Schoifet, and Kitty had her biennial checkup with Dr. Debbie Epstein, her gynecologist; and both went well. We were looking forward to a break from our medical visits, most of which turned out to be false alarms.

Off to the Erie Canal

On Sunday, August 11, we set out for Little Falls, New York, to take part in an Elderhostel called "Boats, Trains, Great Camps." This small city sits at a falls on the Mohawk River section of the Erie Canal. The Best Western Motor Inn was our headquarters, and our room looked out on the river and the big curve of the adjacent main line of the old New York Central Railroad, now CSX. I loved to hear the powerful hum of the multiunit diesel locomotives whenever trains came through town. As an old mill town, Little Falls suffered economically as industry departed upstate New York. As a gateway to the Adirondacks, Little Falls has made a great effort to build a tourist economy. A walking tour of the city center showed that they had done a great job of saving and using many beautiful old homes and buildings. They had also created a really nice park along the river to commemorate the city's importance in the heyday of canal commerce—the traffic is now mostly recreational.

This short Elderhostel took us out on the river and through one of the highest locks on the canal. We heard quite a bit about General Herkimer, who led the German settlers of the Mohawk Valley in the 1777 fight against the British and Indians at Oriskany Creek. We visited Herkimer's home and the Oriskany Battlefield state historic site. While Herkimer was wounded in the first volley, the settlers' strong resistance led the British to abandon the western prong of their three-pronged attack on Albany. The highlight of the Elderhostel was an excursion into the Adirondacks with a boat tour of beautiful Raquette Lake. The camps, homes, and even churches along the extensive lake were magnificent and undoubtedly best seen from the water.

In the Adirondacks, our next stop was a visit to Great Camp Sagamore. Camp is a bit of a misnomer as these rustic lodges were huge and built by wealthy families as summer retreats during the "gilded age." Sagamore was started by W. W. Durant in 1895, but his bankruptcy caused the place to be sold to the Vanderbilts in 1901. Oddly enough both were railroad families, the Durants were Union Pacific and the Vanderbilts were New York Central. Under the Vanderbilts, the lakefront camp was greatly extended by adding guest residences and recreational amenities. This

Adirondack great camp, one of many, is now in the hands of a foundation that offers tours and residential programs at this historic landmark.

During free time one day, Kitty and I traveled to the Utica area and went to the casino on the Seneca reservation. There, for the first time, I walked away with more money that I had when I started. We also drove up to Rome to visit the Fort Stanwix's site but came away disappointed as we could not find parking anywhere near the revolutionary war battlefield. The next morning, the group had breakfast at Utica's beautiful 1914 Union Station. It is said to be the last of the old New York Central's major stations still serving its original purpose. The city is a major rail and road hub and is home to the Utica and Mohawk Valley Chapter of the National Railway Historical Society. Kitty and I had spent sometime in Utica in the early 1960s while I was attending a technical society meeting. Now it seemed to be very different, much less going on.

The Elderhostel wrapped up on Thursday, and we decided to do some touring on our own. We drove into the Adirondacks and made the easy auto climb to the summit of Whiteface Mountain. From the 4,867-foot mountaintop, one should see many of the Adirondack High Peaks, but we did not. As usual, clouds had moved in, and the scenery had moved out. We spent the night not far from Whiteface at Ledge Rock, a motel in the town of Wilmington. Next morning we set out for Auriesville by way of the I-87 Northway south to Albany and then west on the I-90 Thruway to the Amsterdam area. I'd never gotten off the Thruway in that area, but it seemed very familiar to me. In school we had studied the three adjacent towns (real cities, I suppose) of Amsterdam, famous for carpets; Johnstown, for shoes; and Gloversville, for the obvious. This was another old New York industrial area whose economic fortunes had sadly declined.

After a few misses, we finally found our way to our destination, the National Shrine of the North American Martyrs. In school we'd read about the valorous Jesuits, Isaac Jogues and company, who had been tortured and killed by the Mohawks. I'd always wanted to visit this spot where they had worked and died in the 1640s. Then it was a walled Mohawk village and the original home of Kateri, a Christian Indian known as the Lily of the Mohawks. I've forgotten why the Indians decided to kill the priests and brothers living among them, but the event certainly resonated in New York State and elsewhere. They became America's first and only canonized martyrs. The site sits on a bluff overlooking the Thruway, the Mohawk River, and the railroad. The wooded grounds are extensive, with many shrines and a large assembly hall or church. We particularly enjoyed the building dedicated to the sainthood cause of Blessed Kateri Tekakwitha.

On Friday we left the shrine and made our way to the baseball museum at Cooperstown, a very difficult place to get to from anywhere, but particularly so from Auriesville. The town was jammed, so we had to park on the outskirts and take a shuttle bus to the museum. We did the obligatory tour of the museum, which is a very fine and enjoyable museum. I, naturally, had a good look at all the memorabilia associated with my favorite team, the New York Yankees. The town itself is quite pretty and sits on Otsego Lake, the source of the Susquehanna River. From Cooperstown, we drove down through Binghamton and onto the Pennsylvania Turnpike's Northeast Extension.

Home to Wedding Bells

While on the turnpike on Friday evening, we tried to call Chris on our cell phone to let him know we were just a few hours away. Each time, the call was answered, but we heard nothing. After several tries, we gave up. Later, Chris said that he was much annoyed as he had to stop mowing his lawn to answer our calls. He said he could hear us talking, so we figured something was wrong with the phone's loudspeaker. I opened up the phone and fiddled with the loudspeaker connector, and it worked thereafter for the several years I kept the phone. Back in Cinnaminson, a first order of business was to check on tuxedo rentals for the upcoming wedding. Tuxedo rental was so expensive that it seemed to me that, with the likelihood of three weddings in the coming years, buying instead of renting was the better way to go. On September 15, at Sym's in Cherry Hill, I bought my very first tuxedo and accessories. I was ready for anything in the wedding department.

Our Jim had been dating Amanda Schwan, whom he had met at the Jersey shore. She lived with her family in Birmingham Township, near West Chester, Pennsylvania, and had a son, Shawn. Jim and Amanda became engaged in the spring of 2002 and were married on October 7, 2002, at the Schwan home. Bill Hess, pastor of West Chester's First Presbyterian Church, performed a very nice wedding ceremony. It was a beautiful day, and we took some excellent pictures on the back lawn. Afterward we all went to the nearby Dilworthtown Inn for a celebratory dinner. Amanda and Jim lived in the Schwan home for several years thereafter. A formal church wedding and reception at the Concordville Inn was held the following August. Their son Daniel James was born in May.

In the fall, my back started bothering me, and I spent sessions with Chiropractor Gary Noseworthy. His treatments must have worked well for me because we left for Florida by car on October 12 and stayed overnight at the Santee Best Western instead of Clark's. We stayed in Florida until November 22 when we began the drive north with an overnight stay at the Hampton Inn at Lumberton, North Carolina. Chris and Danielle were married in St. Mary's, Pompton Lakes, on November 30, 2002. The reception was held at the Smoke Rise Village Inn at nearby Kennelon. It was a beautiful place in a great setting, and we all had a great time. We stayed the night at the Radisson in Fairfield and many of us went to the De Bow home next morning for breakfast. They had the all-time biggest and best bagels. Chris had sold his house in Riverside, so and he and Danielle rented a row house in the Roxborough section of Philadelphia. It was quite a nice place, with a rear deck that overlooked Wissahickon Park. It had been a very eventful year, and we looked forward to an uneventful Christmas season and an early return to Florida. But it didn't quite work out that way.

Early in December, I was invited to receive a military service medal from the Burlington County Freeholders. It was being awarded to all county residents who had served honorably in the armed forces during wartime. I received my medal from one of the freeholders in a ceremony held at Burlington County College. It was a very nice affair, with a bare minimum of speech-making. Kitty later had the medal framed, and it now hangs in my study.

Our immediate family assembled at Christmas day at the home of Chris and Danielle to exchange presents. From there, we were to go to the Blue Bell home of Susan and Joe Cassidy for dinner. They had bought the house from Susan's parents, Pat and Dick Davis, when they downsized some years before. The Fletcher clan was also expected for dinner. Well as the afternoon gave way to the evening, heavy snow was falling, and I became quite anxious about the drive out to Blue Bell. I followed Chris and Patrick on back roads to Cassidy's, and we did quite well until we got to a sharp turn onto the short hill into their community. My tires were a bit worn, and my Toyota Avalon could not get enough traction to make the hill from what amounted to a standing start. I backed out of the road and around the corner and up the road in the opposite direction from my original turn. With no other cars in sight, I zoomed down the road and around the corner, and my momentum carried me up the hill. Because of the weather, the Fletchers decided not to make the trip to Pennsylvania, but those of us who were present had a very nice Christmas together.

Another Eventful Year

It was our custom to ring in the New Year with Agnes and Jim Fletcher. We'd play pinochle and then watch the ball fall on Times Square in New York City. I believe I mentioned that we played partnership pinochle every Saturday night, if at all possible, but after *The Lawrence Welk Show* that Jim and I enjoyed. We'd take turns hosting cards, and the special New Year session, which usually involved something special to eat. We usually took a day or two thereafter to pack the car and get ready for our trip to Florida. In 2003, we left on the third and made the usual overnight stay at the Best Western in Santee. Then next day, we got to Hampshire Gardens about 4:00 p.m. When we opened the door, we found many baby geckos (lizards) running around inside, along with many dead. They all eventually died because they couldn't find their way out. Over time we'd find the corpses behind the furniture. We asked the exterminator about this surprising infestation. He looked around and found that there was an opening underneath our wall heating-cooling unit. We filled the gap and since have had no more gecko visits.

This winter season was enlivened by the arrival of a bearded and mustached Patrick. His few days with us happened to coincide with the annual luau held around the pool area. This particular luau was the best ever as the weather was good—many get rained out and moved indoors, which ruins the effect. Patrick enjoyed himself and was a big hit with our neighbors and friends. We have a fine picture taken of our table: Patrick, Kitty and I, and Lee Oberschmidt, John Dennen, Jim and Dorothy Ann Taylor, and Linda Asher. It was a fun group. Patrick got to dance with some of the ladies, and we were asked the following year if he was coming back to see us. Our trip north in April was uneventful.

In June, Patrick made a trip to Ireland with Craig Adams, a Holy Cross buddy. Craig wanted to visit his family and asked Pat to come along because he was a veteran of European and Irish travel. In 1999, Patrick had spent twenty-three days in Europe with schoolmate John Woods, mostly at the villa in Tuscany that John's family had rented for the season. But they did make side trips to Germany, Switzerland, Holland, and Belgium. John and Patrick decided to do Ireland in 2002. Over seventeen days, they covered the entire perimeter of the island and then went to see our family in Northern Ireland. They stayed with the Clarke twins, my second cousins Brigid and Margaret, who live in Dungannon. From that base, they visited as many of the O'Donnell, Quinn, and Clarke's relatives as time permitted. Patrick's companion, John Woods, was quite impressed by the scope of Patrick's family contacts because John was unable to locate any of his family.

Back to 2003, Craig and Patrick flew into Shannon Airport and made their way up to Northern Ireland. Craig was able to locate some of his relatives, one of whom was a religious brother who took them on a tour of the "war zone" in Belfast. Patrick again visited our family. On their way west and south to Shannon, they stopped in Kinvarra, County Galway. There in a pub, they met a California family, the Pedersons—mother Betty, son Duane, and daughter Janet—who were also touring and looking up ancestral places. Back in the states, Patrick and Janet stayed in touch and began to exchange coast-to-coast visits. Together they attended the October Cape Cod wedding of one of Patrick's friends. Kitty and I met Janet for the first time later in the year and found her to be a charming young woman. Janet and Patrick's meeting in Kinvarra was as fateful as was the unlikely meeting of Kitty Carlin and I back in 1956.

In August we had Jim and Amanda's formal wedding at First Presbyterian in West Chester with Pastor Bill Hess presiding. While I can't find the program, I do recall that Amanda's son, Shawn, was the ring bearer; and a neighbor's daughter was flower girl. Amanda's brothers Chris and Scott and Jim's brothers Chris and Patrick provided the manpower; Amanda's sister Tara was maid of honor. A reception followed at the nearby Concordville Inn where we stayed for two nights. Although we had rain during the day, it didn't really dampen our spirits. It was an enjoyable wedding for me because I was able to spend some time with my cousin Eileen Quinn Dodd and her husband Joe who also stayed over. I reminded Joe that back in 1956 and working part-time, he sold me a pair of beautiful black Nunn-Bush shoes that I wore for my wedding, but never actually fit me. I gave them away.

Off to Nova Scotia

In September, we took off by car for an Elderhostel given by Mount St. Vincent University, Halifax, Nova Scotia. We drove flat out to Bangor, Maine, where we spent the night. Next day, Sunday, we joined the group at the Halifax Holiday Inn. On Monday we did a Halifax highlights tour with major stops at Citadel Hill and Pier 21. Halifax was a major British naval base through the nineteenth century, and the citadel's guns helped protect the sheltering harbor. The bastion's defenses are formidable and designed to turn away a ground attack. Pier 21 is the Canadian equivalent of Ellis Island and a very interesting place to visit. It's possible that my father was processed there on his arrival in Canada. From Canada he eventually, by hook or crook, made his way to the United States.

On Tuesday we took an all-day train trip to Sydney on Cape Breton Island. In its heyday, the town was an important coal-mining center, but now it's just the end of the line at the far eastern point of the province. This train ride was long enough for me to learn the entire life stories of many of my fellow travelers, but far too long to justify the expenditure of time. For most of the trip, there was nothing to see but trees. After group dinner, Kitty and I walked to the Sydney version of an Indian casino where I again won a few dollars, this time Canadian. On Wednesday we went to the University College of Cape Breton (UCCB) for lectures on the island's history and culture. After lunch, it was off to the coal miner's museum where I foolishly agreed to go into the mine while Kitty stayed topside. Actually it was no problem; the ceiling was high and the tunnels wide and well lighted. That night we went for a two-night stay, including dinner, at the Fortress Inn at Louisbourg.

For years the French fortress at Louisbourg was a threat to Britain's base at Halifax and colonies on the Atlantic seaboard. The fortress was taken in 1745 by New Englanders with British support but was returned to France by treaty. In 1758, some thirty thousand British troops retook the fortress, and the fortifications were destroyed. The British garrison was left in 1768, and the site lay in ruins until it became a National Historic Site in 1928. The site, as it stands today, includes a portion that's been reconstructed since 1961. The fortress and a nearby lighthouse are in a locations that are very exposed to the open Atlantic Ocean. I was very interested in visiting Louisbourg as I'd read about it in my favorite Kenneth Robert's novels of the French and Indian Wars. While the Canadian government has done some excellent restoration, there really isn't that much to see or do at the fortress. However, just being there and getting a sense of the place was good enough for me.

One evening we went to a music program at the playhouse in the town of Louisbourg. The building is the actual replica of Shakespeare's Globe Theater that was used in a movie filmed in the area. The program of song and dance called Lyrics and Laughter was performed by young, attractive, and accomplished men and women. The intermission offered light refreshments at no cost. It was an excellent evening. On Friday, we drove back to Halifax by bus, with a stop at the Alexander Graham Bell Museum in Baddeck. Bell's nearby home is still standing and occupied by family, so we were unable to visit there.

Off to Green Gables House and Homeward

Back in Halifax, we packed up the car and set out for Prince Edward Island (PEI) because Kitty wanted to visit the Anne of Green Gables house. After spending the night in Truro, Nova Scotia, we had a choice of two ways to get across Northumberland Strait to PEI—either go east to a ferry or west to a bridge. Not knowing the ferry's schedule, we chose the bridge, which turned out to be nine miles long with a toll of about C$40, certainly not a commuter bridge. We stopped at the island's very nice visitor's center and picked up a map to speed us on our way to Cavendish and Green Gables, the house author Lucy Maud Montgomery used as a setting for her "Anne of" stories. I enjoyed touring the house and grounds even though I'd never read the books nor did I see the movie or the television series. It's amazing to realize as I write that the first of the "Anne" books was written one hundred years ago. From the house, we took a short drive along the Gulf of St. Lawrence beaches on the north shore of the island. On the way back to the bridge, we had lunch at Tryon Head on Northumberland Strait.

Back in the province of New Brunswick, we made a major mistake. We left the main highway at Moncton in order to drive along the Bay of Fundy to the Hopewell Rocks Ocean Tidal Exploration Site. The narrow road was under construction for miles, and we arrived at our destination as they were closing. An employee was nice enough to let us ride on the tram as he drove down to the water's edge to pick up the guides and other staff. We got a brief glimpse of the amazing mini islands that are left high and dry at low tide. Better yet, the ticket booth was closed, so it was a free ride, but we did tip the driver. To continue south from Hopewell, we had to take the scenic route through Fundy National Park and into the town of Sussex. We stayed overnight at a Quality Inn that had a very nice restaurant. Everyone, staff and guests, were very friendly; and we enjoyed our brief stay in Sussex.

Sussex is on the main road, so in the morning, we set out for Frederickton, the provincial capital and over the border onto I-95 in

Maine. Though we were anxious to get home, we thought we'd visit my cousin Patricia in Niskayuna on the way. This meant that we'd take the Massachusetts Turnpike west toward Albany. It was raining and getting dark when we got to the Berkshires in western Massachusetts, so it seemed like a good idea to spend the night there. We checked into the Best Western in the town of Lee. To my disgust, their restaurant offered strictly Asian cuisine, meaning we had to set out in the gloom to find a more suitable eating place, which we did. As we entered the hotel some men coming in behind us asked if I were an engineer. When I answered yes, I was given the room number for "the party." At the party, I learned that it was arranged by and for engineering graduates of New York University. I was welcomed anyway and commiserated with the group on the death of the engineering major at NYU. I also shared their disgust at the end of football at NYU many years ago. Their team was called the Violets, a soubriquet that hardly challenged the Lions of Columbia or the Rams of Fordham.

Next morning we took a quick drive around Lee and thought that we'd like to come back and spend some time in the Berkshires. But off we went to Albany and west to the Schenectady suburb of Niskayuna. We had a nice lunch with Cousin Pat and Joe Early and talked family stuff, politics, and whatever. We could easily have talked for hours, even days, but we wanted to be home by nightfall. Joe, though quite ill, insisted on leading us back to the New York Thruway, and I'm glad that he did. The roads are very tricky and congested in that area. We took the Thruway south to I-287 and then to U.S. 202, a routing which keeps us off the horrible Garden State Parkway and New Jersey Turnpike. We stopped to eat along U.S. 202, and it was well after dark by the time we got back to Cinnaminson. This ten-day jaunt to Canada and back sorely tested our travel stamina. Somehow I didn't think we'd be driving anywhere of consequence any time soon.

Florida and Vermont, 2004

The winter of 2004 in Florida was extraordinary for two reasons. First of all, Patrick and Janet came down for a visit, her first ever trip to Florida. They had just become engaged. We picked them up at Palm Beach International Airport and did some sightseeing as well as fine dining at our favorite places. The most interesting tourist bit was a water taxi tour of the complex waterways of Fort Lauderdale. The number of high-rise condos going up along the water was absolutely mind-boggling. We learned that Fort Lauderdale had a shortage of berths for 100-foot-plus yachts, so a wealthy Miami investor was buying up small marinas in order to combine and expand them. One night Patrick and Janet took our car and went out on their own for the evening. Patrick really enjoys spending a few days with us in Florida. Of course, it was easy for him to fly down as he accumulates a lot of bonus air miles.

The winter of 2004 was probably the height of the South Florida's construction boom. All of the vacant or underutilized land near Hampshire Gardens was giving way to condos. The city of Boynton Beach even eased its four-story height limit to allow high-rises in downtown areas along the Intracoastal Waterway. Some units at Hampshire Gardens sold for prices over $200,000. All of the nearby construction contributed to traffic problems, noise, and annoying wind-blown dust. Displaced foxes and raccoons started showing up in our community. For some reason, a sick raccoon climbed up to die on the balcony next door. At night our people became a little fearful of these invaders.

As usual, Kitty and I drove back to New Jersey in April but had to leave for Vermont in May. The occasion was the wedding of Pat and Joe Early's youngest son, Joseph, to Angela Corsi, of Vermont. Joe had recently accepted a football coach position with Middlebury College. I might have insulted him when I said I didn't know that the college even had a football team. We stayed in the Burlington Hotel where a rooftop reception was held. The wedding ceremony, however, was performed in St. Mary's Church of the Assumption, Middlebury, some thirty-five miles away. Of necessity, there was lots of driving back and forth through Middlebury. At the reception, we got to meet for the first time some of Pat Quinn's relatives on her mother's side—an interesting group indeed. The rooftop setting offered a very dramatic view of the sun, setting behind the Adirondack peaks across Lake Champlain.

Burlington is Vermont's largest city and a lake port. It is home to several colleges, and we took a drive around the beautiful campuses of the University of Vermont and Champlain College. We enjoyed browsing the Church Street Marketplace, a pedestrian mall right in the center of the city. On departure day, we chose to take the ferry across the lake to New York and drive down to New Jersey via the Northway and the Thruway. However, it was the day of the annual Vermont City Marathon, which brings a lot of runners and lookers to the city. To avoid getting hung up in traffic, we made an early start so as to get the first ferry of the day. The morning was very chilly with a brisk wind, which made being on deck very uncomfortable. Kitty stayed in the car, but I went below to hang out in the comfort of the cozy coffee bar. One of the reasons we had chosen this routing through New York was to take the raft trip through the scenic Ausable Chasm. Our family had made this enjoyable excursion back in 1949. When we got to the attraction, we found that it had become a great deal more formal, expensive, and time-consuming than I remembered, so we reluctantly gave it a bye and lit out for home.

Kitty's eldest sister, Marietta, known as Mimi, was married to Dr. Edward M. Sullivan. When Kitty and I lived in Collingswood, we often played bridge with Mimi and Ed who lived nearby for a time. Ed had a popular and successful obstetrics and gynecology practice in the vicinity of their home in Springfield, Delaware County, Pennsylvania. Mimi, herself, had been a medical technician. Ed had degrees from St. Joseph's University and from the Temple University School of Medicine and served as a medical officer in the U.S. Navy. He was an associate and adjunct professor at Hahnemann, Jefferson, and Temple medical schools. Ed and Mimi were avid golfers and members of the nearby Llanerch Country Club. Around the time of his retirement from his full-time medical practice in 1998, Ed developed heart problems, which led to his death in June of 2004, at age seventy-six. His funeral at St. Dorothy's church in Drexel Hill was attended by a multitude of associates, friends, and former patients. It was quite a testimony to his standing in the community. The service concluded appropriately with the Navy Hymn, always a favorite of Ed's and one that I loved to hear myself.

In August we joined the Schwans and Jim's family for the formal adoption hearing, which brought Amanda's son Shawn Patrick into the O'Donnell clan. It was held on a rainy day in the Chester County courthouse in West Chester. Afterward we had a celebratory dinner in a local restaurant. In September Jack Michael, our third grandson, was born to Amanda and Jim. All of their children were born in the Chester County Hospital, also in West Chester, a very nice place indeed. In October, we drove down to Florida, making our usual overnight stop in Santee, South Carolina. As we got south of Orlando on I-95, we began to see damage from some of the five hurricanes that had passed over Florida a month or two before—blue tarps covered many a roof. Fortunately, West Palm Beach on the South was pretty much unscathed. In November, we flew back to Philadelphia from West Palm Beach for the 2004 holiday season. It was an uneventful flight on the much maligned U.S. Airways.

An Eventful 2005

In January, we flew back to Florida for the winter season. We were on the same flight as neighbor John Dennen who was being picked up at the airport by his brother-in-law Phil Wynne. We readily agreed to share the ride. The first thing that seemed strange was Phil's arrival on foot at the baggage claim area. That meant that we'd have to lug our bags to wherever he'd parked the car. I quickly found a loose baggage cart; and we were able, barely, to put three sets of luggage on board. Phil didn't seem too sure where he had parked, but we all set off to one of the garages. I wheeled the cart onto the elevator and up to the second or third level. As Phil and John searched for the car, Kitty and I guarded the luggage cart and awaited further instructions. Suddenly I noticed John and Phil going up the ramp to the next level. I struggled to push the heavy cart up the ramp after them. When we got to that level, they started up to the next, and I again followed. Off they went again, and I said to Kitty that we would wait where we were as I'd found an abandoned wheel chair to sit in. We'd waited quite a while when a rental car van stopped, and the driver asked if we needed help. We explained our problem, and the driver went off to find Phil and John. Eventually, Phil and John arrived to pick us up, explaining that we'd been in the wrong garage. We'd lost an hour and a half, and I was worn out from pushing the cart up hill.

No sooner had we gotten settled in our unit for the winter when we learned that we had a first granddaughter. On January 23, Katherine Marie O'Donnell was born to Chris and Danielle at Chestnut Hill Hospital in Philadelphia. We flew up to see the beautiful newcomer who has been called Katie evermore. Cassie De Bow, Danielle's mother, had been helping out with the new baby, and Kitty relieved her for several days. As an infant, Katie would prove to be very easy to live with. After about a week or so, up north we flew back to Florida. In 2004, we'd started to upgrade our living room furniture by acquiring from JC Penney a Florida-style cocktail table and matching lamp tables. We picked the furniture up at the store and, with a great deal of help, somehow squeezed them into the Avalon. But there seemed to be no room for Kitty, yet we somehow squeezed her in as well. It was a very uncomfortable ride home, but mercifully a short one.

The new living room furniture was an amazing improvement, so in 2005, we decided to add a wall unit from Ikea. We'd seen it at Amanda and Jim's and thought it would work well in our apartment in place of the rickety, temporary media unit we bought when we first moved in. It was shipped to us, at reasonable cost, from Ikea in New Jersey because we had no Ikea stores in Florida. The unit came to us in several huge boxes, and I had a real workout putting it together. As I built it, I realized that we would not need the topmost section. I called Ikea, and they picked it up at no cost to us. It was a real pleasure shopping from the Ikea catalog and very economical. Best of all, the wall unit is perfect for our apartment.

The summer back in New Jersey was very busy with family birthdays and celebrations. My niece Maureen Maguire and her husband, Jack Kulaga, held their annual Independence Day party at their home in Lincroft, New Jersey. They had added a beautiful and well-landscaped swimming pool, which kept the many grandnieces and grandnephews busy. The adults and older children enjoyed the volleyball competition. We oldsters sat in the shade and chatted until it was hamburger and hot dog time. As usual, we saw fireworks from many of the small towns we passed through on the drive home. Maureen and Jack's party is always a highlight of the summer season.

In September, it was bridal shower time for Janet Pederson, Patrick's intended. The affair was hosted by Ann Sullivan McShane at her home near Media, Pennsylvania. In October, the wedding festivities kicked off with a grand rehearsal dinner at Brittingham's Irish Pub, Lafayette Hill, Pennsylvania. On October 15, Janet and Patrick were married at Epiphany Catholic Church, Plymouth Meeting. It was the third time I got to wear my tuxedo and most likely the last. A reception followed at the Flourtown Country Club. It was a great setting for pictures, both indoors and on the golf course itself. After a Paris honeymoon, Janet and Patrick returned to their apartment near Blue Bell.

Wilma Pays a Visit

After a busy summer and early fall, Kitty and I drove down to Florida after the wedding. We'd been warned that a hurricane was on a track headed for Florida, but we felt that the likelihood of it hitting our area was small. Well, we were wrong. Hurricane Wilma had been in the Gulf of Mexico but turned east and crossed the Florida peninsula, smacking Boynton Beach. With hurricane shutters closed and no power, we were in the dark as the storm raged around us. We heard stuff crashing against our shutters, particularly on the rear windows, but had no idea what it was. It turned out that it was parts of the roof over the nearby shuffleboard courts. As the eye passed over, we were able to go outside for a look, and while there was lots of debris, the buildings didn't fare badly. We scrambled inside as the back edge of the storm approached, and it seemed worse than the front edge.

As the only officer present, I checked the building after Wilma passed. We had a section of concrete balcony railing knocked off by a swinging screen door and some facia and gutter damage. I made a temporary railing repair using two-by-fours and removed the broken concrete. Since we had a $35,000 deductible insurance policy, we wouldn't be collecting anything. After the storm, we faced the possibility that we'd be without power for a week or more. The supermarkets started disposing of all refrigerated or perishable items, and there was no ice or bottled water to be had. Many towns were without tap water because their systems required electricity but not Boynton Beach. We had nothing to eat or drink other than canned food and water. Fortunately, the day before I'd bought a supersized hoagie, which lasted several days. I had also stocked up on bottled water and batteries, so we had a radio and some lighting to supplement the ubiquitous candles. However we faced a bleak future without our morning coffee.

Early next day I decided to take a drive around the area to see what had happened. Trees had fallen across Federal Highway, our street, reducing travel to one lane. Along Congress Avenue, the newer wooden utility poles had snapped off, and the wires lay in the street along with parts of palm trees; it was a mess. At the tennis complex, many trees had fallen, knocking down stretches of fence. The covered pavilions alongside the courts were destroyed. The canvas had been torn off and blown away after bending or breaking the supporting posts. Amidst all this bad news, I spotted a restaurant that was open and serving customers; it turned out that they had their own generator. I raced home to get Kitty and got right back to where we joined the line of desperate people. We had a wonderful breakfast there each morning that we stayed in Florida. I don't remember what we did for the other meals once our hoagie was gone. Some of our year-round neighbors shared an electric generator and were able to do some cooking.

My tour of inspection convinced me that we would likely be without power for some time. I didn't enjoy being in the dark, without television and without decent light to read by, and without many food options. Furthermore, it was obvious that there would be no tennis for a while. Consequently, we decided to leave and quickly made reservation on the first flight to Philadelphia out of West Palm Beach when it reopened. We paid $138 each, one way on Southwest, as we still had our return flights from the original U.S. Airways round-trips we'd booked before we left by car for Florida; or so we'd thought. Just to be sure, I called U.S. Airways and explained to them why we wouldn't be able to use the northbound flights we'd booked. We gave our lantern, batteries, and bottled water to the Oberschmidts before we left. So after only five days in Florida, we headed back to Philadelphia. As he did so often, Jim Fletcher picked us up at the airport and drove us home to Cinnaminson.

After all of the excitement—births, weddings, and hurricane—it was nice to enjoy the fall and holidays at home. Thanksgiving dinner for the O'Donnell side of the family was hosted by my niece Kathleen Crumlish and her husband Rob Merola. They have a big house that can accommodate the forty-odd potential diners and have been good enough to host the dinner for several years. By December, it became clear that brother-in-law Dick Davis was losing his battle with lung cancer. We didn't see Dick very often; Pat didn't encourage visits, so we were shocked at his condition when we finally did see him. He died on December 15 and on the 19 had a quiet funeral, followed by interment at St. John Neumann Cemetery in Chalfonte, Pennsylvania. Dick was a U.S. Regular Army and had served in Korea during that war. He could have had a full military funeral, and I'm not sure why he didn't. He will always be remembered for his hearty laugh and his performance of the French song *Alouette*.

Before long, Christmas was upon us. On Christmas Day, we all assembled at the new (to them) home of Jim and Amanda to open presents. They had recently bought a townhouse in a small community near Kennett Square. The area is very nice but rather remote from the business centers that surround Philadelphia. However, it is close to Amanda's parents and is in the Chadds Ford-Unionville school district that she likes. Niece Susan and Joe Cassidy hosted Christmas dinner for the Carlin side of the family that year. As usual, we spent New Year's Eve with Agnes and Jim. The men won the 2005 pinochle competition 83-78, one of the closest margins in the nine years we've kept score.

After Wilma, 2006

In January, when I tried to get seat assignments for our US Air flight back to West Palm, I learned that our tickets had been canceled since we hadn't used the West Palm to Philadelphia portion. Naturally I went ballistic so that they were forced to fully research our case. They eventually found a file notation that I had called to explain that due to Wilma we'd be unable to use the front part of our tickets. While we got reinstated, I realized that it was a bad idea to book a flight back from Florida until we'd arrived there. This meant that we'd probably have to book one-way flights on Southwest and take our chances on getting the adjacent aisle seats we preferred.

When we got back to South Florida, we saw a sea of blue coverings on area roofs. Colonial Club, just to our south, lost the shingle roofs on almost all of its many buildings. Assessments on unit owners for new roofs and other repairs exceeded $20,000 each. It took a few years, but they got beautiful new aluminum roofs and fascia. We, at Hampshire Gardens, had minimal damage. Our rock roofs were rock solid, but the shingle roof on the clubhouse had to be replaced. Shingle roofs are just no good for Florida and are rarely used except in really cheap construction. Our community of twelve buildings collected some $150,000 for debris removal and repairs. We were able to bank $100,000 of that insurance payment. Even today, there are many area roofs still covered in blue plastic. I spoke to a roofer who was getting into his truck at the supermarket and suggested that he must be making a ton on roof rework. He said he was being run ragged and might just move out of Florida.

We had decided to spruce up our apartment, which had become rather dowdy by contemporary standards. We were still using the original bedroom furniture comprised of twin beds, a double dresser, and a nightstand. We painted all of them, including headboards, in bright Florida green with white trim. With new brushed aluminum drawer hardware and mirror to match, the room looked 100 percent better. We then engaged two young men, perennial visitors to Hampshire Gardens, to paint the whole place, except for the kitchen. They had come highly recommended and did a first-class job. They liked to use bright colors and accent walls, and we let them do their thing. I used a Home Depot gift card to buy a cordless telephone system with two handsets. We priced new carpeting but decided to hold off until the fall for reasons you'll learn later. We had no visitors that winter and drove back to New Jersey in April.

We had a busy spring and summer. Granddaughter Elleanor Susan (Ellie) was born May 11 to Amanda and Jim. She was grandchild number six, second granddaughter, and is an O'Donnell-style redhead. Late in July, my niece Patricia Margaret "Trish" Crumlish was married to Mark Andrew Stillwagon at a grand ceremony and reception at the Mill at Spring

Lake Heights, New Jersey. Both are civilian employees of the U.S. Army in Germany and returned there after the wedding. Mark has a young daughter, Melissa, from a previous marriage. In August, granddaughter Ellie was baptized at First Presbyterian, West Chester. It's a very friendly church, and Pastor Bill Hess always does a fine baptismal ceremony.

After our short stay in the Berkshires a few years back, we looked for an opportunity to go back for an extended stay. Such an opportunity turned up in the Elderhostel catalog, so we signed up for Historic Homes in the Berkshires. In September, we drove to Hancock, Massachusetts, and checked into our base camp, the Country Inn at the Jiminy Peak ski resort. For the next several days, we visited historic homes and museums in the upscale towns of Lee, Lenox, and Stockbridge. We also had a tour of Tanglewood, the historic summer concert venue, followed by a picnic on the grounds. For me the highlight was the tour of the Mount, summer home of Edith Wharton who chronicled the Gilded Age in her many novels. I was inspired to read her 1905 best-seller *The House of Mirth*, which was written at the house. I found the story of striver Lily Bart fairly interesting but much too long. It would have made an excellent short story. I haven't had the urge to read any more of her books but did enjoy the 1993 film of her novel *The Age of Innocence*.

Driving to and from our hotel each day, we rode through the historic city of Pittsfield. This is the capital city of the Berkshires and a once-thriving industrial city. They have done some great restorations to make the downtown area very attractive and welcoming. I wished a detailed tour of the city was on our agenda but no. We did get to see the wonderful Norman Rockwell Museum in Stockbridge. His paintings adorned the cover of the very popular *Saturday Evening Post* (SEP) magazine for years. The original cover art is on display in the museum, and for me, it was a trip down memory lane. I remembered many of them from the 1940s and early 1950s. The SEP (1821-1969) offered a mix of articles and fiction, and many famous writers contributed short stories. Admittedly, I was not a big fan of the SEP in its heyday, but I miss it now when I can't have it. Oddly, we have the May 1907 edition of the *Ladies' Home Journal*, a sister publication of Philadelphia's Curtis Publishing Company. While the magazine itself is huge, the print is so miniscule it's unreadable. One wonders how anyone in 1907 could read it by gaslight or candlelight.

One of the Elderhostel guides was an assistant football coach at Williams College in nearby Williamstown. He suggested that we visit the campus as it had an excellent art museum. On an afternoon when we had some free time, we drove up to Williams, parked, and set out for the museum. It turned out that we'd parked in exactly the wrong place because we had to climb many steep flights of stairs to get to the museum's entrance level. It was indeed a fine small museum, but it had one memorable exhibit that defies all notions of art. In a gallery all to himself, Chinese artist Zhan Wang built a stainless steel model of Beijing using cookware—trays, tongs, servers, pots, coffee pots, and miscellaneous utensils. His Urban Landscape was absolutely wonderful, well worth the drive and all the steps. We learned that the Chinese language is the fastest growing major at Williams College, and consequently, the museum often exhibits Chinese artists.

The last day of the Elderhostel was Friday, September 29. We arranged to visit the Earlys near Schenectady on the way back to New Jersey, so we picked up our box lunches and set out for New York. Though we had told Pat that we would have had lunch before we got to her place, she laid out quite a spread, and we ate again. We had the usual talk of family, politics, and every topic under the sun. Though Pat and Joe, in particular, have serious health problems, we had a good time. Unfortunately, we left without the supply of donuts Pat had gotten to sustain us on the drive south.

To Florida and Back, Autumn 2006

In mid-October, we left for Florida with the idea that we'd finish our projects there before the winter crush. It was very warm when we arrived and found, naturally, that our General Electric air-conditioning unit had died. We replaced it with a similar summer-winter unit from GE. There are cheaper units, but we liked this one because it did not project out onto the balcony like the others. The new unit also was self-evaporating, that is, required no drip pan to collect the condensate. I also got a new La-Z-Boy recliner to replace the well-used chair that was eight years old.

A major project that we'd deferred to fall was the installation of a tankless hot water heater. We felt that we should replace the old tank before putting down new carpeting. Some of our more adventurous neighbors had been going tankless as the unit took up very little space under the kitchen counter, essentially giving us a new storage cabinet. I had bought the unit in January but had difficulty finding anyone to install it. Since the tankless unit required more power, I had to get an electrician to upgrade the wiring to the heater. That much I'd gotten done in March. Finally, in October, I got a local plumber, Jim Peterson, to do the plumbing and final hookup. We were very pleased with the result and asked Jim Fletcher to make a door for the area now accessible under the kitchen counter. With the information I'd given him, Jim had the door and the hardware ready for us to take back or ship back to Florida in January. With the completion of all the nasty jobs, we were finally able to get new wall-to-wall carpeting installed throughout our apartment. The place looked great, and Kitty was mighty pleased. We had become a showplace, and neighbors liked to come by for a look.

Margaret O'Regan was one of our fellow Berkshires Elderhostelers and was also from New Jersey. In chatting with her, my story seemed to match that of her former neighbor and longtime friend John Collier. I was astounded; John and I were classmates in grammar school and high school and often partied together during our college years. John dated Rosemary Dugan for many years, and they eventually married. John didn't make our wedding as he was in the army at the time. We had lost touch over the years and both ended up in Alcoholics Anonymous. I think alcohol played a significant role in keeping us apart. It turned out that the Colliers had a place in Hobe Sound, not far from us in Florida. With the contact information I received from Margaret, I called John, and we arranged to meet for lunch at Panama Hattie's in North Palm Beach, midway between our places. I had last seen John in New York City sometime in the 1970s and Rosemary back in the 1950s. We pulled into the restaurant parking lot at the same time and immediately recognized one another; Rosemary hardly seemed to have changed at all. I learned that they were avid golfers and that John was still active in Alcoholics Anonymous. It was really great to see them both after such a long hiatus. I had hoped that we could get together again back in New Jersey, but it never happened.

In November we flew back for the holidays on U.S. Airways, but we booked our flights well after our arrival in Florida. As usual, Fletch picked us up at the airport. Our first order of business was to get over to Mercy Suburban Hospital, Norristown, Pennsylvania, to see Ryan Patrick, our new grandson. He was born to Janet and Patrick, who had recently bought a house in nearby East Norriton. To me, Ryan has a round Quinn-like face, where O'Donnells have a more angular face. Nonetheless, he was and is a very handsome lad. Betty Davis, Janet's mother, came down from Massachusetts to help out for a few days. That Christmas, the opening of presents for the immediate family was held at our house, and then we were off to dinner with the extended family, but I know not where.

Winter 2007

In 2006, we had shipped some bulky bedding, clothing, and miscellaneous stuff to Florida via the postal service in two large boxes, both insured for a minimal amount. Only one of the boxes arrived, and we filed claims with the postal service and with our homeowner's insurance carrier. We eventually recovered $600, but Kitty was very unhappy at the loss of some favorite things. Thereafter, we shipped anything of value by UPS through their agents at the Staples stores. With UPS, you get a tracking number and can follow the progress of your package on their Web site. Before flying back to Florida, we had to ship the cabinet door that Fletch had made; it was just too big to carry to the airport. I packed the door and its hardware in a large flat box that Fletch had found and sent it out via UPS. When it didn't arrive on the scheduled date, I checked the UPS Web site and found that it was hung up in Jacksonville.

When the door arrived a few days later, it was in an entirely different box—a box for the ages. I assume that my box had fallen apart and that UPS had repacked it in a double-walled box of much-thicker cardboard than any boxes I've been able to buy at Staples, Lowe's, or Home Depot. The box was so well packed that I had difficulty extracting the door and its hardware and disposing of the box and packing material. I'm guessing that a similar fate had befallen our missing box, but that the post office just put it aside rather than repacking it. I put a final coat of paint on the door and easily installed it with the materials Fletch provided. If you ignore the faulty packing job, it was indeed one of my more successful projects.

In January, Janet and Patrick brought Ryan down for a short visit. I think Patrick wanted to get the little guy tuned into to the good life in Florida. Our next-door neighbor, Katie Minneker, offered the use of her apartment during their visit, so Kitty and I spent the night in her lush bed. While they were with us, our new Toshiba thirty-two-inch LCD TV arrived, and Patrick helped me unpack and set it up. Our ten-year-old nineteen-inch Sony was still in working order but just too small for its place in our huge wall unit, and even worse, I could no longer read the ball scores on it. I couldn't sell or give away the bulky old set, so I donated it to Faith Farm, our prominent local charity. I also gave them a lot of computer accessories I was no longer using. Anyway, our guests had a fine time as the weather was perfect. We dined one night at the nearby Prime Catch; it's located on the Intracoastal Waterway and is one of Kitty's favorite restaurants.

Sadly, in January, we lost Liz Oberschmidt, at age seventy-five, to cancer. She and her late husband Leo had been good friends for over thirty years, going back to early days at the Tenby Chase Swim Club. Once she and Leo settled alongside us in Florida, they became our pinochle partners. Knowing that her health was failing, Liz had done something truly remarkable. She threw herself a wonderful party at the Delran, New Jersey, home of her son Leo. Everyone she knew was invited, and it was a grand affair. When she went into Samaritan Hospice in Mount Holly, we realized that the party had been Liz's farewell to the world. Before leaving for Florida, we got in a final visit and found her comfortable and in good spirits. She regretted that she never had a chance to work on my bad attitude toward religious faith. We didn't fly up for the funeral; we had been there for Leo's and that had to suffice.

Over the years, I had kept in touch with Tommy O'Rourke, a childhood neighbor and close friend. I had lost touch but got reconnected through a cousin who had met Tom at an Alcoholics Anonymous meeting near their mutual home town of Santa Clara, California. His mention of Riverdale or Yonkers triggered in my cousin a notion that Tom was a pal of mine that she had likely seen at our house. She was right, and through her we got reconnected. Tom and Diane, his second wife, eventually moved from Santa Clara to the farm town of Modesto. Tom's first wife mysteriously dropped dead in broad daylight on a street corner in Yonkers a few years after they were married. Tom was an electronic technician and traveled extensively on government contracts, eventually settling in the Golden State. You may recall that he and I had been briefly in the same electronic training unit at Keesler Air Force Base near Biloxi.

Out of the blue, I got a call from Tom, saying that he was visiting his sister, Eileen, in Fort Lauderdale and would like to get together. We invited them to our apartment, and it was great to see Tom whom I'd last seen briefly in California in the 1960s. I had sent him pictures of John Collier and I together, and Tom said he would not have recognized either of us had he passed us on the street. We went to lunch at the Banana Boat on the Intracoastal, and Diane really enjoyed the ambience and the passing boats, a far cry from hot and dusty Modesto. After lunch, I took our guests over to the Wakodahatchee Wetlands to view the birds and other wildlife. We walked out on the boardwalk to a pavilion near a nest under construction by a great blue heron. We watched the male flew over to the woods to pick up branches, which he used to expand the nest already occupied by two hungry chicks. Diane was fascinated by this scene—we were so close to the nest, we could almost touch it—and she took many pictures. Diane loved everything about our get together, and we hated to see her go.

We hurried back from Florida to meet Meghan Elizabeth, our third granddaughter. She was born on April 10 to Danielle and Chris at Grand View Hospital, Sellersville, Pennsylvania. In June, we had a grand celebration of our fiftieth wedding anniversary. Patrick and Janet arranged a brunch for sixty people at the nearby Riverton Country Club. The room looked out upon the golf course and opened out onto a lovely terrace. Background music was provided by a pianist who knew all of the show tunes and big band hits loved by people of our generation. It was wonderful to celebrate with family, relatives, and friends on a beautiful day and in a beautiful place. Thanks to Patrick and Janet for putting it all together. Some of our guests came back to the house afterward to keep the party going.

The rest of the year was full of family doings. Ryan was baptized in May at Epiphany Church, Plymouth Meeting. In November, Meghan was baptized at their parish church, Corpus Christi in Upper Gwynedd, Pennsylvania. In between, we spent a week at the shore, Sea Isle City to be exact. Chris, Patrick, and their families shared with us a huge four-story duplex on a beach block around Seventy-eighth Street. Kitty and I got the beautiful master suite on the top floor, with king-sized bed and truly glamorous bathroom. But best of all, the place came with an elevator. We had a lot of fun playing a very competitive card game called Phase 10. Danielle and Kitty were the big winners. Kitty and I discovered a wonderful bakery in Somers Point. We brought back over a dozen donuts; and they seemed to disappear very quickly, especially the jelly donuts, which seemed to be everyone's favorite. I went on the beach one day but mostly just hung out and read while watching Meghan, of course. One night we went to Ocean City and walked the boards. At Gillian's Wonderland Pier, we did some of the rides with Katie and Ryan. I enjoyed the Ferris wheel and the carousel, perhaps others. I found that the wheel had changed since I had last been on it. The seats used to be open, and you could actually slip out and fall. Now you are locked into a cage with no escape. It is much safer, but no longer a thrilling ride. Sitting in her stroller, Meghan was very bright and alert, drawing lots of oohs and ahs from passersby.

We got bad news in August. Uncle Jim Fletcher had been behaving erratically for several months and couldn't shake a sinus infection. He was being treated for the sinus infection, but it wasn't getting any better. Finally, one of the staffers in his and our family practice decided that he should have some tests. I don't recall the exact progression of the tests, but the conclusion was that he had a cancerous brain tumor and some cancer in his colon. This last was really surprising as he'd been undergoing regular tests for colon cancer. It was decided that he would have the brain tumor removed; and then, after recovery, the cancerous section of his colon. It was all presented as routine.

Jim's brain tumor was removed at Thomas Jefferson University Hospital by Dr. Andrews, chief of neurosurgery. Jim seemed to be recovering nicely and was sent home a few days later. While he had some trouble walking, he seemed to be in good spirits over the first weekend at home. However, he crashed a day or two later, and his son Mike rushed him to Virtua Memorial Hospital in Mount Holly, where I had my knees replaced. There he was treated for a serious infection in his lungs, which he may well have picked up at Jefferson. Jim never seemed to fully recover from the combination of brain surgery and infection, so there could be no operation to remove the colon cancer. By year's end, Jim was able to get around by himself with a walker but with some difficulty.

As usual, the O'Donnell family Thanksgiving Day dinner was hosted by niece Kathleen and Rob Merola. As I probably said before, their spacious home is perfect for a large family gathering. There suddenly seems to be a lot of youngsters in the family, and they can be out of sight in the furnished basement or upstairs. Of our boys, only Patrick and family made the scene. Chris and Jim usually go to their mother-in-laws on that day. After dinner, we played a fun game, and the prize of $40 was won by one of my nieces. On Christmas Day, we assembled at the new (to them) home of Chris and Danielle to open presents. They had bought a big colonial-style house near Lansdale, Pennsylvania, near the Merck facilities where Chris works. It's a little farther to Danielle's office at Medco Health Solutions in Blue Bell. After presents, most of us went to Christmas dinner at the home of Trish and Bob Keen in nearby Harleysville.

We played our usual Saturday night pinochle with Agnes and Jim Fletcher right up through New Year's Eve. In spite of Jim's failing health, the men won 68-57 for the year 2007. We rang in the new year with shrimp and champagne, knowing, without saying so, that it could well be our last pinochle session together.

The End of the Story, 2008

We drove down to Florida in January, and it was an unevenful passage, just the way we like it. Neither of us was sick this winter—I now blame the usual January bronchitis attack to the plane ride south. Kitty resumed her golf with the ladies, and I rejoined my Jewish tennis group. I was saddened to learn that Lou Shapiro, everyone's favorite, had died over the summer. I continued to play Tuesday, Thursdays, and Saturday, and had only one rainout the entire season. Actually, the winter weather was cooler and windier than we consider normal, so Kitty missed a lot of pool days. But she kept in shape walking around Boynton Beach Mall and other emporia. We had a scare when our new Toshiba LCD television set started going bad by losing color. Fortunately, it happened a few days before our one-year warranty ran out, so we were covered, but were without the set for about ten days. We were told it would have cost us $500 had the set been past its warranty. Ouch! We've had so much trouble with electronic equipment in Florida; the summer heat in the closed up apartment seems to be the culprit, as heat is the bane of electronic equipment.

I had decided not to continue local phone service with Southern Bell and use our Verizon cellphone for all calls. This proved to be a bad decision as we ran well over our allotment of cellphone minutes in January. Furthermore, Kitty's tennis captain objected to making a long-distance call to reach us. As I was watching the local television news one evening, an ad appeared for something called MagicJack Phone Service. I logged onto their Web site and learned that it offered unlimited nationwide calling for $20 per year, using your high-speed Internet connection. There is, however, a one-time charge of $40 for the physical MagicJack unit. I signed up for the service, got a local Florida phone number, and it has worked out very well for us. Although I have little need to do so, I can even use it for long-distance calls while I'm here in New Jersey.

While in Florida, we closely followed Jim Fletcher's cancer treatments. Over time it became apparent that his course of chemotherapy was eroding his ability to get around, without limiting the spread of cancer from his colon to his liver. We left Florida on Friday, April 18, so as to be at Meghan's first birthday party on the following day. As an antidote to the boring drive on I-95, we elected to drive up through Charlotte and onto I-81 for the pleasant ride through the Shenandoah Valley of Virginia. From I-81 at Harrisburg, Pennsylvania, we drove directly to Meghan's home near Lansdale. This western routing adds about two hours to the drive home from Florida but is more scenic and interesting. Anything is better than driving through I-95 time after time.

Once back in New Jersey, we were able to visit Jim Fletcher and continue our pinochle sessions, something that Jim looked forward to. In just a few weeks, it became almost impossible for him to arrange and hold his cards. Agnes helped him to play his hand, and we took turns dealing for him. We had our last pinochle session on May 17, and Jim died ten days later in the early morning of Tuesday, May 27. His death ended, among many other things, our long years of four-handed partnership pinochle. I kept scores for the last twelve years with results as follows: men 947, ladies 698. A funeral mass for Jim was held at St. Charles, with music provided by in-laws and granddaughter Maggie. His ashes were later interred at Mount Carmel Cemetery in Moorestown. Jim was many things to family and friends: father, Mister Fix-it, pinochle partner, fellow traveler, and bowler are some of them. I'll really miss his coming by mornings for coffee and man-talk. No one comes by any more.

As I conclude this saga, I realize that I have written far more than I ever intended. The ending may seem abrupt, but it's very likely merciful. I am in good health, but back pain is limiting what I can do. I cannot walk very far, and my tennis game is going badly, as I tend to focus on the pain rather than the ball. I'm trying to play once or twice a week, if anyone will have me. We have been blessed with seven grandchildren—four boys and three girls—and expect another boy in a few months. However, Fletch's death has me thinking that it's time to downsize again and move to Pennsylvania where we'll be closer to the children and grandchildren. We don't want to wait too long as we may find ourselves unable to cope with a move—it seems like now is the time to go. As I write, the painter is doing his thing on the outside to get the house ready for sale.

Flash! I can't let this book end without recognizing the arrival on July 31 of Matthew Peter O'Donnell, grandchild number eight. Mattie was born to Patrick and Janet at Mercy Suburban Hospital, East Norriton, Pennsylvania. He is a real cutie.

In the winter of 1997, my sister Pat and her husband, Jim Maguire, enjoyed a trip to Key West with us.

Kitty and Marion Dennen loved this restaurant on the ocean in Palm Beach. Marion was Kitty's duplicate bridge partner in Florida and in New Jersey. Sadly, Marion died of lung cancer in a year or so.

In June of 1997, Jim graduated from Rutgers with a degree in computer science.

In 1998, we moved back to Sacred Heart Parish, Riverton. We had helped to build the church in the 1960s.

In 2000, Kitty and the Hampshire Garden ladies league played at the Palm Beach Golf Club. The course lay along Route A1A between the ocean and the Lake Worth lagoon.

Kitty and Katie Minneker, our next-door neighbor, are off to lunch, winter 2000.

In 2000, the Minnekers spent a year or so in a rental house near Louisville, Kentucky. Here Kitty and Katie are making friends with some neighbors.

We pick up Kitty's new Mazda in Philadelphia in May of 2000.

In July of 2000, we had a visit from some Clarke relatives, Patrick and Bridget Clarke Boyle, second cousins. Bridget was over with her daughter for a visit to Patrick, her brother, who has a construction business in New York City.

Here we are celebrating Kitty's birthday at Braddock's in Medford. Danielle De Bow joined us.

Kitty and my sisters, Maureen on left and Pat on right, celebrate at the 2000 family Independence Day cookout at Mo and Jack Kulaga's in Lincroft.

The Carlin homestead at 240 Linden Street, Moorestown, was substantially renovated after its acquisition by Andy and Barb Fletcher.

Chris and Danielle, 2001.

Kitty's sister Agnes and Jim Fletcher at their fiftieth wedding anniversary celebration in 2001.

Kitty and I in Western regalia (?) at a Hampshire Gardens party, winter of 2002.

In the 2002 Hampshire Gardens show, I played Father Jim with Alice Tuttle as Mother Superior.

Prospective bride and groom with their mothers—Kitty at left and Cassie De Bow at right—at a 2002 bridal shower for Danielle.

Our Jim and Amanda were married at the Schwan home in October of 2002. The bride's parents, Susan and Bob, complete the picture.

Chris and Danielle leave the church after their wedding in November of 2002. The wedding was held at St. Mary's Church in Pompton Lakes, New Jersey.

In 2003, Patrick joined us for a luau at our place in Florida. Liz Oberschmidt is in red at the left end of the table. John Dennen, also formerly of Tenby Chase, has the white cap on the right. Leo Oberschmidt's head appears just beyond John. Everyone expected Patrick to be back for the 2004 luau. He came back but didn't make the luau.

Jim and Amanda Schwan had their formal wedding at First Presbyterian Church, West Chester, Pennsylvania, in August of 2003.

Kitty and I enter the wedding reception at the Concordville Inn.

**Our house at 1014 Cherry Lane, Cinnaminson, New Jersey.
It looks as though the shrubbery is taking over.**

Janet Pederson and Kitty at Hampshire Gardens, winter 2004.

Bridget Hackett Cotter is the daughter of a first cousin. Her late mother, Mary McVeigh Hackett, was much older than I, and we never met.

We celebrated grandson Danny's first birthday at the home of Amanda's parents in 2004. That's Shaun standing next to Amanda.

Longtime friends Liz and Leo Oberschmidt, our Florida pinochle partners, winter 2005. Death would claim both within two years.

This post-engagement portrait of Patrick and Janet Pederson, of Burbank, California, was taken in the spring of 2005. Pat and Janet met near Galway, Ireland, in the summer of 2003.

Patrick and Janet leave the altar after their November 2005 wedding at Epiphany Church, Plymouth Meeting, Pennsylvania.

Kitty and I with the bridal couple.

The five Carlin sisters at Patrick's wedding. Joanne and Agnes are seated. Standing are Pat and Marietta who is next to Kitty.

Taking a break in the shade at the Boynton Beach Tennis Center, winter of 2006.

In 2006, Jim and Amanda's growing family. *From left to right:* Shaun, Ellie, Danny, and Jack.

Saturday pinochle with Agnes and Jim, c. 2006. Unbelievably, death would claim Jim within two years.

In the fall of 2006, we met John Collier and his wife Rosemary for lunch at Panama Hattie's in North Palm Beach. John and I were classmates through high school, and he was a close pal and bridge partner. I had met Rosemary at some of the dances we attended during college days.

In the winter of 2007, I received a call from Tom O'Rourke; he & his wife were visiting in Florida from California and wanted to get together. Tom, on left, was probably my closest pal through high school, and it was good to see him and meet Diane.

Kitty holds granddaughter Ellie while brother Danny looks on.

Kitty and I celebrated our fiftieth wedding anniversary with a grand affair at the Riverton Country Club, June 2007.

Granddaughters Katie (big sister) and Meghan, 2007.

Granddaughter Meghan keeps her eye on cousin Ryan, Sea Isle City, September 2007.

APPENDICES

APPENDIX A

Donegal and the O'Donnell Clan

(Excerpted from the Donegal Web site)

Donegal

Donegal town is set in a valley girdled by Barnesmore Mountains and Donegal Bay, and overlooking the town can be seen the remains of several earthen forts. There is a record of an early Danish fortress being destroyed in the town by Murtagh Mac Lochlainn, High King of Ireland in 1159. The O'Donnell castle in the town was built by the first Red Hugh and his wife Lady Nuala in 1474. There were two Red Hughs, the last being the most colorful. He was captured by the English and thrown into Dublin Castle. He eventually escaped and managed to make his way back to Donegal. It was said that he suffered frostbite during his escape and as a result lost a big toe and had to ride into battle from then on.

The O'Donnells were noted for their patronage of the church and learning. Hugh O'Donnell and Lady Nuala brought the Franciscans to Donegal, also in 1474. They built an abbey and established a community that contributed greatly to the spiritual needs of the area. It was there that Brother Michael O'Cleary with Peregrine O'Cleary, Peregrine O'Duignean and Fearfasa O'Maolconry worked on their famous "Annals of The Four Masters," a full account of Gaelic Ireland, since what they took to be its birth until the Flight of the Earls. It was compiled during the period 1632-1636

The last great battle in which the O'Donnells were involved was the Battle of Kinsale in 1601 where they were badly defeated. Red Hugh went to Spain to seek help to resume the fight but he died there. Other chieftains, including the O'Donnells and the O'Neills, were forced into exile. This became known as the Flight of the Earls and it took place from Rathmullan in 1607. Before going, they partially destroyed the castle to prevent the English from using it. Their flight led to the "Plantation of Ulster." The O'Donnell castle and lands were given to an English captain, Basil Brooke, who carried out major reconstruction work and added a wing to it, known today as the manor house. Basil Brooke eventually moved to Lough Eske where he built a house.

The O'Donnell Clan

Surnames were not in use in Ireland until about the tenth century. The O'Donnells take their name from Domhnaill, son of Eighneachain (d. 905) and they are sometimes called Clann Dalaigh from Eighneachain's father (d. 874).

The O'Donnells were a leading branch of Cineal Conaill (race of Conall), formed by Conall Gublan who established the Kingdom of Tir Conaill (the county of Conaill), which almost corresponds to the present county of Donegal. Conall Gublan, so called on account of his fosterage or schooling at Binn Gublan (now Ben Bulben) in County Sligo, was a son of Niall of the Nine Hostages, the High King of Ireland in the fifth century.

Niall of the Nine Hostages was one of the last pagan High Kings of Ireland. On one of his foreign military expeditions he captured, brought to Ireland, and sold as a slave a young boy who afterward became St. Patrick, the patron saint of Ireland. Niall was an ancestor of eight of the most illustrious families in Ireland and of the most lasting, most important, and most powerful Irish dynasty. His descendants shared the High Kingship, the Cineall Conaill having ten High Kings up to the Anglo-Norman invasion in the twelfth century.

St. Colmcille belonged to the Cineall Conaill, the same family as the O'Donnells and he, too, was close in line for the High Kingship. Other important members of the Cineall Conaill were the O'Cannon's, the Maeldory's, the O'Boyles, the O'Doherty's and the O'Gallagher's. Many of these families lost the "O" over the years.

As leading members of the Cineal Conaill, the O'Donnells were very highly regarded in other countries, as well as Ireland, and were accorded the designation of Princes, Chiefs, and Kings of Tir Conaill by the rulers of England, Scotland, France, and Spain. Keating, the Irish historian, describes the inauguration ceremony of the O'Donnells as follows: "the ceremony of inauguration of the Kings of Tir Conaill was thus: The King, being seated on an eminence, surrounded by the nobility and gentry of his own country, one of the chief nobles stood before him, with a straight, white wand in his hand: and on presenting it to the King of Tir Conaill, used to desire him: 'To receive the sovereignty of his country, and to preserve equal and impartial justice in every part of his dominions.'" The reason that the wand was straight and white was to put him in mind that he should be unbiased in his judgment, and pure and upright in all his actions.

Tir Conaill

The passing of the name Tir Conaill into distant history has an unfortunate story behind it. The change commenced in the year 1584 when Sir John Perrott assumed the deputy Lordship of Ireland. Within a year, he had assembled a parliament in Dublin which, according to the Four Masters included the following septs: Hugh O'Neill, chief of Tir Eoin; Hugh O'Donnell chief of Tir Conaill; John Og O'Doherty, chief of Inishowen; Turlough O'Boyle, chief of Boylagh and Owen O'Gallagher, O'Donnell's marshal.

The parliament passed a decree, which stated that the area has to be known henceforth as Donegal (Fort of the Strangers). O'Donnell who refused to accept any interference in the running of Tir Conaill strenuously resisted this measure. He forbade the entrance of the English sheriff into his territory. However, by the time the Flight of the Earls had, some twenty years later, the name Donegal started to prevail in the vernacular.

APPENDIX B

O'Donnell, McDonnell and Donnelly

By James D. Ryan, *Irish American Magazine*

These three names and others of a similar sound, such as Donnellan and Donlon are sometimes confused as spelling variations have occurred among emigrant families. They are, however, totally distinct, and indeed both the McDonnells and O'Donnells are made up of several distinct septs. The main branch of the O'Donnells, based in Donegal, is the most eminent of the Gaelic families, leading back to Niall of the Nine Hostages, but owing their name to one Domhnaill. Another branch was from Thomond (Limerick and Clare areas), and a third from Galway/Roscommon. The Donnelly or O'Donnelly family is also from the same stock as the O'Donnells, that is they too are descendants of Niall of the Nine Hostages and are originally from what is now County Tyrone. The McDonnell family is more complicated, having several separate roots, one of which is the McDonnells of Antrim who were a gallowglass (mercenary soldier) family. The roots of the other McDonnell families are in Donegal and Clare.

Because of the frequent omission of the 0 and Mac from Gaelic names, it is sometimes difficult to distinguish the origin of a person named Donnell. Among the prominent Donnells whose specific kinship is difficult to determine is James C. Donnell (1854-1927) who was brought to the U.S. at age two and became a crude-oil hauler in Titusville, Pennsylvania, later rising to become a vice president with Standard Oil, and later president of Ohio Oil. He was a classic oilman who, even at the height of his executive career, liked to work at the wellhead. He is estimated to have drilled 42,000 wells during his career and at his death was one of the last oil giants.

Among the prominent Donnellys is Ignatius Donnelly (1831-1901) an entrepreneur of Philadelphia who went to Minnesota to found a town and, on its failure, farmed the site instead. He became a Republican politician and lieutenant governor of the state, and later a congressman from 1863-69. Dan Donnelly (?-1820) born in Dublin became a very successful boxer whose fights were attended by thousands and won him very large purses, which he used to set himself up as a publican. One of the reasons for his success was the fact that he had extremely long arms. Unbelievers can visit Kilcullen, County Kildare, where one of his mummified arms is on display!

Perhaps the most famous of the O'Donnells was Red Hugh O'Donnell (1575-1608) who, at age 18, was kidnapped and imprisoned by the English Deputy in Ireland who feared the power of the clan. In 1591, along with the famous Hugh O'Neill, he escaped, and in the following year he was inaugurated as Chief of the O'Donnells. He joined with O'Neill and the other Ulster chieftains in the Nine Years War against the English, which was finally lost at the Battle of Kinsale. Hugh O'Donnell was sent to Spain to enlist further help and died there, apparently from poison administered by an English spy. Other famous kinsmen of Red Hugh were Manus O'Donnell (?-1536) under whose guidance a book on the life of St. Columcille was completed in 1536 and Niall Garbh O'Donnell (1569-1626), who was imprisoned as a rebel in 1608 and died in the Tower of London.

Of the Donegal O'Donnells there was Frank Hugh O'Donnell (1848-1916), the politician and journalist who left the Parliamentary Party due to his opposition to its leader Parnell, but later wrote a major history of the party. Peadar O'Donnell (1893-1986), also of Donegal, was a socialist republican, writer, and IRA commander. Possibly the most prominent O'Donnell in Ireland at the moment is the very popular Donegal-born country singer, Daniel O'Donnell.

In the U.S., prominent members of the family include Thomas Jefferson O'Donnell (1856-1925) who started life as a journalist in his native New Jersey, but later became a lawyer, and a pillar of the legal profession in Denver, Colorado; and James O'Donnell (1774-1830) who was born in Wexford and trained as an architect. He went to the U.S. in 1812 and rapidly established a name for himself and was elected to the American Academy of Fine Arts in 1817. One of his greatest buildings was the Church of Notre Dame in Montreal, which was completed until after his death.

The O'Donnell, McDonnell, and Donnelly families were prominent in the American struggle for independence. There were 155 Donnellys, 139 McDonnells and 36 O'Donnells in the American Revolutionary Army. The list of soldiers also illustrates the range of spelling variations that occur in the U.S., which includes Ensign John O'Donnell, of Washington County Militia, Pennsylvania; Captain Nathaniel Donnell of Steven New York Artillery; Colonel Andrew Donnally of Botecourt County Militia, Virginia; Lieutenant Patrick Donnelly of 7th Regiment, Maryland Line; Lieutenant Moses Donley of Bedford County Militia, Pennsylvania; and Captain John McDonnol of Shee's Regiment, Pennsylvania.

APPENDIX C

Dungannon's Castle Hill

The Dungannon Observer (Tyrone)

March 31, 1951

The streets of Dungannon ascend steeply to a green plateau which crowns the town. This is Castle Hill. Here, in the heart of Ulster, the O'Neills, kings and chieftains, built their home and fortress.

They chose historic and strategic ground. In the early centuries it was the residence of Genann, friend of Cuculain, and one of the leaders of the Red Branch Knights. In its commanding position, Genann found it a natural and easily defended Rath or Dun. So did the O'Neill's in the later centuries when they made it their headquarters and built their castle upon it.

Of their castle nothing remains; but in the long years of its proud existence the Red Right Hand flag which flew from its battlements stood for the power, and might, and justice of one of the most famous of all the Chieftains of Ireland.

THEY CAME 600 YEARS AGO

The O'Neill's appear to have come to Dungannon about the year 1300, though many claim they were established in Tyrone [was Tyrowen] much earlier. Be that as it may—and there is some evidence of it—the first written testimony that links them to Dungannon is dated 1329. Writing from Dungannon in that year, Donald O'Neill sent a spirited message to Pope John in which he styled himself the "King of Ulster and the true heir to the whole dominion of Ireland." He assures the Pope that, since the days of St. Patrick until the landing of Henry II in 1172, no less than sixty monarchs of his race had been true Kings of Ireland.

Whatever about the place of residence of these O'Neill's, there is no doubt whatever that from the year 1300 at least chieftains of this blood have ruled from the Castle Hill of Dungannon. They come valiant and inspiring from the pages of history. From the parent house they have spread and founded lines in many corners of Ulster. We can easily bring their lives and times to the forefront of memory.

There were the O'Neill's of Clannaboy or Antrim branch. There were the O'Neill's of the Bann and the O'Neill's of the Fews—and they all sprung from the stock that had its main root in Tyrone.

PROGRESS OF CENTURIES

From that hill they have watched the progress of centuries, and when, in 1607, the great Hugh himself burnt down the Castle and joined the "Flight of the Earls" to France, it seemed the Red Hand had been lowered forever. It was the greedy day of the Chichesters and the Caulfields and they quickly set about the work of gathering in their spoils.

Through the brief period of the 1641 rebellion, the victorious Sir Phelim O'Neill returned, the new era and tenure of the foreigner had begun.

Standing on the high Castle Hill today and surveying the encircling wide sweep of Ulster, it is easy to conjure the picture that saddened the O'Neill Chieftain in those dark days when the country he loved was in the sordid market for adventurers from England.

FORMIDABLE FIGHTERS

To the north there were allies in plenty. There were the staunch O'Neill's of the Bann. There were the O'Neill's of that other branch of the family in Killeiter in the south of the present County of Derry. All along those friendly plains that foot Slieve Gallon dwelt hundreds of valiant soldiers, proud, every man and boy among them, to serve under the banner of their Chief. Further north were the clansmen of O'Cahan, kinsman and tributary to the O'Neill.

Over the Sperrins lay the land of Tyrconnell [now Donegal] where the O'Donnells were the traditional allies of Tyrone. Together they had fought many battles and the watcher on Castle Hill may well recall many occasions when their united might had disposed of many of the enemies of their country. There were happy social reunions between these two Ulster families as well. And there was the famous visit to Dungannon by the tired and wounded Red Hugh from his prison in Dublin on his way back to his home in Tyrconnell.

AN EVIL PLOT

How the heart of the Earl of Tyrone must have gladdened on welcoming this young chieftain and how the resolution and purpose of every O'Neill of the North must have hardened at the tale of his treacherous capture. For what boy of 15 could have seen in an invitation to board a fine vessel at Rathmullen the evil plot to take him off his own shore and carry him prisoner to Dublin?

All around the O'Neill's the story had gone, from the clan Clannaboy, to the east, to the furthermost O'Neill's of the Fews, south from Castle Hill and among the jagged outline of the peaks of Armagh. These latter O'Neill's had early established themselves there and for generations resisted the northern penetration of the English of the Pale.

In the last years of the struggle for Ulster the Chief of the Fews, Henry O'Neill shared the fortunes of the great Earl of Tyrone.

The Clannaboy O'Neill's lay over the water of Lough Neagh— formidable fighters they were who had defied the English in their strongholds in Antrim. By the early fourteenth century they had pierced into the heart of the English garrisons of Down and Antrim and their strongholds of Castlereagh and Edendubhcarig (later Shane's Castle) were hopeful signals on the horizon from Dungannon that the Irish were still masters of their own soil.

SPIRIT STILL LIVES

Lying nearer and crowded in memory are Benburb, the Friary of "The Brantry," and the Hill of Armagh.

Benburb was where O'Neill had another stronghold and castle; and the city of Armagh was the scene of so much English murder and sacrilege. Armagh was where the poor Franciscan monks were whipped out of their convent, but were befriended securely by the forces under the command of Hugh, the redoubtable Earl of Tyrone. To "The Brantry" he led them, to thickly wooded fastness where without fear they could pursue their self-denying way in peace to God. Safe and unmolested, they remained while O'Neill ruled in Dungannon, but orphaned and in danger when he himself was driven out.

NOTHING REMAINS

Nothing of O'Neill's Castle remains on the Hill. The tower ruins that are now shown are of a later mansion erected in 1790 by T. Hannington, a local magnate of the new order.

But the spirit of the O'Neills is not dead in Tyrone. And men of the old race and the new feel proud that it was to Dungannon the Volunteers of Ulster came in 1782 and put on record sentiments of liberty that were the honour of Tyrone and of all Ireland.

That stands forever to the pride of Tyrone and Dungannon; and in a fast changing world, who can say what a sentinel on Castle Hill may yet see and remember?

("J.C.M." writing in the Irish News.)

APPENDIX D

Residences

This listing excludes short-term or extended residence in hotels.

Address	From	To	Comment
294 Cypress Ave., The Bronx, NYC	Birth	Sep 1934	
675 E. 135 St., The Bronx, NYC	Sep 1934	Aug 1935	
34-43 34th St., Astoria, Queens, NYC	Aug 1935	May 1936	
195 Valley Rd., Clifton, NJ	May 1936	?	
127 Valley Rd., Clifton, NJ	?	Nov 1938	
683 E 138th St., The Bronx, NYC	Nov 1938	Sep 1942	
503 W 175th St., Manhattan, NYC	Oct 1942	Sep 1943	
437 W 261st St., Riverdale, NYC	Oct 1943	Jun 1956	
MacDill AFB, Tampa, FL	May 1951	Sep 1952	incl
Keesler AFB, Biloxi, MS	Oct 1951	Jun 1952	incl
White Horse Pike, W. Collingswood, NJ	Jun 1956	?	
111B E Coulter Ave., Collingswood, NJ	?	Jul 1959	
409 Kathleen Ave., Cinnaminson, NJ	Jul 1959	Jul 1969	
Coral Gables Apts., San Vicente Blvd., Santa Monica, CA	Apr 1962	Nov 1962	incl
174 Westover Ct., Delran, NJ	Jul 1969	Jul 1998	
2460 S Federal Hwy., Boynton Beach, FL	Jan 1996	Present	incl
1014 Cherry Lane, Cinnaminson, NJ	Jul 1998	Present	

APPENDIX E

Schools Attended

Grade	Schools	Start	Finish	Notes
K	PS#5 (Clifton)	?	Fall 1937	
1B	Same (Ms. Spangenberg)	Winter 1938	Spring 1938	
1A	Same (Ms. Dunn)	Fall 1938	N/a	NJ schools reverse order –
1B	PS#63 (Bronx)	N/a	Winter 1939	do B before A
2A	St Luke's (Bronx)	Winter 1939	Spring 1939	
2B	Same	Fall 1939	Winter 1940	
3A	Same	Winter 1940	Spring 1940	
3B	Same	Fall 1940	Winter 1941	
4A	Same	Winter 1941	Spring 1941	
4B	Same	Fall 1941	Winter 1942	
5A	Same	Winter 1942	Spring 1942	
5B	Incarnation (Manhattan)	Fall 1942	Winter 1943	
6A	Same	Winter 1943	Fall 1943	
6A/B	St Margaret's (Riverdale) Sr. Marie Rita	Fall 1943	Spring 1944	Left back
7A/B	Same (Sr. Bernadette)	Fall 1944	Spring 1945	
8A/B	Same (Sr. Maureen)	Fall 1945	Spring 1946	Graduation
HS	Cardinal Hayes (Bronx)	Fall 1946	Spring 1950	Graduation in St. Patrick's Cathedral

1st Year	Manhattan College Liberal Arts Major	Fall 1950	Spring 1951	
N/A	USAF 1. Basic Electronics and 2. Radar Bombing Systems	Spring 1951	Fall 1952	Early release With GI Bill
2nd–5th year	Manhattan College Engineering Major	Fall 1952	Spring 1956	BEE degree
Part-time Masters	Rutgers Camden Business Administration	Fall 1983	Spring 1986	MBA degree
Part-time Masters	NJ Inst of Technology Computer Science	Fall 1987	?	Did not complete

APPENDIX F

The New York Street Game of Skully

Chalk and a Bottlecap: Skully, a game that has its roots on the streets of New York goes by many names, including skelly, skelsies, and deadbox. The game is played by first drawing a large square in chalk on the pavement. Along the borders of the square, there should be twelve boxes numbered one through twelve. In the very center of the square is a box numbered 13 and surrounded by a "dead man's zone."

The Fine Art of Skully Tops: Bottle caps, also known as "skully tops," act as playing pieces, and are carefully chosen for their smoothness and glide. Some kids modify their bottle caps by putting crayon wax or pennies or even bits of orange peel in them to make them heavier so that they glide better.

How to Play: Skully usually works best with three to six players. If you're going first, you slide a bottle cap into the square marked number one. If you miss, you leave the bottle cap where it lands and let the next player take a turn. If you are successful, you get another turn to slide the bottle cap into square, number two. The goal of the game is to slide the bottle cap into each numbered square until you reach square 13. When this happens, you must slide the bottle cap into each square again, but this time in reverse, from thirteen to one. After this is accomplished, you can shoot your cap into the "dead man's zone" around square 13, and then "hunt" the other players by "blasting" their bottle caps out of a square with your own. If you hit a player's bottle cap three times, he is out of the game. The last player remaining on the "skully board" wins.

APPENDIX G

The Toals and the Taxi Business

There were two Toal brothers from the old country (Carrickmore, County Tyrone), Peter (called Pete) and Terence (called Terry). Pete was the elder of the two and a good IRA man; Terry probably was as well. They would have been born around 1900. Carrickmore is rather far from my father's home near Brantry Lough, but much closer to my mother's home near Pomeroy. My father would likely have met the Toals while he lived with his uncle Rt. Rev. (Msgr.) Francis Donnelly, pastor at St. Columcille Church, Carrickmore until he died in 1940.

When the Toals came to America they probably worked in the small food store chains that were owned & operated (as sweatshops) by those who came before. Somehow they got into the taxi business before World War 2. As the business went thru the wartime boom they became very well off so that after the war they had fleets of hundreds of cabs. Pete was the more easygoing of the two and really didn't like to work. Terry worked harder and thus became the wealthier of the two.

After struggling during the depression, my father got into the taxi game around 1940, probably as a driver for the Toals. With their help he then bought and operated his own cab during the later years of World War 2, hiring another driver for the night shift. He then went on to buy a house in Riverdale, add additional cabs, and prospered in the postwar boom. Somewhere along the way Pete Toal engaged my father to manage his large taxi fleet as well as some cabs owned by Pete's brother-in-law, Bill McFadden. By then my father had his own small fleet, Prospect Cabs, to run along with Pete's, Pomeroy Cabs, and Bill's. They operated out of a garage on West 61st Street on the far west side of Manhattan. The garage was owned as a cooperative by the Irish taxi owners, themselves. The cooperative was known as IOTA (Independent Owner's Taxi Association). IOTA also had a garage on West 56th Street. It was there that Terry Toal ran his cabs. Because of crowding and for political reasons, these operations have since moved out of Manhattan to more open space in Brooklyn.

Pete and his wife Peg McFadden lived in a spacious two-family house on Liebig Avenue in Riverdale. I'm told that Peg was born in England of Irish parents. Pete and Peg lived downstairs with their daughter Jane. Peg's bachelor brother Bill lived upstairs and worked for the telephone company. Peg was a bit of a recluse and didn't even attend the County Tyrone Ball, the year's big event. I got the feeling that I was supposed to date Jane Toal, but I don't think that I ever did. Jane married Roger Daly and they live in New Rochelle, Westchester County, New York. Roger recently retired from running the Pete Toal taxi fleet. I don't know who is running it now. Sadly Pete died some years before my dad. In September 1965 he was remembered in the County Tyrone Society's 75th Anniversary Banquet Souvenir Journal as "a friend to all" and I believe that was indeed the case. This banquet was otherwise a great occasion. At our family table we had Mom and Dad, my wife Kitty and me, my sister Patricia and her husband Jim Maguire, and Aunt Nell O'Donnell. Jane and Roger were present, but Peg Toal did not attend. Neither did Terry Toal nor his wife.

I have many fond memories of Pete Toal who I often saw at the taxicab garage and around Riverdale. I remember riding home from work with him when we were stopped for speeding on Riverdale Avenue. As the cop approached Pete turned to my dad and asked "Should I slip him a pound note?" meaning a $5.00 bill, equivalent to a British pound in those days. The cop saw the fiver among Pete's license and registration papers and said "Are you trying to bribe me?" Pete quickly grabbed the fiver saying "How did that get in there?" He got the well-deserved speeding ticket as he had a very heavy foot. One day Pete and some other men came to the house to help my dad put up a new clothes line pole, as the original had rotted out. It was a very big pole as the laundry window at the time was 2½ stories up from the backyard. As the men struggled to set the new pole Pete said "You can hire men to do this kind of work." No one paid him any attention and the pole was placed where it needed to be. Obviously, I've remembered Pete's words some sixty years or more later; and I've been guided by them.

I was much less familiar with Terry Toal, but believe he and his wife Marion Early also lived in Riverdale. They had two sons, Jackie and young Terry who was called "T" Toal. He was considered to be a spoiled brat. Jackie ran the Terry Toal fleet until his premature death in an auto accident. He was riding in his convertible and talking on his mobile phone at the time. Young Terry, about my sister Maureen's age, then took over the taxi operation. He lives somewhere in Westchester County and is a member of the prestigious Winged Foot golf club. Marion Toal had a brother, Jack Early, who also ran cabs at the 56th Street garage.

Terry's wife Marion has a home in Boynton Beach's Leisureville section, which is quite near us. I visited her there a few years ago. One of the many Kelly cousins was there at the time. The curious thing was that Terry Toal had courted my Aunt Mary Quinn back in the Old Country. I guess he biked over from Carrickmore to Cornamaddy. The story was that Mary, who was very fussy, was not taken with Terry. For a time my Aunt Mary and Terry's wife Marion were living in Florida not far from one another. Mary died in 2001, but Marion lives on.

Marion's sister, Catherine Early, married Dan O'Neill who had a very good position with the telephone company on West Street, Manhattan, a long way downtown from his Riverdale home. Dan also had some cabs that he ran with Terry Toal's. The O'Neills had two daughters and a son Danny, Jr. The elder of the two daughters was Peggy who was in my sister Patricia's class at St. Margaret's School. Dan O'Neill was a leader in the parish and ran the annual Minstrel Show. He arranged for me to handle the coat checking concession for the shows, a modest moneymaker. He was also a factor in local Republican politics. He hired me during the 1944 and 1948 presidential elections to stuff mailboxes and hand out flyers on Election Day. He later arranged for me to be the paid precinct election chairman during the 1952 presidential election. I worked for Eisenhower, but, disloyally, voted for Stevenson. Young Danny became a lawyer via the Fordham School of Law.

Dan's sister, Maria Theresa, was married to the labor leader Mike Quill, a Kerry man. Quill headed up the New York Transit Worker's Union (the TWU). His politics were considered somewhat "pink" in a time when full-blooded Communists were "reds." Quill usually called transit strikes on New Year's Day to disrupt our return to work or to school after the holidays. The city usually settled fairly quickly. Mike was unique in that the longer he lived in the USA, the thicker his Irish brogue became. And it was very thick indeed! I did meet Mike Quill once, at a wake, possibly Pete Toal's.

APPENDIX H

Favorite Bagpipe Tunes

O'Donnell Abu!

(The Clan Connell War Song)

"O Donnell Abu!" is a song about the wars between Protestant Elizabethan troops and Ireland's native and Catholic clans, led by Hugh O'Donnell, in the 1590s. The words were composed by a young Galway man, Michael McGann, in 1843, to an older tune by Joseph Haliday.

"O'Donnell Abu!" Lyrics

Proudly the note of the trumpet is sounding
Loudly the war cries arise on the gale
Fleetly the steed by Lough Swilly is bounding
To join the thick squadrons on Saimiers green vale!
On every mountaineer! Stranger to flight or fear!
Rush to the standard of dauntless Red Hugh!
Bonnaught and Gallowglass, throng from each mountain pass!
Onward for Erin! O'Donnell abu!

Princely O'Neill to our aid is advancing
With many a chieftain and warrior clan!
A thousand proud steeds in his vanguard are prancing
Neath the Borderers brave from the banks of the Bann!
Many a heart shall quail under its coat of mail,
Deeply the merciless foeman shall rue
When on his ear shall ring, borne on the breezes wing
TyrConnell's dread war cry O'Donnell abu!

Wildly o'er Desmond the war wolf is howling
Fearless the eagle sweeps over the plain
The fox in the streets of the city is prowling
And all who would conquer them are banished, or slain!
On with O'Donnell then! Fight the good fight again!
Sons of TyrConnell are valiant and true!
Make the proud Saxon feel Erin's avenging steel!
Strike! For your Country! O'Donnell abu!

Garryowen

"Garryowen" is a 1800s Irish tune that originated in the Irish town of Garryowen. It was a favorite drinking song of the British Army and was later adopted by the U.S. Seventh Cavalry headed by General George Armstrong Custer. The unit had many Irish troops, one being Captain Myles Keogh, who came from the town of Garryowen and introduced the song to Custer. "Garryowen" was played by the Seventh Cavalry as they entered the Little Big Horn Valley on that fateful day. Only Keogh's horse (Comanche) survived the battle. "Garryowen" is still the song of the present-day Seventh Cavalry. There's a town called Garryowen, population 200, near Custer's last battlefield.

Original Lyrics

The most accurate origin of the song "Garryowen" is from a small Irish town outside of Limerick, where the Fifth Royal Irish Lancers made their home. The town's name, Garryowen, means Owen's Garden; but the old tune soon came to become associated with the Lancers' drinking. The Irish poet Thomas Moore wrote the words around 1807.

> *Lyric 1*
> Let Bacchus' sons be not dismayed
> We'll break windows, we'll break doors,
> But join with me each jovial blade;
> The watch knock down by threes and fours;
> Come booze and sing, and lend your aid,
> Then let the doctors work their cures,
> To help me with the chorus.
> And tinker up our bruises.

Chorus

Instead of spa we'll drink down ale,
We'll beat the bailiffs out of fun,
And pay the reck'ning on the nail;
We'll make the mayors and sheriffs run;
No man for debt shall go to jail
And are the boys no man dares run,
From Garryowen in glory.
If he regards a whole skin.

Lyric 2
We are boys that take delight in
Smashing the Limerick lights when lighting,
Our hearts so stout have got us fame,
Through the streets like Sporters fighting,
For soon 'tis known from when we came;
And tearing all before us.
Where're we go they dread the name
Of Garryowen in glory.

Chorus

U.S. Seventh Cavalry Lyrics

In 1905, Chief Musician J.O. Brockenshire, of the Seventh Cavalry Band rewrote the music to "Garryowen" and composed these stanzas and chorus before the regiment was deployed to the Philippines.

Lyric 1
We are the pride of the army,
And a regiment of great renown,
Our name's on the pages of history,
From sixty-six on down.
If you think we stop or falter
While in the fray we're gin'
Just watch the steps with our heads erect,
While our band plays "Garryowen."

Chorus

In the Fighting Seventh's the place for me
It's the cream of all the cavalry;
No other regiment ever can claim
Its pride, honor, glory and undying fame.

Lyric 2
We know no fear when stern duty
Calls us far away from home,
Our country's flag shall safely o'er us wave,
No matter where we roam.
'tis the gallant Seventh Cavalry
It matters not where we're goin'
Such you'll surely say as we march away;
And our band plays, "Garryowen."

Chorus

Lyric 3
Then hurrah for our brave commanders!
Who lead us into the fight.
We'll do or die in our country's cause,
And battle for the right.
And when the war is o'er,
And to our home we're goin'
Just watch the step, with our heads erect,
When our band plays, "Garryowen."

Chorus

APPENDIX I

About Riverdale

(Mostly from Wikipedia, the free encyclopedia)

Riverdale is about three square miles in area. It is bordered on the north by the city of Yonkers, Westchester County, New York; to the east by Van Cortlandt Park and the Kingsbridge section of the Bronx; to the west by the Hudson River; and to the south by the Harlem River. The five subsections of Riverdale are Spuyten Duyvil, South Riverdale, Central Riverdale, Fieldston, and North Riverdale. The leafy, scenic enclave of Fieldston, a private community, was designated as an historic district in by the New York City Landmarks Preservation Commission in 2006. Fieldston Hill at 250th St. and Grosvenor Ave is the highest point in the Bronx.

Housing in Riverdale ranges from multi-story apartment buildings dating from the 1950's and 60's to large, architecturally distinguished houses built in the early 20th Century, mostly in Georgian- and Tudor-revival styles. In 2005-06, Central Riverdale experienced a building boom with the addition of over a dozen mid- and high-rise condominium buildings.

Riverdale is in a sense the closest northern suburb of New York City although it is not its own municipality. Administratively, Riverdale is part of Bronx Community District 8. Two weekly newspapers, the *Riverdale Press* and the *Riverdale Review*, focus on news of interest to residents.

Wave Hill, a combination botanical garden and outdoor art gallery, is located in the so-called Estate Area overlooking the Hudson River. Wave Hill House was built as a country home in 1843 by jurist William Lewis Morris. From 1866-1903 it was owned by William Henry Appleton, who enlarged the house in 1866-69 and again in 1890. A publishing scion, Appleton brought to Wave Hill such pioneering natural scientists as Thomas Henry Huxley and Charles Darwin. Huxley was astounded by the site, declaring the Palisades across the river one of the world's greatest natural wonders. It's probably no coincidence that North Riverdale's north-south avenues are named after scientists: Delafield, Liebig, Tyndall, Huxley, Spencer and, to the south, Faraday. Wave Hill has since been the home of many well-known people including Arturo Toscanini and his daughter (1942-1945).

Riverdale is home to three prominent private schools: Horace Mann, Riverdale Country, and Fieldston; and two Roman Catholic colleges: Manhattan and Mount Saint Vincent. The public elementary schools are P.S. 24 and P.S. 81. The public middle school and high school is P.S. 141, the Riverdale-Kingsbridge Academy. Many Riverdale residents attend Bronx High School of Science and the High School for Environmental Studies. Riverdale is also home to SAR Academy, a private Jewish day school near the Riverdale train station, and SAR High School. Catholic elementary schools in the area are in the parishes of St. Gabriel and St. Margaret of Cortona.

The New York City subway system serves the area along its eastern border as far as 242nd Street at Van Cortland Park. The Hudson line of the Metro-North commuter railroad runs along its western edge, the Hudson River. This was originally the New York Central main line to Chicago and still carries freight and Amtrak long distance trains. There are two Metro-North stations, Riverdale at West 254th Street and Spuyten Duyvil. Commuting time from the latter to Grand Central Terminal in Midtown Manhattan is around 25 minutes. The Mount Saint Vincent station at West 261st Street has apparently been abandoned. Manhattan can also be easily reached by the MTA Bus Company's (formerly Liberty Lines) express service. By car, Riverdale is commonly reached from the south via the Henry Hudson Parkway (Route 9A), which bisects much of the neighborhood. This major thoroughfare connects to Manhattan over the Henry Hudson Bridge, which was completed in 1936.

Many residents of Riverdale are reluctant to be associated with the Bronx, and use "Riverdale, New York" as an address. The U.S. postal service once did not consider this acceptable and hence letters addressed as such were often returned to the sender. Now it has no problem sending letters to Riverdale residents. However, further complicating matters, some residents insist on using the "Bronx" address so as not to appear politically incorrect. Riverdale's zip codes are 10463 and 10471; while

10471 is entirely in Riverdale, 10463 largely covers the less prestigious neighborhoods of Kingsbridge and Marble Hill, leading some to mistakenly call those areas "Riverdale."

Riverdale is home to large Jewish and Irish American communities. Greeks and Koreans are also a presence. The exteriors of many of Riverdale's mansions have been used in movies, most notably the "Corleone House" located on Palisade Avenue (across from Wave Hill). It was used for the exterior shots in the film, *The Godfather*.

Partial list of celebrities who have lived in Riverdale

- John F. Kennedy, U.S. President, lived at 5040 Independence Avenue, across the street from Wave Hill.
- Eliot Spitzer, New York State Attorney General and candidate for governor, was born in Riverdale.
- Lou Gehrig, Yankee baseball immortal, lived at 5204 Delafield Avenue.
- Willie Mays, baseball star, lived at The Whitehall, a residential co-op at 3333 Henry Hudson Parkway.
- Carly Simon, singer songwriter, is a Riverdale native.
- Ed Sullivan, television personality, also lived at the Whitehall.
- Greta Garbo, actress, lived at 2475 Palisade Avenue.
- U Thant, U.N. Secretary General.
- Arturo Toscanini, conductor, lived at Wave Hill.
- Ella Fitzgerald, jazz singer, lived at the Riverdale Children's Association.
- Mark Twain, author, lived for a time at Wave Hill.
- John Kieran, New York Times writer and TV personality, lived along the parkway.
- Madame Chiang Kai-Shek, wife of the Chinese leader who preceded Mao Tse Sung, lived near Wave Hill during World War 2. She was one of the famous Soong sisters. Chiang was forced to convert to Christianity in order to marry her.

APPENDIX J

A Night in Coligan's Barn

By James O'Donnell
CARRICKMORE 1923

NOTE: This is one of a number of poems to be sung that were composed by my father during the time he lived in Carrickmore, County Tyrone, in what is now Northern Ireland.

Oh God save you all I'm welcome here this minute I'm after landing,
And lucky enough to get a seat, for often I'm left standing
I was tossed about in Coligan's Barn, like a ship upon the ocean.
It was all bad luck put it into my head, that ever I took the notion.

Since my boyhood day I did attend all dances, wakes and caleidhs
And have spent good nights amongst the Currans, the Duffs, the
      Slanes and Dalys.
This night I had no peace or ease, twas far more like a shambles,
Than anything that I have seen since I began my rambles.

It's in the townland of Granna boys this barn is situated,
It's owned by Barney Coligan, and it's newly renovated,
So Biddy Coyle, she got the barn, to give her friends a spree sir
They first arrived at Currans ye know, and there they had their
      tea, sir.

The girls that were invited sure they numbered near a score, sir.
But none were in the style with those from Carrickmore, sir.

Miss Donaghy, the Tailoress, and Tanny they sang gailey
For they have boys in Granna you know Frank Boyle and Paddie Daly

Miss Rafferty was also there; she's the Lord Mayor's daughter,
Her father is a merchant, that sells everything but water,
Along with her came Sarah Freel, on courting she was bent, sir
Her dress was very scanty, but she had on lots of scent, sir.

You may think that I am telling lies when I say that she was stylish.
Believe me, for she would not dance a dance that was plain Irish.
She would not dance the jigs or reels, barn dance or waltz so
    plain, sir.
Her heart was set on jazzing with her sweetheart Packey Slane, sir.

The Duffs and Slanes they all were there, and Ellen Tommy Daly,
And Mikey Paddie Willies wee doll, the Currans and Coyles I'll
    bail ye
The Coyles from Lough McGrory too, did come to help the fun, sir.
And Maggie May McCrory that's in Foxes of Crookdun, sir.

Now the boys that did attend that night by me they won't be
    named, sir
On account of having been there myself I really am ashamed, sir
For there were the most disgraceful crowd if I described them fairly
Except a few from around the rock and they departed early.

Sure the jazzing got so very bad that Frank Coyle took the floor, sir
And asked them to conduct themselves and not to jazz anymore sir.
For Granna always had a good name and I hope you won't
    disgrace it
And give the Carmen poets the chance for quickly they'd embrace it.

I mean to keep good order now, no matter who's offended.
For the Carmen lads were still good sports wherever they attended.
I don't want to be talking again or they'd have me in the poem, sir,
So if you don't conduct yourselves I'll chase you all right home, sir.

The barn it would have held many more if it wasn't for the walls, sir.
But many that were loose that night should have been tied in stalls,
    sir.

I thought there would have been a fight as the night it did advance, sir.
Frank Anthony hit Jim Foster for asking his girl to dance, sir.

Dan Tucker came from Ballybrack, and no one could him blame, sir
He brought a gang from around the Forge and they were very
   game, sir.
I think they did enjoy themselves; they sang and danced quite
   merrily,
But as none of them could find a girl they left and went home early.

I do not like this jazzing boys, so with them I departed.
I very soon got clear of all for my cabin home I started.
To tell the truth I ran my best until I crossed the river,
I got safe out of Granna but I thought it would be never.

When I got my senses back again and thought the whole thing over
I bade farewell to dances boys, tho I am still a rover.
But I cannot stand those Belfast boys or bottle men at all, sir,
While they were courting all the girls I was holding up the walls, sir.

So now my friends I must conclude just with a word of warning,
For if I told you all I saw 1'd keep ye here to morning.
Girls if you're wise take my advice and to such dances don't go,
For you're safer to waltz with the fellows ye know,
Than jazz with the devils ye don't know.

Original Appendix

Tune: "The Leitrim Switches"

To be sung lively, night and morning, before and after meals, and
particularly passing Mike Rafferty's.

N. B. not to be shown to anyone except the Quarrymen, and those who
frequent Mr. Daly's and Mr. McKiernan's and the congregations of both
churches at Carrickmore.

Please read it over slowly for those who cannot read themselves.

APPENDIX K

Broadway Shows 1955-56

Show (type)	Theater	Cast When I Saw It
Cat On a Hot Tin Roof (drama)	Morosco	Barbara Bel Geddes, Burl Ives, Mildred Dunnock, Pat Hingle
The Diary of Anne Frank (drama)	Cort	Susan Strasberg, Joseph Schildkraut, Gusti Huber
Fanny (musical)	Majestic	Ezio Pinza, Walter Slezak, Florence Henderson
A Hatful of Rain (drama)	Lyceum	Shelley Winters, Ben Gazzara, Anthony Franciosa
The Lark (drama)	Longacre	Julie Harris, Boris Karloff, Joseph Wiseman, Christopher Plummer, Theodore Bikel
My Fair Lady (musical)	Mark Hellinger	Julie Andrews, Rex Harrison, Stanley Holloway
The Pajama Game (musical)	St. James	Eddie Foy Jr., Stephen Douglas, Helen Gallagher, Julie Wilson
The Ponder Heart (comedy)	Music Box	David Wayne, Una Merkel, Will Geer, Juanita Hall
The Righteous Are Bold (drama)	Holiday	Dennis O'Dea, Irene Hayes
Will Success Spoil Rock Hunter (comedy)	Belasco	Jayne Mansfield, Orson Bean, Walter Matthau, Martin Gabel

BIBLIOGRAPHY

References

Cinnaminson Township. *Cinnaminson Centennial.* 1960.

Gonzales, Evelyn. *The Bronx.* New York: Columbia University Press, 2004.

Hodges, Graham Russell Gao. *Taxi!.* Baltimore: Johns Hopkins University Press, 2007.

Johnson, Deidre. *Edward Stratemeyer and the Stratemeyer Syndicate.* NY: Twayne, 1993.

Ultan, Lloyd and The Bronx County Historical Society. *The Beautiful Bronx (1920-1950).* New Rochelle, NY: Arlington House, 1979.

Ultan, Lloyd and Unger, Barbara. *Bronx Accent.* New Brunswick, NJ: Rutgers University Press, 2001.

Periodicals

Nuti, John G. "The Jablonski Years." *Back in the Bronx,* 2000 (Special Issue), 36-37.

Readings

Charyn Jerome. *Bronx Boy*. NY: Thomas Dunne, 2002.

Cunningham, Laura. *Sleeping Arrangements*. NY: Knopf-Random House, 1989.

Freedman, Samuel G. *Who She Was*. NY: Simon and Schuster, 2005.

Waters, Maureen. *Crossing Highbridge*. Syracuse, NY: Syracuse University Press, 2001.

Wolfe, Tom. *Bonfire of the Vanities*. NY: Farrar, Straus, 1987.

INDEX

M

MacDill Air Force Base 180, 183-6, 190-2

Madrid, Spain 435, 439, 472, 480

Maguire, Jim 236, 249, 457, 506, 510, 572

Maguire, Patricia Margaret (née O'Donnell) 35, 47, 71, 103, 139, 151, 226, 280, 310, 457, 506, 572

Mahoney, Ursula 103

Manhattan, New York 18, 21, 23, 31, 37, 49, 57, 84, 113, 140, 202-3, 207-11, 354-6, 605-6, 610-1, 619

Manhattan College 180, 194, 222, 280

Mara, Wellington 146

Martin Marietta 443, 499

Maurin, Peter 175

McAleer, Mary (née Gallagher) 53

McCreesh, Arthur 69, 119

McCreesh, Barney 51

McElroys (neighbor) 22-3, 51

McEvoy, Bill 208-9, 238, 241

McGann, Joseph (Rev.) 120

McGarrity, Tom 367

McGlone, Jim 349

McMahon, Mary Jeanne 103-4

McNamee, John 130

McSwiggan, Terrence 116

Memphis, Tennessee 506-9

Minneker, Katie (Florida neighbor) 566

Montreal, Canada 17, 137-8

Moorestown, New Jersey 236

Moorestown High School 223, 241, 364

Morgan, Henry 127, 177

Morocco 434

Moses, Robert 35

Mott Haven, New York 21-2, 26-7, 32, 39, 43, 46, 53, 69, 72, 76, 81, 119

Mulligan, Jim 273-4, 277, 279

Murray, Jack 526

N

NASA (National Aeronautics and Space Administration) 420-3

Nassau, Bahamas 247, 308

National Aeronautics and Space Administration 420-3

New Jersey Institute of Technology 424

New Orleans, Louisiana 188, 222, 440-1, 481-2

New York City 11-2, 17-8, 20, 22, 31, 33, 36-7, 100, 102, 112, 123, 207, 214-5, 240-1, 605, 618-9

New York Connecting Railroad 31

New York Giants 299

New York Jets 299

New York Yankees 37, 50, 80, 123, 138-9, 191, 202, 254, 549

NIR (Northern Ireland) 15, 19

Nisteruk, Chester (Prof.) 208

NJIT (New Jersey Institute of Technology) 424

Normandie 45, 86

Nova Scotia, Canada 17, 552, 554

NSA (National Security Agency) 254, 262, 406-8, 413-4, 452

NY Thruway 207, 548, 556

NYC (New York City) 11-2, 17-8, 20, 22, 31, 33, 36-7, 100, 102, 112, 207, 214-5, 240-1, 353-4, 605, 618-9

NYCRR (New York Connecting Railroad) 31

O

V

Valhalla, New York   140, 295, 430
Valley Road (Clifton, New Jersey)
    24-6, 38, 52, 71, 80, 240-1
Van Nuys   270, 285, 290, 299
Vancouver, Canada   512-3
Vella, Desmond   124, 166
Vermont   555

W

Walt Whitman Hotel   235
Ward, Hugh   116
Ward, Mike   146
Ward Baking Company   22, 70
Washington Heights, New York   23,
    47, 50-4, 84, 140
Waterfield, Bob   146
Wave Hill   110, 156, 618-20

Weil, Robert (Prof.)   199
West Coast   268, 274, 423
Whelan, Billy   102
Whitey (cop)   29
Wiener, Ferdinand P. "Fritz"   95,
    103, 118, 145, 148, 179
World War I   15, 19, 40-1, 119,
    169, 190-1, 194, 371, 433,
    609, 620
World War II   15, 40, 45, 89, 112,
    169, 199, 371
Wyeth, Andrew   360, 363

Y

Yonkers   90, 125, 176, 209, 214,
    618
Yurko, Mike   367, 369, 381-2

www.ingramcontent.com/pod-product-compliance
Lightning Source LLC
Chambersburg PA
CBHW051218050326
40689CB00007B/732